MACHINATIONS IN COAL MINING

Machinations in Coal Mining

CHARLES ROUND, C.ENG. FIME

The Pentland Press
Edinburgh – Cambridge – Durham – USA

First published in 1996 by
The Pentland Press Ltd
1 Hutton Close
South Church
Bishop Auckland
Durham

ISBN 1-85821-403-3

Typeset by Carnegie Publishing, 18 Maynard St, Preston
Printed and bound by Antony Rowe Ltd, Chippenham

Contents

Illustrations

Photographs

Figures

Plans

Charts

Others

Acknowledgements

I express my grateful thanks and appreciation to Ken Mackie, IE, FIME, MME, Engineering Director, Anderson Group Ltd., Motherwell, for the interest, help and updated information kindly given.

The support and encouragement I have received from my nephew and his wife, Charles and Dorothy Round, residing at Burncross, Chapeltown, Sheffield, has been unstinted, invaluable and more than generously given. Both prior to retirement were valuable members of the West Riding Educational Authority, operating as Headmaster and Headmistress at schools within its administrative environs of South Yorkshire. To Dorothy Round (née Grimes) there is a special added acknowledgement. Her expertise in the checking of manuscripts, galley and page proofs, saved me many days of tiring effort, not to forget the kind hospitality extended over long periods and the driving of long motorway distances to ensure my freedom from stress and risk. To me she is accepted in all aspects as a natural daughter.

To John and Dorothy Fordham, my son-in-law and daughter, Bishops Hull, Taunton, I am again incalulably indebted. Since the tragic loss of Mary my wife on 12 July 1987, they have done much to fill the traumatic vacuum so created. Their understanding and devotion has done much to create a situation where I could rehabilitate myself and continue living with that zest for achievement so characteristic of my make-up. Dorothy has inherited all the fine attributes and devotion her mother possessed which are unselfishly given to enhance my daily welfare.

My son and daughter-in-law, John and Sue Round, of Killiney, Dublin, although living at a distance, are ever near. They have built a fine family and have established themselves with great respect in a foreign country – a situation which gladdens the hearts of most caring parents in a similar situation.

Finally, these Acknowledgements would be incomplete were I not to make a reference to two of The Pentland Press Limited's publishing staff.

Mrs Jill Rowena Cole, Executive Editor, is a fine lady with the disposition, charm and patience of no lesser person than the Duchess of Kent for whom I have great regard. Her support in the selection of my manuscript was followed up with all the care and help one would wish to have when venturing into uncharted waters. My sincere and grateful thanks are publicly expressed herewith. She is one of the naturally nice persons one is privileged and glad to have met and will be remembered for a very long time.

Mrs Mary Denton. Publishing Manager, a Durham miner's daughter of great ability and enterprise, handles very effectively a position of tremendous importance within the production processes. Her generous help and courtesy is thankfully appreciated.

Preface

The title of this book might suggest that it is essentially a technical treatise beyond the ken and interest of the average reader. Such fears are rapidly seen to be unfounded as the author, in clear non-technical language and easy style, accompanied by photographs and computer-assisted line drawings, weaves a rich tapestry covering the evolution of the British mining industry from the late 19th century to the present day. It has the added bonus of being a most interesting autobiography of a man intimately involved at shop-floor levels, progressing from an underprivileged position beginning in south-west Yorkshire, with no secondary education; starting work at the age of fourteen as an underground pony driver; progressing by sheer hard work, tenacity, and honesty of purpose by night school and technical school education to qualify at an early age as a mining engineer and manager. His experience in these, his promotion through general and area managements posts to the highest area production and planning managerial posts, together with the innovations and engineering improvements he instituted, are chronicled along with biographical details of the associated social activities involved. As I was resident and at school in the same area during the initial period covered by the book, I was well acquainted with many of the characters and conditions described and reading the book was similar to watching a video recording of that time.

This book will appeal to a wide audience, both technical and non-technical, and should take its place in posterity as an invaluable and essential historical record of the evolution of the British mining industry from the pick and shovel, steam engine, belt driven era to the modern fully mechanised British longwall mining system with its large electrically powered coal-cutting, road driving, roof-supporting, underground coal and personnel transport systems with their vastly improved productivity and safety.

Throughout one is impressed by the constant drive by management, mine machinery manufacturers and designers (the author played a large part in all these aspects) to achieve this fully mechanised scenario which resulted in increasing output from 5,000 tons in the very early days to currently over 100,000 tons per week, per individual mechanized unit. Unfortunately this came too late to save the industry from the savage effects of the many facets and the economics of 'the second industrial revolution' which has appertained since 1945.

The part played by the author in this process, the difficulty experienced overcoming backward-looking practices, 'management by fear' and 'small minds in big jobs' are vividly portrayed. An interesting episode is the author's three week stint as an able-seaman on a coaster, taken as a well-needed anti-stress therapy provided by an observant higher management.

Excellent vignettes of many well known personalities he encountered augment the narrative which is enriched by discourses on the important and rightful place of the long-standing close relationship of the mining communities with brass bands and their conductors.

Early retirement from Coal Board work after a period as a 'troubleshooter' for the industry presented the depressing prospect of enforced leisure, but fortunately provided him with a career change into mining consultancy. This involved advising on the alleviation of coalmining mining engineering difficulties in Spain and the USA mining machinery development, and advising on claims for property damage due to mining subsidence within the UK.

Ruminations on the present day contraction of the coalmining industry and its social implications complete a highly important work which should be preserved for its historic and biographical note.

Arthur Naylor, CRM, MD, MSc, FRCS
Retired Consultant in Trauma and Orthopaedic Surgery
Bradford Hospital's Unit founder 1945.

The Round Family

Introduction

During the past sixty years, the UK Coalmining Industry has been the subject of intense political, organisational and technical change. On 1 January 1947 the Coal Industry was placed under public ownership, being directed and controlled by a newly created National Coal Board responsible to the then Minister of Fuel and Power. He in turn had similar responsibilities to Parliament for the Board's conduct and operation. During its first year of operation the Board's 980 collieries, employing a total labour force of 711,000 men, produced 187,000,000 tons at an average productivity rate of 1.075 tons and 263 tons per manshift and per man-year respectively. During that first year 608 miners were killed and 2,447 seriously injured.

Prior to and from this datum point the Coal Industry's fortunes are traced in terms of an autobiographical account, and the historic mine mechanisation development which has led to the current 1995 UK advanced longwall mining technology, the fruits of which are being gathered on a world wide scale by the major coal-producing countries: America, Australia, China, France, Poland, and South Africa.

In simple progressive terms, mechanised handloaded longwall units formerly extracted coal at the rate of some 3,000 tons per week. These were later superseded by fully mechanised coal-faces which, in their early days, yielded 5,000 tons per five-day week with a reliability factor (i.e. freedom from breakdowns) of about 60 to 75 per cent, taking in all aspects of delay from a wide variety of causes.

The current situation is such that modern mechanised longwall coalfaces with the Electra shearer coal producing machine and matching infrastructure in the form of hydraulic supports, face and outbye transport conveyors and communications have now come together so that weekly outputs are of the order of 30,000 to over 100,000 tons per week with a reliability factor of over 95 per cent being obtained as a matter of course. Certain current installations in America are producing at rates of over 3,000 tons per hour. The present situation is mind boggling.

Following a restructuring of the National Coal Board's initial organisation provisions in 1967 and the advent of nuclear electrical power generation - which provided for one such installation at Seaton Carew within the Durham Coalfield – the seeds were set for a progressive decline in the Coal Industry's fortunes. Following the miners' strike in 1984 this decline was accelerated by the indiscriminate closure of collieries such that the situation at the end of December 1994 led to the pits being privatised. At this point 16 deep mines were producing at the rate of 28.6 million tons per annum, with 7,400 men on colliery books producing at the rate of about 3,865 tons per man-year. During that year, the industry sustained just two fatal surface accidents and had an impeccable safety record.

It is felt that the basic cause of the Coal Industry's rapid decline was political, it being sacrificed on the altars of political power, deceit, manipulation, greed and party interests, for which a future price will ultimately be levied. Both Government ministers and the National Union of Mineworkers' leaders cannot be considered blameless. Intelligent co-operation as to the Industry's role in the national scheme of things, by allowing it unfettered to develop the vast economic potential of the immense coal reserves of incalculable value we are fortunate to have, had much greater rewards to offer than periodic highly disturbing confrontation. This is particularly distressing in that, over the years in question, the highly technological direction to achieve these rewards had been determined and funded by the British taxpayer.

I have been closely associated with the Coal Mining Industry for some sixty-eight years, and during active life therein took a leading role, along with many others of my kind, in the Coal Industry's technical development, raising its status from backward primitive endeavours to the highly sophisticated achievements of the modern factory, in which hard work has been eliminated and safety standards enhanced to the highest levels of those which prevail in very much less arduous and risky enterprises.

It is against the foregoing background that the experience of a lifetime is narrated as a human story in which countless people in many walks of life may readily identify, particularly miners, mining communities, engineers and managers.

The prospective reader is taken simply step-by-step through my exposure and achievements, together with the manner it affected me and my family, providing plain lay explanation and valuable historic illustrations along the way.

Starting out as a humble pony-driver underground at the age of fourteen and progressing through other subterranean activities to become a coalface worker, I concurrently undertook a long course of evening class and technical college studies, during which I obtained the basic mining qualifications for higher endeavours at an early age. Studying was somewhat second nature to my work and interest and, being naturally endowed with an infectious enthusiasm, I obtained considerable enjoyment from such efforts.

This led on to my appointment to the post of Underground Junior Official, further into Colliery Management posts, and later to the highest levels of appointment within the Mining Industry. The problem of human relationships and internal political pressures and harassment are not neglected, being dealt with in the direct manner of the Yorkshireman, without ill-feeling, animosity or rancour, and are covered by an honest expression of my factual experiences, both adverse and stimulating. Such experiences did much to mould my character and develop an extremely wide range of human understanding, personal confidence and general recollection.

Three major mining disasters are referred to: the Sutton Colliery explosion; the Silverwood underground locomotive manriding accident which involved six fatalities and the Aberfan Colliery waste tip disaster with its awesome cost of the lives of over four hundred children, teachers and adults. The last had a spill-over effect with an associated American colliery waste tip disaster of a similar nature, the author being asked and subpoenaed to meet American attorneys in New York and London respectively.

The Sutton explosion was a major disaster involving five fatalities, two of whom were young boys, and which inflicted serious burns to twenty-three miners.

Further reference is made to major winding engine breakdowns, the first an 'overwind' at Gedling Colliery, Nottingham, which resulted in extensive damage to the shaft headframe and pit bottom shaft fittings. The second was at the No.1 coal winding shaft, Clifton Colliery, Nottingham, involving an irreparable crack in one of the very old steam cylinders. Here, some 1,200 miners were redirected to temporary work at adjacent collieries in a matter of hours.

Being caught up in social welfare activities and old age pensioners' associations, particularly in the South Western Division No.9 (Neath Area), an event is recorded including the time spent with Lady Megan Lloyd-George MP, daughter of the First World War Prime Minister, David Lloyd George, Lord Halifax, Minister of Fuel and Power, and Sir Grismond Phillips, Lord Lieutenant of Carmarthenshire.

Personal interests in the brass band movement in both West Wales and South Yorkshire, together with activities within the gramophone recording field and radio broadcast field, are chronicled particularly with reference to some twenty-five tape recorded radio programmes featured by the Public Radio WGTS-Tacoma Park (Broadcast Service of Columbia Union College, Maryland. (USA) Authorities.

I describe a situational experience when I undertook an eighteen day round trip sea voyage as a Merchant Seaman between Troon (Scotland) - Casablanca (Algeria) - Cork (Ireland) and Fishguard (Wales): an unusual undertaking for a practising mining engineer.

Following a premature retirement from the National Coal Board's Headquarters staff, the circumstances of which are explained, detail is provided relative to work undertaken as a mining consultant with particular reference to the American and Spanish mining situations. In the latter case reference is made to visits to Spanish collieries involving thick and highly inclined seam workings. The experiences are simply explained and illustrated.

I undertook several visits to the American coalfields of West Virginia, Virginia, Pennsylvania, and Kentucky. One particular visit, authorised by Lord Robens, the National Coal Board Chairman, had special significance and resulted in the development of Riddings drift mine which achieved American standards of performance in terms of productivity, some twenty years ahead of its time, within the UK.

Account and discussion is provided relative to mining-subsidence damage claims in which bona-fide claimants were shabbily treated by National Coal Board area officials.

My human account is much more than an autobiographical record. It provides an historic record with photographic and sketched evidence of events within the coal industry over the past sixty-six years or so.

I have spent some two years writing up and completing this account of my personal and historic experiences before and after the State's control of the UK Coal Mining Industry, and during the current situation of privatisation. Much of this is common to, and will readily be identified by, all former and current working miners, miners' families and communities, mining engineers and mine management. All will no doubt recall their personal memories and the great comradeship within the Industry that existed during their working days.

Those who are being introduced to mining for the first time will surely find great interest in this fascinating account, showing the sort of difficulties encountered by anyone attempting to emerge from the very lowest to reach the highest levels of operation within a profession. Intellect and imagination are not necessarily confined to the birth accident of privilege and academic prowess with their associated labels, as has been demonstrated throughout history.

I

Educational and
Academic Development

Early Days

Having been born of working class parents on 13 January 1914 at 29 Elm Street, Hoyland Common, Barnsley, South Yorkshire in an almost exclusive mining environment, I suppose my future under the circumstances was more than likely to be tied within the Coalmining Industry. The city of Sheffield, with its many iron and steelworks, had within its precincts a number of collieries including Nunnery, Handsworth, and Tinsley Park, some ten miles south of my village on the A61 turnpike road. Chapeltown and Ecclesfield, some two and four miles respectively to the south on the same road, were also established in the manufacture of iron and steel. They too had common coal supplies from Newton Chambers Co.'s Thorncliffe and other collieries located adjacently.

Employment was obtained generally within the mines and steelworks and miners' sons tended to follow the same pattern of employment as their fathers, brothers and cousins. Steel and iron workers' sons similarly sought their employment in the steel and iron works of Rotherham, Sheffield, Ecclesfield and Chapeltown. Coal and steel were the basic contributors to the local economy.

Travel within the wider area involved the London Midland & Scottish Railway, London & North Eastern Railway, horse and trap and the reliable bicycle, not to forget 'shank's pony', the most common form of getting from one place to another over short distances at that time. Barnsley, situated approximately five miles to the north on the same turnpike road, was the virtual centre of Yorkshire's coal activities, but with the passage of time and the near exhaustion of the rich Barnsley coal-seam, the centre moved eastwards with the construction of the more modern and larger collieries in and around Doncaster.

The coal mines were taken into national ownership on 1 January 1947, by which time Doncaster had been developed into an important large coal producing area with an annual output of some 8.2 million tons, the total output for the Yorkshire coalfield for that year being 38.31 million long tons, with a labour force of 138,500 average manpower employed in the North-Eastern (Yorkshire) Division. Nationally some 1,000 collieries had produced 184.4 million saleable tons, with an average labour force of 703,900 men.

FIGURE I (page 24) : The rough sketch shows the location and disposition of the village of Hoyland Common in relation to the current M1 London-Leeds motorway. It forms part of the large conurbation of Hoyland Nether, then the administrative centre of a large populated urban area which embraces Hoyland Common, Platts Common, Hoyland, Milton, Elsecar, Jump and Hemingfield. The whole area is rich in the history of early industrial progress, with old bell pits and deep mined collieries. Iron and coal were first mined in the mid/late 1700s, with the manufacture of iron and steel being undertaken locally at a later date.

The village's economic activities were formerly centred upon this group of collieries dotted round the villages to north-east, south and west, together with an agricultural contribution in that appreciable areas of land were under culti-vation, embracing wheat, oats and a wide variety of root crops and grassland. A substantial area of land was cultivated by the villagers (predominantly mine and steel workers) which took the form of Local Authority annually rented allotments. Probably this was their most practical and needful pursuit of the time, being necessary to eke out the low wages earned in the pits. In many cases they emerged into sizeable smallholdings in which pigs and hens were husbanded.

Within a radius of three miles of Allotts Corner, a distinguished landmark, Tankersley, Barrow, Rockingham, Wharncliffe Silkstone, Hoyland Silkstone, Skiers Spring (Earl Fitzwilliam's Ownership) and Skiers Spring (Newton Cham-bers Ownership) Collieries were all readily accessible. A number of coal-seams: Barnsley, Lidgett, Flocton (thick & thin), Fenton (high & low), Parkgate, Thorn-cliffe, Silkstone, and Whinmoor outcropped or were close to the surface on the southern and western fringes of the village.

Socially the religious lives of the inhabitants were bound up in two churches, three chapels (Wesleyan Primitive Methodist denominations) and a Salvation Army Corps, all of which actively catered for the people's spiritual needs and promised that the people's hard lives would be compensated for in after life. Whitsuntide processions were of great importance with the parents having made great sacrifices the preceding year to ensure their children were immaculately turned out with new clothes for the occasion. Music flourished in the form of two brass bands (Wharncliffe Silkstone Colliery Band and the Salvation Army Band), small orchestras, dance bands, school choirs, choirs and choral societies and vocal quartets. The Musical Festivals of Barnsley, Hoyland, Wombwell and Mexborough were the annual highlights of the musical calendar.

Sporting activities took in all forms of athletic event - soccer and rugby football, cricket, walking, crown green bowling, tennis and golf, the last two mostly the province of the middle class: doctors, teachers, solicitors and shop-keepers. The miners had their own pursuits including gambling in the local Bellground and Spring Woods. These were Nipsy and 'Knur and Spell', some-what in principle akin to golf. The two implements used in the former were a standard miner's wooden pick shaft. This was specially shaped with a flattened hitting head at one end. The shaft was reduced and a handle formed at the opposite end. The 'nipsy' was usually a short length of hard wood (oak and lignum vitae were popular choices) about 1½ - 2" in diameter and 3" in length, one end being shaped to a rough point. In playing the game the nipsy was positioned on top of a vertically placed building brick. The players struck the pointed end with the converted pickshaft head end causing the nipsy to rise vertically or fly off in a random direction according to the degree of skill applied. They played the game on the basis of maximising distance between competitors. The players would estimate the achieved distance with joint acceptance or disagreement according to the circumstances. In competitions and gambling often a 'strider' was used and his determination by actual rough measurement was final. The game of 'Knur & Spell' adopted the same principle but the striking medium was more sophisticated, being longer with a degree of 'whip' or flexibility somewhat akin to that of a fishing rod. A small very hard 'spherical clay 'potty' about 1.25 inches in diameter, kiln prepared and hardened, was the driven element. It was hit in two ways: statically whilst suspended in a loop of string or dynamically whilst rising or falling vertically from spring ejection type mechanisms. Whether or not these games have passed into oblivion I do not know but they were extremely popular some seventy years ago, and absorbed a great deal of gambling stakes.

Pigeon flying, involving both 'milers' and 'homers', had a large following and the breeding of competitive birds for both pursuits was widely undertaken. The 'milers' were released at the LMS railway bridge near the bottom of Stead Lane (PHOTOGRAPH 3) being timed as the first arrival at its respective pigeon loft. Homers were entered annually in the San Sebastian competition, Spain, where the birds were released and checked on the time the birds took to fly home. Birds were often lost in this event.

Another popular pastime amongst both miners and steelworkers was that of coarse fishing. They were attracted over the weekends to the local lakes, rivers, reservoirs, ponds and canals and to the Fens of Lincolnshire and Yorkshire.

In those early days a substantial area of common land was vested in the village people over the wider area. I recall my father telling me on a number of occasions that the mineral rights under these lands were transferred to the then Earl Fitzwilliam who in exchange built the inhabitants a Town Hall located in the centre of Hoyland. Our village was quite active with an engaging social life. It had its fair share of churches, chapels, public houses and working men's and social clubs, with a wider interest in music, centred about its brass bands. Wharncliffe Silkstone Colliery Band and the Salvation Army Band undertook regular practice twice weekly in the Hare & Hounds public house and Salvation Army Tabernacle respectively, and there were choirs, male quartets, glee clubs and vocal talent in fair profusion competing at the main annual local musical festivals which flourished at the time. Sport in all its forms was widely practised and encouraged with an annual event in the village sports field, the location of the Yorkshire Council Cricket team.

William Alfred Round, my father, as a small boy of nine years, was moved from Dudley, West Midlands, to Caton near Lancaster into the care of his grandmother, having lost both his parents. His elementary schooling consisted of three hours per weekday, four days per week, in reading, writing and arithmetic, with English grammar as a very important part of his learning. During the weekday afternoons he worked at a brickyard involved in the manufacture of building bricks, earthenware pipes and tiles. Here he developed such skills as to enable him, in later years, to set up brickyards in open fields and thereafter manage them on behalf of their owners. He was a remarkable man who had, by ruthless determination, extended his education by intensive reading and study. He had also become proficient as a brass instrumentalist and music arranger. Both his parents had been orchestral musicians which had involved them in travel within various parts of the country.

Harriet Elizabeth Norton, my mother, was of Irish descent but born in Liverpool. She was a gentle and kind lady who for a number of years prior to her premature death (following an internal operation at the Jessop's Hospital, Sheffield, at the age of fifty-three years), undertook voluntary nursing of the sick within the village of Hoyland Common. There is little doubt in our minds that the cause of her death was really obtained by excessive hard work and by sustaining her family during the very difficult and hard times of the First World War, and later in the 1920s and early 1930s. Following their marriage, thirteen children were born into the family circle, nine of whom died as a result of the ravages of the First World War, poor food, diseases such as meningitis and influenza and by fatal accidents. The four of us remaining were:

John Edgar Round, my older brother, who rapidly developed at an early age as an excellent solo and principal cornetist with Wharncliffe Silkstone Colliery Band, and in local orchestral and dance band combinations. During his musical career he won countless gold and silver medals competing in instrumental solo competitions which flourished over a wide area at the time. John at the time of his premature death was senior overman (underground official) at the colliery. He married Maud Housley of Warren, Chapeltown, near Sheffield, and they had two boys, William and Charles. The former became a miner and the latter a teacher, ultimately headmaster of Worsborough Junior School. John died instantaneously from a stroke whilst following his underground duties at the age of forty-eight.

Doris Eva Round is my oldest sister, currently in her ninetieth year. At the age of fourteen she became a domestic servant to the spinster sister of one Joshua Rawlin, at the time manager of Wharncliffe Silkstone Colliery. She remained with Miss Rawlin throughout her single and married life, in the later years as a companion until the demise of this fine lady. Doris has had a very hard life, particularly following the death of her husband Albert Sykes at the age of forty-two, two years after I went to live with them, through pneumoconiosis (dust disease of the lungs) arising from his underground coalface employment at Cortonwood Colliery. They had a daughter Margaret who became a shop assistant and a son Frank who was trained and worked as an underground mechanic. Despite all her difficulties she has always maintained a very cheerful contented approach to life and found time to assist older neighbours who lived close by.

Marion Round, was my younger sister who spent several years in domestic service in various parts of the country until she married James Ratcliffe, an underground junior official at Elsecar Main Colliery. She died of a stroke at the age of sixty-five whilst on a caravan holiday in Scarborough. Marion was an excellent cook and taught the subject at the Hoyland Kirk Balk evening classes for many years. Jim and Marion had two daughters, June and Jennifer, who both excelled academically. June was a highly skilled nurse and married Dr Gerald Bickler of Leeds. Jennifer, following her London graduation, married David Anderson, a Civil Servant. He later was closely involved in the United Nations discussions during the Falklands crisis, for which he was awarded the Falklands Medal.

Charles Round, currently aged eighty-two years, narrator of the experiences which follow, was the youngest of the surviving four. Married to Mary Greaves, a miner's daughter, they had two children, John and Dorothy. The family is shown in the Frontis Photograph, the children being quite young at the time.

Living in Elm Street, Hoyland Common, at the time of my birth were quite a number of old miners who were knowledgeable about the former working of the ancient bell pits before they themselves were absorbed into the deeper mines, Lidgett (Pillbox) and Tankersley Collieries. The former was adjacent to the LMS railway station for Wentworth and Hoyland Common and the latter about a mile to the east of the village. Both of these mined the coal at greater depths, the latter at over 750 feet.

Emmet Walker and Mrs Walker with their family of five boys (all of whom started out their careers in the local mines) and four girls lived next door to us as neighbours for many years. He had been a miner all his life and in his later years (he lived into his early or mid-nineties) used to delight us youngsters with accounts of his experiences. It was from him, during my early student days, that I was made aware of the nature of bell pits, which were profuse in the woods and fields to the south of the village. Nature however over the past seventy years or so has sought to conceal their origin and location to great effect, much in the same way as in fifty years it has turned the Old Pillbox pit-tip into a very large foliaged green and wooded hillock.

Bell pits and bell pit locations are indicated in FIGURE 2. They lay along the western fringe of the village of Hoyland Common, somewhat parallel to the present London-Leeds M1 motorway. Quite a number of the upper Yorkshire coal seams outcropped or were at shallow depths, including the Barnsley, Swallow Wood, Lidgett Coal, Joan Coal, Flockton thin and thick seams, High Fenton, Low Fenton, Parkgate, Thorncliffe and Silkstone seams. Some sixty years ago these easily accessible deposits were worked by Bell pit mining systems (Figure 2). However, nature has greatly restored the former disturbed environmental situation, over the intervening period.

In the latter years of the eighteenth century, it was customary to sink a small shaft seven feet or more in diameter though the sub-soil to the coal seam lying within a few feet of the surface, maybe down to twenty feet or more. On reaching the seam the coal was immediately extracted and, by working circumferentially round the shaft perimeter, the coal was removed by wicker baskets or boxes (to which wheels were later added - followed by steel rails, all of which made the work involved less arduous). As the overburden above the seam was exposed, it naturally sought to collapse. To prevent this, sprags and props cut from the timber growth within the adjacent woods were obtained, trimmed and set between the exposed roof and floor, in approximately concentric rings. As this circular area of extraction was extended, the natural weight of the overburden was such that it tended to subside in a saucer-like manner, breaking the supporting props until a point was reached where the whole mass would collapse right through to the surface, often with fatal results to the men working thereunder. The mine would be abandoned and another started a few feet away, and worked until that too succumbed to the same natural phenomena.

Firedamp and blackdamp were problems in that ventilation was primitive. Early lighting below the surface was with wax candles and later with acetylene lamps, progressing in later years to the Humphrey Davy safety lamp and accumulator electric hand lamps. Finally electric cap-lamps became the most convenient form for general underground application, by which time 'bell pits' were of past historic interest.

The early local timber was used for supporting the excavated bell pits workings, but was of poor quality, generally lacking strength, so better quality timber came to be imported from the Scandinavian countries. I recall undertaking a surface 'timber stock check' during the early 1940s at Elsecar Main colliery and counting upwards of 500,000 props of various lengths and diameters and some 300,000 wooden bars of similar variation, probably about three months' colliery supply at the time.

It was not until the early 1940s, that steel began to be applied to the support of mine workings, both at the coalface and for the lining of underground roadways, these ultimately being superseded by mechanical/hydraulic support systems.

The old Lidgett (Pillbox) Colliery, closed before the 1900s, was a source of fascination to us as children. We would slide down the face of the bare tip on strips of canvas belting and play around dangerously in the abandoned winding engine house. With the introduction of the motor car during the early 1920s the old Pillbox winding engine house was converted into a petrol station and garage. It first started out by selling ROP (Russian Oil Products) at 11*d*. per gallon. It is presently operating quite well although the present brands of petrol are considerably higher priced at some £2.50p per gallon!

PHOTOGRAPH I is of the old Lidgett (Pillbox) Colliery Tip. During the 1926 miners' strike hundreds of miners turned the tip over into a grey mountainous mass in their search for coal formerly contained in the tipped dirt. Natural environmental cladding of these old tips takes place in stages: first mosses and sparse foliage, followed by coarse grass and shrubs which tend to recede with the further growth of small trees and bushes. In the final stages come trees and large bushes. The process can take many years. The original Pillbox tip, as can be seen, has passed through all the natural environmental stages of rehabilitation and is now densely clad with trees and large bushes.

Of recent years, particularly with opencast mining, the process of site rehabilitation has been planned and the timescale involved substantially reduced. Drainage, grass seeding and importation of trees in an advanced stage of growth ensure that within a very short time the working site has been fully restored, in many situations to better standards than formerly obtained.

PHOTOGRAPH 2 shows the north face of the winding enginehouse looking south towards the old building in its current form as a motor garage and petrol station. The rough road to the left upon which the breakdown vehicle stands leads to the old Newton Chambers Skiers Spring Colliery. On either side of this road a great deal of illicit shallow mining was undertaken during the 1926 miners' strike with methods but little more advanced than the old bell pits shown.

My earliest recollection of fatal road accidents involves two which occurred around 1926 within the one-mile stretch of the A61 turnpike road from this garage (currently called Lidgett Garage), north towards Allotts Corner, Hoyland Common. The first involved a Foden's steam wagon which was travelling south towards Sheffield. Approaching the north-west corner of the old Pillbox tip, the road curves and has a fair gradient. Together with mates I was sliding down the tip and was able to see the steam wagon careering down the gradient out of control. It struck the corner of the tip and turned over. Burning coal and scalding water was ejected across the road and steam was noisily issuing into the atmosphere shrouding our view of the calamity. When we got to the scene of the accident a number of workmen from nearby were present. The driver had been thrown out of the vehicle, fatally burned and scalded.

The second fatal accident involved a small child of about eight or nine years of age and occurred about half a mile further north opposite what was locally called the Second Steps, a series of green fields with public footpaths through them in which we undertook sports of all kinds, being driven out from time to time by Johnny Mathewman, the local farmer. The child had been playing in the fields and whilst crossing the road was struck by a new Lagonda open car. Her injuries were fatal. At the time this was probably the only car on the road for miles around, other than the pony and cart, solid tyre charabancs and steam driven vehicles. There was very little traffic about on the roads in those early days.

PHOTOGRAPH 3 shows the bottom section of Stead Lane, Hoyland Common where it joins Milton Road leading to the A61 road. The LMS Hoyland Common and Wentworth Railway station stands at the T-junction of the two roads. Throughout this area we as youngsters from Elm Street and Chapel Streets and the 'foreigners' from Milton Village, about two miles distant, spent many hours playing in Spring Wood and on the adjacent dirt tips. In these woods gambling schools were set up by men from the villages, broken up from time to time by the police. Pigeon fanciers with their 'milers' and 'homers' used to release their birds along this stretch of road, clocking times and making heavy bets. In addition, for many years it was a very popular haven during the summer months for romantic couples.

Spring Wood, to the left, formerly housed a large explosive magazine which supplied Skiers Spring Colliery (Newton Chambers Ltd) with explosive cartridges and detonators. During the late 1920s, in the early hours of the morning, it exploded, killing the magazine attendant. The noise was horrific over a radius of some three miles and gave rise to great concern throughout the adjacent villages including Milton, Wentworth and Harley. Apparently the cause of the explosion was assessed as the nitro-glycerine having become unstable owing to the low temperatures of that particular winter night.

The woods were exceptionally beautiful, being rich in a wide variety of birds, flora, and other living things. With the advent of spring each year, a fabulous array of bluebells delightfully scented the whole environment. In those early days the farmers grew and sustained a wide variety of crops which contributed colours and scents to give life real meaning after a particularly hard winter. With the changes that have taken place over intervening years however, much of nature's former rich endowment has disappeared.

Dotted throughout the whole locality were a number of open ventilation mine shafts about 12 feet diameter at depths of 750 feet or more. On the surface they were surrounded by a circular wall about 8ft. high. These shafts were associated with the early methods of ventilating the deeper mines. As ventilation systems improved and more efficient ventilation fans were designed the practice of sinking such facilities was abandoned. They were often a source of vandalism and claimed a number of human lives. Ultimately these old shafts were filled in or capped at the surface in the interests of public safety

Education

Early elementary education from the age of 5-6 years, at the Hoyland Common Infants School, was followed by transfer from 7-14 to the Junior and Senior Boys School. Between the ages of nine and eleven, I had health problems, and being considered anaemic was excused school attendance for eighteen months continuous duration. This really was an educational setback, particularly in that in sitting the 11-plus Examination for Secondary Education at the Barnsley Grammar School I recall being totally lost in the subjects I had missed; consequently I failed the examination. However, I did have one consoling feature in that I did quite well in the intelligence tests by scoring 40 out of a total of 50 which was above the class average at the time. It is at this point that one of my old school teachers, Hurvey Holden, enters this narrative. Hurvey was a bachelor whose love for his musical instrument, the violin, greatly outshone many men's love for their wives. He carried it with him everywhere and played it on the slightest whim, spending all his spare time playing as both a paid and a voluntary musician with orchestras, at religious festivals and with local dance bands. He was an excellent player and a fine teacher. During his various musical engagements he was very often associated with my brother John on trumpet and so he got to know my family extremely well. Following my failure in the 11-plus examination, he took me in hand, projecting a great deal of his effort in my direction, setting me additional tasks not given to others and ensuring that these were completed satisfactorily (sometimes by a heavy stroke of his violin bow on my backside). Fortunately for me we were together for three consecutive years as we jointly moved classes upwards with the result that final class grading was achieved and I left school along with the best of the pupils at the higher levels.

On 14 January 1928 my elementary school days were behind me with no thought of further education, having against my father's wishes got work underground at Wharncliffe Silkstone Colliery (Davy Oil-Lamp and Check Number 884) as a pony driver. However Hurvey Holden was not finished with me yet. In August of that year together with a mate I was walking down the main road in Birdwell (the adjacent village about a mile to the north of ours) when a loud voice from across the road yelled:

'Round! Come over here at once!' The voice was unmistakably that of my former teacher Hurvey Holden. Meekly I went across the road alone to face him. Such was my respect.

'Tell me,' he asked, 'Have you enrolled for night school?' I'd never even considered it but couldn't say so.

'No, Sir,' I replied.

'Then you must go and enrol for the Industrial Course next Wednesday and tell your Dad I will have a word with him.'

As future events unfolded, that meeting was probably one of the most important of all face to face meetings I ever had before or since. It directed the way I had to go. Dutifully Dad was told what had happened and that my enrolment would take place. He was delighted and at this point I believe he forgave me for not having persevered with learning to play the cornet and joining the band as he had wanted. He had told me many times previously that it would be something I would regret in later years, particularly as it was another avenue leading to the availability of employment. As things worked out how right he was, but for a different reason. Many were the times during heavy periods of future managerial stress that I would have found great solace and relief by pouring out my soul through a musical instrument.

During my early school days Father was a great inspiration, being a man of very wide experience developed in many parts of the country, keen to take up new things and to look for ways to enhance one's interest and development. I recall at the age of nine having a No.5 Meccano set for Christmas at, I now realise, great personal sacrifice to both my parents. At the age of eleven, being quite proficient in building catalogue models and developing original ones, I was encouraged to compete in the Meccano model building competitions which were held annually.

It was under the following circumstances that I entered and won a major competition in the mid 1920s. Dad was employed at Wharncliffe Silkstone Colliery as a boiler fireman, generating steam for the two colliery winding engines, surface and underground plant, etc. He worked alternate shifts, days and afternoons. On the Friday afternoon shift I had to go with him to work to collect his wages for handing over to Mum. It so happened that an old Clarke & Stevenson disc-type coalcutting machine was awaiting repair for a number of weeks on the surface. It was lying adjacent to the Lancashire Boiler Range, father's place of work. This took my interest and within six weeks a true replica was built with Meccano which worked precisely as the original one did underground, although I had never before seen one at work. The model was duly entered for the next major competition and won first prize; the sum of £5. In the same competition Dad entered a brickmaking machine, an excellent working model which really made small bricks. He won a prize of £2. This money was used to purchase a Stuart & Turner's No.2 Model steam engine with a bore of 1.25 and stroke of 2.5 inches respectively and rated at one-eighth horsepower. Within six months we had a complete small power plant consisting of a 14 inch diameter, 22 inch long, gas fired Lancashire boiler fitted with lever safety valve, pressure gauge, fusible plugs and preheated water pumped feed (the water being drawn from a heat exchanger heated by exhaust steam from the steam engine), and driving a small dynamo. This to me, at the time, was real magic and gave me an excellent understanding of the principles of such systems in later years.

In the mid 1920s radio entered our lives. Dad bought a Fellows 3-valve light-emitter set with head-phones, run by a 6 volt accumulator, a 90 volt high tension and a 9 volt grid-bias dry battery. With the advent of the 2 volt dull-emitter valve Dad was encouraging me to help him build radio sets. The John Scott Taggart ST 200/400 series were our forte. Dad built them for a number of people and, with the coloured wiring system he had developed, made it comparatively easy for me to carry out simple maintenance where and whenever required.

The people for whom I undertook maintenance often remarked, 'What a bright young boy you must be!'

They did not know that Father had but taught me to carry out only simple tests and in some cases where failure occurred I took the receiver home to Dad on the basis that it required some special part exchanges. I suppose in those early days a youngster of about eleven or twelve walking around with a loops of wire, fuses, pliers, a screw driver and a voltmeter might well look impressive. Today a youngster of that age has mastered the computer with all its complexities.

My personal pride was in the construction of a crystal set in a cigar box, (winding my own variometer tuning device) and later the 'Midget One', a one valve set, followed by the 'Sidney Two', a two-valve shortwave radio, both of which were publicised by the radio magazines of the day. I recall very vividly my first American conquest Schenectady W2XAD and W2XAF, around the early 1930s, and a beautiful contralto voice singing 'Softly Awakes my Heart' from Saint-Saens' *Samson and Delilah*. In retrospect I believe her to have been Marion Anderson. Guy Lombardo and his Royal Canadians, Paul Wightman and his Orchestra, Fred Waring and his Pennsylvanians quickly followed. Later, coming home from work about 11.30 p.m. off the afternoon shift greatly facilitated my searching through the short-wave bands (they weren't so crowded in those early days) until about 3 or 4 a.m. each morning.

It is remarkable how one remembers forever long past isolated incidents. In this connection whilst searching through the shortwave bands during those early years, when I tuned in to Pittsburgh KDKA, a dance band, Hal Kemp and his Orchestra, was playing a delightful number 'In an Eighteenth Century Drawing Room'. The melody was firmly implanted within my mind. Little did I know then that this was a popular arrangement of the Mozart Piano Sonata in C Major, K525, which led on to a love of all Mozart's sonatas and concertos.

The recollection does not end there. Whilst undertaking National Coal Board duties in Kent in the early seventies I was having dinner at the Castle Hotel,

Dover, and listening to an accomplished lady pianist doing her best to entertain the diners who showed little apparent response, I went across to her and asked her if she could play this sonata.

'I'm a little rusty but I'll do my best,' she remarked.

She played the whole sonata exquisitely from memory, almost bringing tears to my eyes. Before I had finished dinner, the lady came across to me and said,

'Sir, I hope you enjoyed listening to the sonata as much as I did playing it, for it brought back to me many happy childhood memories. Many thanks for asking me to play the piece. It isn't very often I am asked to play such beautiful music.'

Naturally my thanks were more than profuse. I don't think however that the other diners noticed or even showed any interest.

Melbourne 3L0 with its Kookaburra call sign appeared on the scene a little later and this was followed by Peking and Tokyo. Such were my formative early years prior to settling down to serious study in the pursuance of objective mining qualifications, during which I made a number of false starts.

As a matter of duty to my father and Hurvey Holden, I enrolled for the evening classes at the Hoyland Common Boys School and spent one year under the sponsorship of the Hoyland Nether Urban District Council, taking the County Council of the West Riding of Yorkshire Education Department's Industrial Course 2. This was a general course which embraced feudal history, maths and English. It was quite interesting, but not sufficiently so as to fire the degree of enthusiasm needed to continue, particularly in view of the apparent enjoyment my young mates were having in other directions. However, being persuaded to continue, I enrolled in a Joint Mechanics/Mining Course at the Wombwell King's Road Evening Institute through the courtesy of the authorities indicated. I was motivated in this choice, undoubtedly, by my former Meccano interests.

My choice was an erroneous one in that there were no prospects of getting the practical experience necessary to develop a career as a mechanical engineer. One had to be privileged or articled to get such training in the colliery fitting shop, or indeed in any of the relevant situations in the electrical shop or surveyor's office or management training. However I went ahead. As main teacher we had an engineer, Mr Mellors from the village of High Green, who had his own manufacturing workshop doing sub-contract work for companies such as Newton Chambers and other large Ecclesfield-based specialised enterprises. I recall quite vividly that at a point about halfway through the course, whilst I was attempting to draw nuts, bolts and threads of different types, he looked over my shoulder, shook his head and said,

'If I was you I would try something different. You won't be able to do machine drawing and this course will certainly be beyond your capabilities.'

Naturally I was crushed and almost threw in the towel. I knew he wasn't being nasty and was trying in a way to be helpful. Discussing the incident with Dad, in his usual way he was most positive.

'You're not going to let a little thing like that put you off; you've got to prove him wrong.'

And so I continued and worked really hard. The examinations came and I remember that the main question, which carried some 50 or 60 marks, was making a scale drawing of a shaft pedestal in plan, front, side elevation and sectional elevation. It looked formidable, but getting 'stuck-in', somehow things started to emerge and I managed not only to finish the drawing but also to answer the remaining questions. The examination results were awaited by us all with deep apprehension but did not arrive until after Christmas 1929.

It so happened that, shortly after that Christmas, I was travelling from Hoyland Common to Wombwell by the Yorkshire Traction Bus Service when the teacher referred to moved from his seat to sit alongside me and his first words were,

'Please accept my sincere apologies for what I said to you earlier in the term. In the examination you made 84 marks (distinction) in machine drawing and have averaged over 80% for all subjects. You obviously have ability and potential.'

Whatever I had felt after what he had said earlier paled into insignificance by what had just been said and did much for my future resolve. However the choice had been erroneous and for the reasons given I decided to follow a pure mining course augmented by the practical experience being obtained irrespective of the need to study. Again the Hoyland Nether Council and the Yorkshire West Riding Educational Authority were of great support with the provision of fees, a limited book allowance and appropriate courses.

PERFORMANCE RESULTS

Session	Course	Subjects	Pass Level
1928-29	2nd Yr Industrial (Hoyland Common Evening Institute)	Practical Maths Drawing English & Medieval History Experimental Science	2nd Class
1929-30	1st Yr Senior (mixed course)	Mining Mathematics* Handsketching, Properties of Material & Simple Machine Construction* Mining Science	1st Class
1930-31	2rd Yr Senior Mining	Mining Heat Engines Mining Heat Engines & Electricity Mechanics	2nd Class
1931-32	3rd Yr Senior Mining	Mining Surveying Mathematics	1st Class
1932-33	3rd Yr Senior Mining	Mining Heat Engines & Electricity* Mechanics* Ventilation*	
1933-34	4th Yr Senior Mining	Mining* Economics & Legislation	1st Class

Facilities at the Wombwell Kings Road Evening Institute were really primitive consisting as they did of corrugated iron clad sheds heated by primitive coke stoves. In winter we boiled in the vicinity of the stoves and froze at all other points in the classroom. What we lacked in facilities was more than made up for with an abundance of teaching ability and high quality teachers led by Mathias Clarke (Mining Principal of the Schofield Technical College, Mexborough and the Wombwell Kings Road Evening Institute). These people were highly dedicated and their efforts were unstinted in attempting to bring out the best in each one of us.

Now, at the age of about twenty, I was at a crossroads. I couldn't sit for the Board of Trade Mines Department Examinations Second Class Certificate of Competency (Colliery Undermanagers) or the First Class Certificate of Competency (Colliery Managers) under Section 10 of Coal Mines Act 1911, as I was under the age of twenty-three, the specified legal requirement. I couldn't go any further at the Wombwell Kings Road Evening Institute.

The Mexborough Schofield Institute was too far away and badly served by bus or rail transport. Barnsley Technical College, being much nearer and more modern, was the obvious option. There was, however, a real snag, Advanced mining courses were for day-release only. One had to be a privileged choice of the colliery management to get such release with pay, or one had to get the permission of the same management to have a day release without pay, which was invariably denied or too costly for us to contemplate. Wharncliffe Silkstone Colliery had one such day release student named Sidney Wroe. Generally understood to be a nephew of a former colliery manager named Jonathon Wroe, he occupied a sort of apprenticeship position at the colliery serving the colliery manager. I got to know him very well later when he qualified by obtaining his First Class Colliery Manager's Certificate, following which he migrated into the West Midlands Coalfield.

Dr J.S. Penman was Barnsley Technical College's Mining Departmental Principal at the time and he agreed to take me on the day-course if I could make the necessary arrangements to attend full day on Thursdays, with a supplementary evening class, each week during the operating session. The only alternative to the sponsorship approach referred to was to work the nightshift (10 p.m. to 6 a.m.) regularly which I was able to do. Thus I would work the Wednesday night, dash home after work and snatch an hour and a half's sleep before catching my bus to Barnsley about 8.30 a.m., complete the full Thursday session until about 4.30 p.m., dash back home, and try to get some three or four hours sleep before going to work that night. This procedure I followed for three annual sessions. It was tough but it worked.

Dr J.S. Penman (his brother David Penman was for a number of years Chief Inspector of Mines, India) was a true mentor in all senses of the words. He thoroughly understood our individual needs and catered for them with real dedication. His teaching staff at the time were William Birch (Deputy Principal); J. Woodyatt; W.G. White, electrical engineer (Houghton Main Colliery); J.D. Tilley; Cyril Weaver (Manager North Gawber Colliery). The last, sadly, was later fatally injured in an underground rock fall. They were all excellent teachers well versed in their respective subjects, able to communicate the most complicated matters in simple terms we could understand.

The day classes were supplemented by an evening class each week. I found it tough and more than once fell asleep in the class. It was on one occasion whilst taking electrical engineering that I dozed off, for how long I don't know - but I do recall being brought back to matters in hand by a voice saying,

'We will now move on to alternating current motors - perhaps Mr Round will find this to be more interesting.'

I looked up at him. He smiled, knowing full well my situation.

I found the day-release classes were much better than the former evening studies I had taken. It was a privilege to study under such fine people who enabled me to make excellent and enthusiastic progress.

PERFORMANCE RESULTS (PART TIME) DAY RELEASE STUDIES

Session	Course	Subjects	Pass Level
1934-35	A.2 Mining	Mine Machinery II* Ventilation & Lighting Electrical Machinery Mine Working *	First Class
1935-36	5th Year Part Time Day	Ventilation & Explosions* Mine Machinery* Mining Mine Surveying* Electrical Engineering	First Class
1936-37	Advanced 2nd	Geology Sinking & Prospecting Mine Machinery* Mine Surveying* Colliery Engineering *	First Class

*distinction

During the above period I was very active in a prize competition run by Thomas Wall & Sons Ltd, Publishers of Wigan, Lancashire, in their bi-weekly publication *The Science & Art of Mining*, an amazing student journal. The competition required submission of written answers to set questions in three interim stages. The best answers were chosen for publication in the journal with the student receiving a postal order prize of 2s. 6d. The Mining Department of Wigan Technical College undertook to mark the submitted answers and allotted marks accordingly, providing also published general comment as to the overall performance and general weaknesses of the material examined. The student submitting the most published answers during the twelve months period in each section received a book prize to the value of £2 10s. The three sections were:

Firemen's Section: The lowest section embracing the requirements of shotfirers, deputies and overmen, and others working for the appropriate qualifications.
Undermanager's Section: Those undertaking studies towards obtaining an undermanager's qualifications.
Colliery Manager's Section: The highest section catering for those preparing for their First Class Certificate of Competency as colliery managers.

The last two often required research and practical investigation. During the summer vacation Dr Penman let me have the keys to the college and library for this purpose. The practical aspects I met by fixing up visits to local collieries at weekends by arrangement with the respective local colliery managers. In the course of three consecutive years, 1935, 1936 and 1937, between the ages of nineteen and twenty-one, I won the three stages together with book awards for each section, achieving a total of 73 published answers (which I still have).

Early in 1934 the Board of Trade amended the Mining Industry Act 1920 to the effect that one could sit for the Second Class Certificate of Competency (Undermanagers Certificate) at the age of twenty-one with the proviso that it would not be issued to successful candidates until they reached their twenty-third birthday. Along with three other persons I sat the examination in November 1935 and was one of the two that passed; I received my Second Class Certificate No 3660 issued on 13 January 1937. Upon reaching my twenty-third birthday the following May I sat the First Class Certificate of Competency Examination (Colliery Manager's Certificate) and again was successful, receiving my final qualification, Certificate No.2996, on 30 July 1937.

One of the statutory requisites concerning the issue of both a First and a Second Class Certificate of Competence was the possession of St John's First Aid Ambulance Certificate (or its Scottish equivalent). Concurrently with the rest of my studies I received training in that field under Herbert Bottomley, Wharncliffe Silkstone Colliery's first aid ambulance superintendent in charge of the ambulance room. Between the ages of sixteen and twenty-four I obtained my First Aid Certificate, 2nd Year Voucher, Silver Medallion and five silver bars and over a period of four years took part in first aid competitions as No.2 member of the Colliery's second ambulance team.

At the age of twenty together with my brother-in-law James Ratcliffe I joined the Workers Educational Society (WEA), a political organisation with a left-wing political bias, and attended classes over a period of three years. The subjects taught embraced economics, capitalism and political dogma, and was run by a tutor formerly victimised for his views and beliefs. Comrade Randle was a mild gentle person and quite a good lecturer. During his victimisation ordeal the local people did much to support him and his family in many ways. The classes were held on a Sunday morning in the Miners Union Hall, Central Street, Hoyland Common, in the large conference room which served as the local cinema for silent films. On Saturday mornings children's matinees of Tom Mix, Lilian Gish, Charlie Chaplin and Jackie Coogan kept us enthralled and in a perpetual state of excitement. Talkies made their appearance with Al Jolson at a Wombwell cinema during the 1930s.

Although I finished the WEA course I couldn't get interested in the political field, and concentrated on my mining studies, taking no further part in subsequent courses.

From 1934 until I got married on Christmas Eve 1938 circumstances dictated that I had to leave home, which was then at No. 57 Springfield Road, Hoyland Common, where I lived with Mum and Dad. My father had been rendered redundant and unemployed from his job as a Lancashire boiler fireman by the introduction of the modern pulverised fuel 'Water Tube' steam raising plant at Wharncliffe Silkstone Colliery. The government of the day introduced a means test whereby unemployment benefit was by taking into account other sources of family income. The effect was that my total earnings as a young underground coalface worker were virtually deducted from Dad's state unemployment benefit.

The upshot on the household budget was catastrophic particularly as at the time Mother was in poor shape and receiving hospital treatment. However, the family rose to the occasion. I left home and went to live with my sister Doris at No.4 Wharncliffe Cottages, Tankersley, Barnsley, some two and a half miles away. What arrangements Mum and Doris came to I never knew. However, it sorted out the problem, although the damage done to Mother was irreparable for we had very close relationship.

Shortly after I left home and went to live with Doris and her family Mother was committed to the Jessop's Hospital, Sheffield for internal operations, eighteen days before Christmas 1935. On Christmas Eve her condition deteriorated such that Doris and I spent some forty-eight hours continuously at her bedside up to her death at about 5.25 p.m. on Boxing Day.

Before she died she told Doris of the steps which she had taken to promote a simple twenty-first birthday party seventeen days later for myself and committed her to follow it through. We were all shattered. She had suffered great pain for some considerable time stoically without complaint, and had been a wonderful mother to us all, including her first grandson William Round, my brother's oldest boy who, as a youngster, went with her on all church going and other outings regularly.

Doris, some eight years my senior, took over the role of mother, at the time being married to Albert Sykes and with two young children, Margaret and Frank. Albert died at the early age of forty-two (two years after I went to live with them) with lung dust congestion contracted at the coalface underground at Cortonwood Colliery, whose authorities denied any liability and made no compensatory payment. In those days dust respiratory disease was difficult to prove and invariably claims were denied.

Doris and Miss Rawlin, the spinster sister of Joshua Rawlin, manager of Wharncliffe Silkstone Colliery and formerly her domestic employer, were now companions and together they ensured we were all well looked after and that we had good food. I recall Miss Rawlin used to make the children and myself a milk pudding, of a different ingredient daily, to augment our meals after coming home from school and work, a situation which prevailed right up to my marriage.

At this point I must introduce my future wife, Mary Greaves, the daughter of a miner who lived close by. She was from a family of five sisters and three brothers and had formerly lived in Ecclesfield near Sheffield, some five miles or so distant. Joe Greaves, her father, walked between Ecclesfield and Wharncliffe Silkstone Colliery, and upon descending the mine shaft walked a further mile and a half underground, a round trip of some twelve to sixteen miles daily for twenty-five years or more. It was only when two of the boys, Arthur and Harry, started work underground at the colliery that he agreed to move nearer to his place of work.

Mary worked as a domestic servant for the colliery surveyor for the princely sum of about 2s. 6d., the going rate at the time. Going to work daily, she used to pass the isolated row of houses of which Wharncliffe Cottages forms an extension. She was just over fifteen years, and I at the time was almost 17½. Our paths crossed regularly, and a few 'good mornings' soon got us on talking terms, followed by an occasional Saturday night out at the Hoyland Kino and a Sunday night walk of about two miles in and about the beautiful Tankersley countryside.

Within twelve months we were both seriously devoted to each other and I took her home to meet Doris with whom she got on with remarkable ease. Although Mary and I tried to see each other as much as possible I had difficulties: those of studying intensely and trying to have as much time with her as possible. Mary, in her practical way, resolved it for me in the following manner. She would call in two or three nights during the week and chat with Doris, leaving me to study as appropriate, and when I had finished for the night I'd walk her home before going to work or retiring to bed.

On Saturday nights she would come to my home at about 5.30 p.m. We would greet each other and then she and Doris would catch a Yorkshire Traction Bus to Barnsley and spend the evening at the Theatre Royal variety show, returning home about 10.30 p.m. With Doris's two young children Margaret and Frank in bed I was able to work without hindrance. This went on for some five years. What a sacrifice Mary made at that time!

However, whenever we could we really did have a break and during her holidays she would bus to Scarborough and I, with her brother Harry, would in the early days cycle the journey of eighty-four miles (and later would pillion ride on his belt-driven two stroke Douglas motorcycle), to stay with her married sister for a few days. Actually we were both staying with Mary's older sister Gertie in August 1937 when Doris sent me a telegram to say I had passed my Colliery Manager's Certificate. Mary was as thrilled as I was. Matters had worked out really well for us both.

We were married at the Old Tankersley Parish Church at 8.30 a.m. on 24 December 1938 and travelled together by bus to Scarborough, spending our three day honeymoon with Gertie and her family, and returning home to get back to work immediately after the colliery's vacational holidays. At that time we had obtained a two-bedroomed cottage, No 40 Cherrytree Street, Elsecar (rented at 10s. per week inclusive of rates) having previously attained underground employment at Earl Fitzwilliam's Elsecar Main Colliery.

Within a year of our marriage, Mary's brother Harry, at the age of twenty-five, suffered a very serious accident at Wharncliffe Silkstone Colliery whilst working at the coalface as a collier. He sustained a fractured spine as the result of being struck by a fall of roof stone, following which he intermittently suffered heavy pain for over forty-five years until he died in his early seventies. He had great difficulty obtaining compensation and after about three years received the sum of £300. Despite his problems he married and, for quite a period, worked at the local steelworks in Ecclesfield, Betty, his wife, carried and sustained him with great devotion for many years. She was and is a very noble lady who, together with her two children by her former husband, gave him great affection. A second son resulted from the marriage. He later developed and still follows a highly successful career in business and has given great support to the family.

On 20 June 1941, John, our first child, was born. We were really happy. Mary was busy helping her younger sister Doris run her father's household some three miles distance at Hoyland Common, their mother having died. Dorothy, our second child, came along on 6 May 1943. I was beginning to make progress as a shotfirer at Elsecar Main Colliery and was teaching at the Hoyland Kirk Balk Evening Institute under the West Riding County Council Authority.

Doris, Mary's youngest sister, married an Elsecar Main Colliery workman, Harry Elliott, engaged in the coal preparation plant as operator and maintenance fitter at weekends. Following the birth of their son Derek, Doris had a post natal relapse such that Mary took the child in her care for upwards of eighteen months. No sooner had the baby been returned to his mother than Gertie had a serious fall through the stair landing and was hospitalised for over two years so that we looked after her young teenage daughter Jean for a further two years.

PHOTOGRAPH I. LIDGETT AND SKIERS SPRING COLLIERIES: OLD DIRT TIPS

This is a view of the Old Lidgett and Skiers Spring Dirt Tip boundaries. The fully tree-cladded area represents a section of the Old Lidgett (Pillbox) Colliery Dirt Tip. In 1926 this tip's surface was fully turned over to depths of 6 feet or more by the miners on strike in the search for loose coal mixed into and tipped with the original pit dirt. After the end of the strike the spoilheap took the form of a grey mini mountain devoid of environmental growth other than very sparse areas of rough grass and gorse shrubs. The current situation represents one of some 70 years or so natural growth.

There is quite a clear line of demarkation between the more recent environmentally clad Skiers Spring Dirt tips and the Old Pillbox Colliery tip, in which the former has its first stage of natural cladding with coarse grass whilst the latter, more advanced, is fully wooded and embraces some of the former bird, animal and insect life originally associated with such situations.

Whilst the above represents natural growth, with the extraction of shallow seams down to 60 feet or more in this particular locality by Opencast Mining operations, the restoration time interval is very much shorter and is normally planned as such. Proper drainage provisions and re-soiling, seeding, and tree planting or replacement within the exploited working areas ensures a speedy return to what are often better environmental and more enjoyable circumstances. Around this particular area and the outskirts of Wentworth and Harley a substantial amount of opencast working has been undertaken, although I doubt that unless told current generations would by looking at the beautiful landscape have any idea such work had formerly occurred.

(OPPOSITE) FIGURE I. HOYLAND COMMON – VILLAGE &
COLLIERY LOCATIONS 1914

Hoyland Common formed part of the larger urban conurbation of Hoyland which also included other mining villages, Platts Common, Hemingfield, Elsecar and Hoyland, the last being the administrative centre within the wider West Riding Authority. During my early days it was part of the Penistone Parlimentary Constituency. On one occasion the seat was held by a Hoyland Common personage.

Consistent with the Co-operative Movement the village had its co-operative store which contributed to the provision of cheaper food during those difficult days. The yearly dividend was an important contribution to the household coffers. Dividends used to run about 1s. 10d. and 2s. 6d. per £ spent at the store.In those early days housewives did their own baking and bought their food, flour and other ingredients at the Co-op. Flour varied in price between 1s. 11d. to 2s. 6d. per stone (14lbs). Yorkshire pudding was an important part of the miner's family diet. It was sustaining and filling, and cheaper than alternatives. Legs of New Zealand lamb often cost less than 2s. each, if I recall correctly.

Tom Avil was the village blacksmith and generally was kept busy repairing farm implements and shoeing farm horses. Doctor Richie, with his surgery located adjacent to the Senior Boys School, was a very important and respected member of the village community. He employed a collector to receive his fees by instalments and took in countless daily visits to patients, together with his morning and evening surgeries. He worked extremely hard. In addition to his normal health care he dealt with colliery accidents other than those in which the injured man was hospitalised at the Sheffield Royal, the Sheffield Infirmary, or Barnsley Beckett Hospital, all of which were run by voluntary contribution.

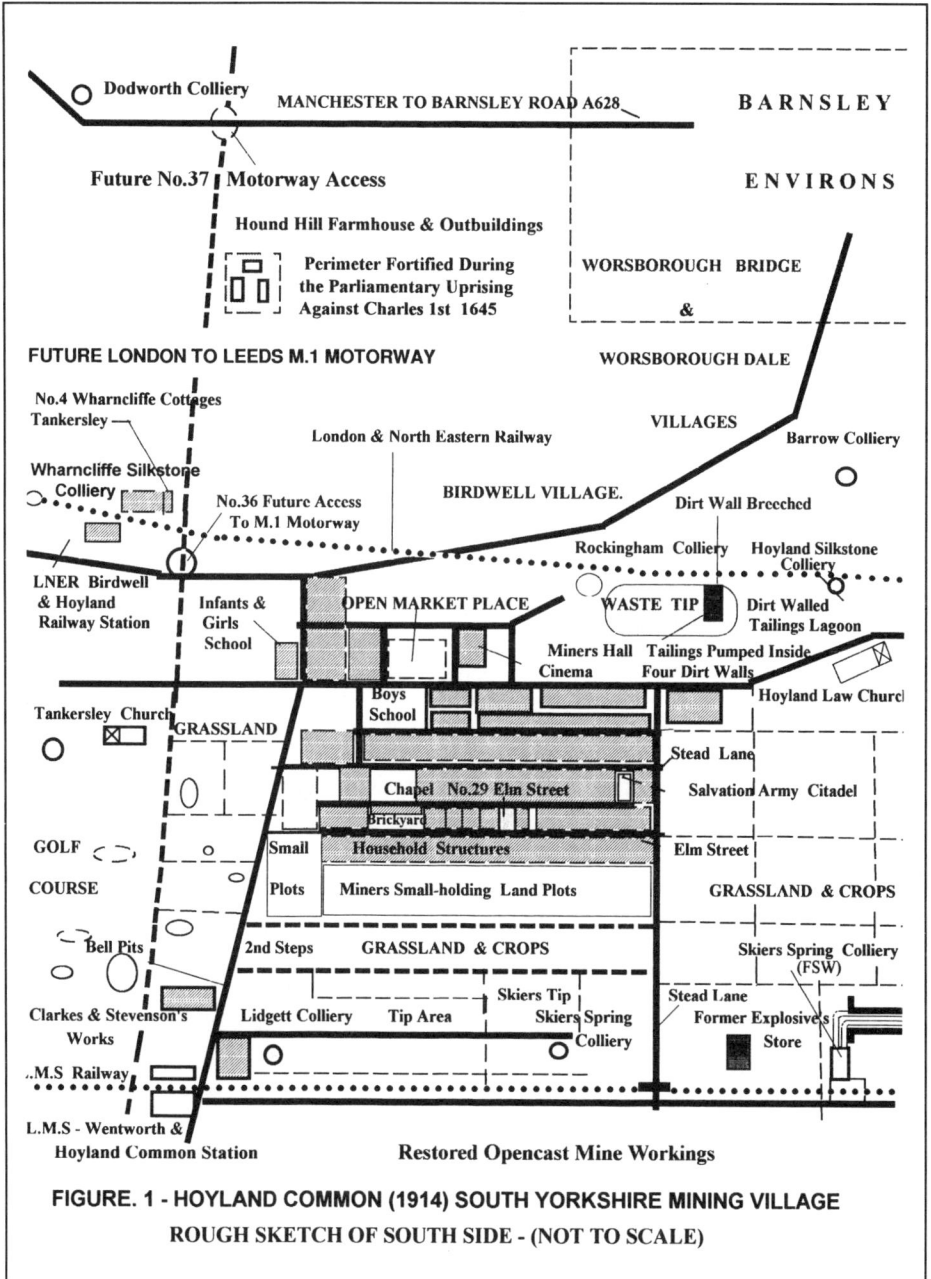

Dodworth Colliery

MANCHESTER TO BARNSLEY ROAD A628

BARNSLEY

Future No.37 Motorway Access

ENVIRONS

Hound Hill Farmhouse & Outbuildings

Perimeter Fortified During
the Parliamentary Uprising
Against Charles 1st 1645

WORSBOROUGH BRIDGE

&

FUTURE LONDON TO LEEDS M.1 MOTORWAY

WORSBOROUGH DALE

No.4 Wharncliffe Cottages
Tankersley

London & North Eastern Railway

VILLAGES

Barrow Colliery

Wharncliffe Silkstone
Colliery

No.36 Future Access
To M.1 Motorway

BIRDWELL VILLAGE.

Dirt Wall Breeched

Rockingham Colliery

Hoyland Silkstone
Colliery

LNER Birdwell
& Hoyland
Railway Station

Infants &
Girls
School

OPEN MARKET PLACE

WASTE TIP

Dirt Walled
Tailings Lagoon

Miners Hall
Cinema

Tailings Pumped Inside
Four Dirt Walls

Boys
School

Hoyland Law Church

Tankersley Church

GRASSLAND

Stead Lane

Chapel No.29 Elm Street

Salvation Army Citadel

Brickyard

GOLF

Small

Household Structures

Elm Street

COURSE

Plots

Miners Small-holding Land Plots

GRASSLAND & CROPS

Bell Pits

2nd Steps

GRASSLAND & CROPS

Skiers Spring Colliery
(FSW)

Skiers Tip

Stead Lane

Clarkes & Stevenson's
Works

Lidgett Colliery

Tip Area

Skiers Spring
Colliery

Former Explosives
Store

..M.S Railway

L.M.S - Wentworth &
Hoyland Common Station

Restored Opencast Mine Workings

FIGURE. 1 - HOYLAND COMMON (1914) SOUTH YORKSHIRE MINING VILLAGE
ROUGH SKETCH OF SOUTH SIDE - (NOT TO SCALE)

(OPPOSITE) FIGURE 2. BELL PITS

A great deal of mining experience was obtained during the working of these early Bell Pits, despite the primitive nature of the operations involved.

One fundamental discovery was that the direction of working had a profound effect upon the miners' efforts in cutting down with the primitive tools of the time – coal, not unlike timber, has a 'grain' – termed 'cleat'. Coal fell from the seam more easily with hand-cutting along the 'cleat' rather than across it. This had quite an effect on the design and layout of the deep mines which replaced the former bell pits.

The long coal faces, of upwards of 1 mile in length, (see FIGURE 4) were arranged to take advantage of this situation. It maximised production for the same degree of human effort. Prior to the introduction of the early coal-cutters to further assist the coal to break down, it was customary for the miners to cut slots at the foot of the seam by hand. The top coal was cut, prized down, or blasted with gunpowder. This was the origin from which the primitive coal-cutting machines were developed.

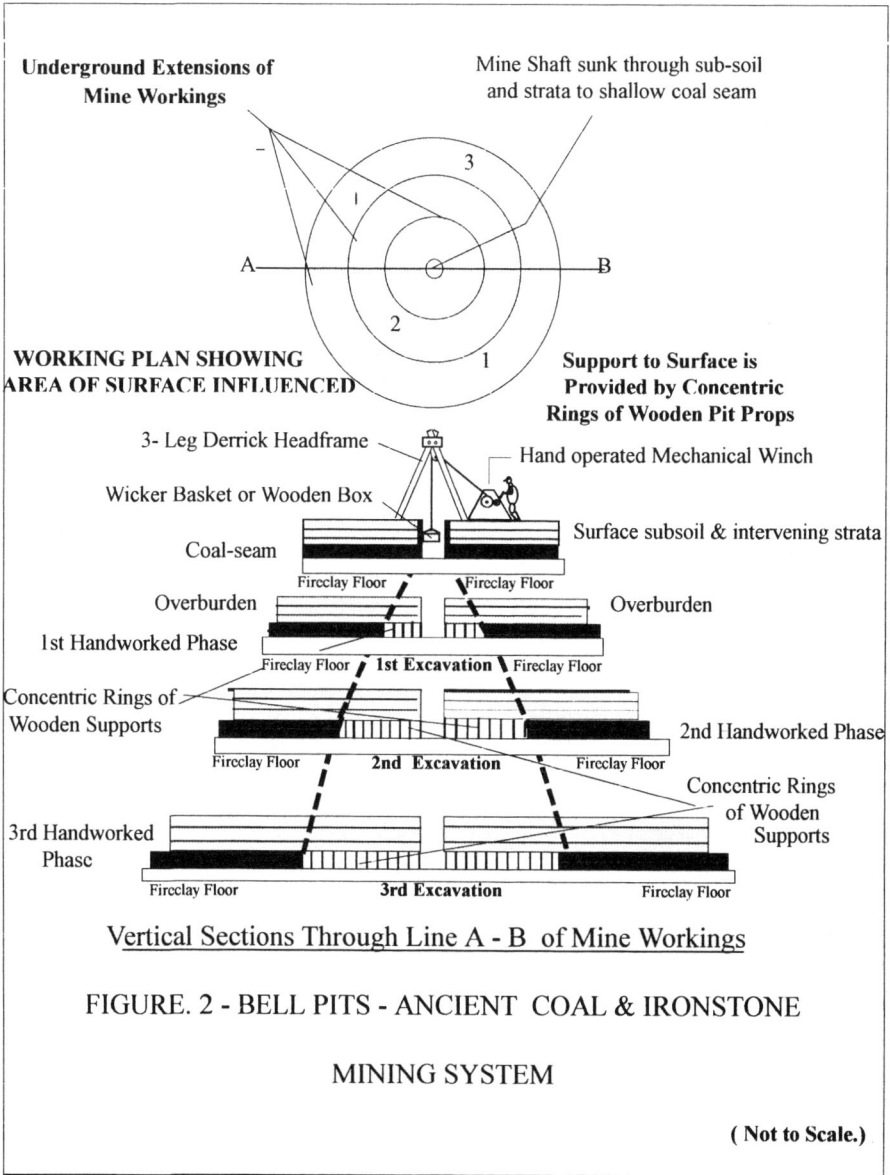

Figure 2 Labels:

Underground Extensions of Mine Workings

Mine Shaft sunk through sub-soil and strata to shallow coal seam

3

1

2

1

A————————B

WORKING PLAN SHOWING AREA OF SURFACE INFLUENCED

Support to Surface is Provided by Concentric Rings of Wooden Pit Props

3- Leg Derrick Headframe

Hand operated Mechanical Winch

Wicker Basket or Wooden Box

Surface subsoil & intervening strata

Coal-seam

Fireclay Floor   Fireclay Floor

Overburden   Overburden

1st Handworked Phase

Fireclay Floor   **1st Excavation**   Fireclay Floor

Concentric Rings of Wooden Supports

2nd Handworked Phase

Fireclay Floor   **2nd Excavation**   Fireclay Floor

Concentric Rings of Wooden Supports

3rd Handworked Phase

Fireclay Floor   **3rd Excavation**   Fireclay Floor

Vertical Sections Through Line A - B of Mine Workings

FIGURE. 2 - BELL PITS - ANCIENT COAL & IRONSTONE

MINING SYSTEM

(Not to Scale.)

PHOTOGRAPH 2.

OLD LIDGETT COLLIERY ENGINEHOUSE (CONVERTED GARAGES)

This view of the Old Lidgett (Pillbox) Colliery Enginehouse takes in the north-west, north-east and south-west corners of the Enginehouses. Two winding engines were installed in the buildings, one for the coal-winding shaft (Intake) the other for the return shaft in which the men were lowered into and raised from the mine. The seam worked originally was the Lidgett Seam, 3'6" in thickness, at a depth of some 287 feet.

The dirt track road to the left upon which the breakdown lorry is standing leads to the Skier's Spring (Newton Chambers) Colliery about 500 yards distant. It is along this track road that an undefined seam of coal some 3'3" in thickness outcropped or was within 10 ft. of the surface and during the miners' 1926 strike was worked illicitly.

Formerly many old bell pits were worked in this particular area, the locations of some being overtipped by dirt from the Lidgett Mine as it worked and developed.

It was one of the very early petrol stations and motor garages set up on the outskirts of Hoyland Common and during its early inception sold ROP Petrol (Russian Oil Products) at 11d. per gallon. This photograph was taken in September 1994, thus the buildings may well have been in operation apart from a short interval for some 120 years.

PHOTOGRAPH 3. UNDEVELOPED VILLAGE SECTION: STEAD LANE, HOYLAND
COMMON, BARNSLEY, SOUTH YORKSHIRE

This shows the current bottom section of Stead Lane approximately a mile from my place
of birth. The rail crossing is the Old LMS Branch Line Sheffield to Barnsley and is
currently in operation. One quarter of a mile through the woods on the east side, Skier's
Spring (Earl Fitzwilliam's Collieries) was located. The Westside Skier's Spring (Newton
Chambers Ltd) was located within the same woods. During our school days and at
weekends as children we spent hours playing and roaming in these woods.

Pigeon fanciers with their 'milers' and 'homers' used to release their birds along this
stretch of road, clocking times and making heavy bets as to the best performance over
the specified distances.

Formerly located with the woodland to the left was an explosives store which fed
detonators and explosives to Skier's Spring Colliery (Newton Chambers Ltd). One winter's
night in the late 1920s, during the early morning hours, it was destroyed, killing the
storeman and frightening a large number of people in our part of the Hoyland Common
and those within the adjacent Milton Village. The noise was horrific and everyone felt
there had been a major colliery explosion. Apparently because of the cold the nitro-
glycerine based explosive had become unstable and had exploded, creating great damage
to the adjacent woodland area.

Early and Actual Mining Experience Underground

Early Experiences

Du
uring our elementary school days as youngsters we obtained a fair degree of mining experience without really appreciating it. Hoyland Common had a long history of very early ironstone and coal production by ancient methods. During the 1926 miners' strike which lasted upwards of a year, modifications of the 'bell pit' mining practice was used by the miners on strike to exploit seam outcrops as a means to obtain the required funds to continue their protest. The police and military constantly patrolling the area shut down these operations as they were discovered.

During the year 1919, Skiers Spring Colliery (Newton Chambers) was under construction with its mine shaft in the process of being sunk to a depth of approximately 1050 feet. When I was about five years old my brother John took me to see the shaft being de-watered, I recall we sat on the spoil heap about eighty yards away, watching the large water barrel being lowered into the shaft, filled and hoisted to the surface where it emptied into a flume which diverted the water into a small stream running through the adjacent woodland (PHOTO-GRAPH 3). Some two or three years later we were able to watch the small 10-cwt. wagons (tubs or corves) carrying stone being hauled by rope to the top of the spoil heap. The contents were tipped and slid down the face of the tip revealing pieces of loose coal mixed with the dirt. These were quickly pounced upon by coal-pickers and loaded into sacks to be sold later at 12*d*. per bag for pocket money.

Half a mile to the north-east of Skiers Spring (Newton Chambers Ltd) Colliery, in a field adjacent to Spring Wood, was the second Skiers Spring (Earl Fitzwilliam's Collieries), Colliery which was mining the rich Barnsley seam from a depth of about 60 to 90 feet. The endless rope system of haulage simultaneously hauled the empty tubs into the workings and the full tubs from the workings to be cleaned and marketed. Often the underground Barnsley seam workings would collapse and a huge cavernous void would appear in the green fields at the surface stretching down to the workings below. In one such incident during the late 1920s, Mr Rupert Simpson, undermanager of the colliery, was killed.

Later in my career I had the task of filling some thirty or more of these large surface voids. These were formed in the fields adjacent to and alongside the Elsecar Reservoir and occurred randomly very many years after the coal had been mined, creating large holes taking up to thirty or forty lorry loads of washery dirt to fill and level. Supervising the task was one Willet Palmer from the village of Milton nearby, formerly a collier, a fine workman and a remarkable man of many talents. In 1950, as manager of Elsecar Main Colliery, I had a serious problem with a blocked water drainage duct at a point where it ran under the waste tip. The duct was under quite a large lake with a maximum depth of about four feet of water. It was mid-winter and very cold. Willet stripped off naked, showing a fine muscular physique of about sixteen stone, and taking a considerable amount of time he dealt with the situation and successfully cleared the obstruction. He really was tough, hard and had no fear in any situation; I respected him both as a person and a workman. The colliery engineer and I would allocate him to special jobs where he could work alone under his own supervision. He appreciated this and would accommodate us in whatever problems we had to deal with.

Whenever I hear the American mining ballard 'Big John', Willet Palmer always crosses my mind. Willet was famed throughout South Yorkshire for his skill at the miners' games of 'Nipsy' and 'Knur and Spell' and for his ability to make the high quality implements associated with both games. Most of the top players would go to his home to discuss their needs and ideas for incorporation into improved designs. In those early days, the game was played over a wide area of South Yorkshire. Not unlike football, leagues were formed and matches played on a systematic basis. Tankersley Park Golf Club had an eighteen-hole golf course attracting company executives from Newton Chambers Ltd. and other middle class members. In those days, golf clubs, not unlike the early miners' tools, had a primitive construction with wooden shafts. Some of these were identified as 'machie niblicks', 'wedge irons' and 'spoons'. Young miners were used as caddies, some of whom became very competent themselves as amateur golfers.

As children during the period of the 1926 strike we attended 'soup kitchens' run by the Hoyland Common Working Men's Club committees. These clubs were great institutions, particularly during periods of industrial hardship. No matter how severe the situation, we as children were provided with a minimum of sustaining food, boots and clothing. The Wharncliffe Silkstone Colliery Band did miles and miles of busking in the unaffected towns and cities, collecting and contributing to the funds of the various Club committees.

With the advent of the solid tyre charabancs, during trips organised annually to Cleethorpes, Scarborough and other places by the Children's Outing Committees of Hoyland Common Working Mens Club, Fitzwilliam Street, we obtained our first view of the sea.

When our fourteenth birthdays arrived we had no fears of going to work down the mine, although our older colleagues already at work fed us with imaginary tales of life underground. Very few of our parents wanted us to go into coalmining and constantly stressed the need to obtain high levels of education and qualifications, such that better employment might be found elsewhere in other walks of life. In those early days there were no pithead baths and, after a shift underground, we had to walk home with dust and dirt permeating our whole body and the clothes we wore.

The old kitchen range with its open top cast iron boiler and miners concessionary coal met our bathing needs, with the heated water being poured into the tin bath placed in front of the kitchen fire. Bathing was far from private with our sisters, neighbours and others often milling around during the operation. With the development of the Yorkshire Traction Bus Services centred on Barnsley in the mid and late 1930s, travelling on the early buses, covered in pit dirt, was always an embarrassing experience. It was not until about the mid 1930s that the Wharncliffe Silkstone Colliery owners provided pit head baths for the men (the underground officials having had them much earlier during the 1920s), and for the privilege of using these baths in the early days we made a weekly payment of eighteen pence.

Although money was short and life in the village was quite tough, there was always plenty of activity. As children we worked on the family's garden allotments; growing vegetables to eke out the low wages, and, for recreation, we joined in with the young miners' sporting enterprises in the street. In the middle of Elm Street there was a sizeable area of open ground which could accommodate football, cricket and other games. This was locally referred to as the 'brick pond', as formerly it had been an old brickyard in which thousands of raw bricks had been stacked and buried. In the actual street area, rounders, similar to American baseball (with no baseball bat), was played extensively by young and old alike.

On the brick pond we had a local game which the young miners played frequently, often in their pit dirt, into the late hours of summer until the light had fallen. This we called 'duckstone' and it had the following format: A building brick was stood vertically on its end about a yard from a boundary wall (in the actual situation this was the back wall of an outside primitive midden toilet) bordering on the ground. On the top face of the building brick a white rounded stone (duck) of about four inches diameter was placed. From a distance of twenty-two yards a stone was thrown in turn to try and displace the duckstone off the building brick (rather like the coconut shy at the travelling fairgrounds, probably from where it was derived). The people who owned these toilets used to get really angry, especially when they were using them at the time the game was being played. In the course of time the constant stone-throwing broke a hole through the midden wall exposing its contents to public view. In such situations the game had to be suspended until the wall repairs were carried out, or played elsewhere.

Actual Mining Experience

On the Friday following my fourteenth birthday I left school to start work underground in the No 1 Pit at Wharncliffe Silkstone Colliery at 5.30 a.m. on the succeeding Monday. I was excited but somewhat nervous, for my older mates, who had been working underground quite some time, as was usual with new starters, painted pictures which weren't too reassuring. I remember very vividly starting work on the Monday morning, and obtaining my personal Check No.884 from the check office, guided by a helpful miner who realised that I was a new starter. I presented my check to the lamproom attendant at an appropriate window who immediately handed me an old battered Davy oil lamp of the same number. Struggling with my snap-tin (containing four slices of bread and lard and a two-pint bottle of cold water) I was helped across the colliery yard to the steps leading to the No.1 Pit bank. Men were in orderly queues awaiting their turn to enter the cage which held sixteen men top and bottom deck, thirty-two in all, in order to descend either to the mid (East Fenton) landing or the pit bottom Warrendyke landing. At an appropriate time I was chaperoned along with the East Fenton consignment into the pit cage. The loud noises as the safety jacks holding the cage were withdrawn, together with ringing bells indicating to the winding engine man the projected destination of the wind, contracted my nerves. I was in near state of panic. The cage slightly lifted and suddenly dropped and my heart felt as though it was moving upwards ready to burst through my head. Just as suddenly I felt the floor of the cage pressing upwards against my feet and my composure returned. It was somewhat frightening on this first occasion of descent into the mine, but I never again had the same experience. Ascending the shaft at the end of the shift was quite pleasantly contrasting.

As a matter of historic interest, the No.1 winding engine was a very early 'vertical single cylinder, 2' 9" diameter, stroke 5' 6", and was mounted below the winding rope drum. When it was installed, the valves (inlet and exhaust) were hand operated with a mechanised valve operation being designed and installed later.

Stepping off the cage into the light of the shaftside landing we moved forward about a hundred yards into pitch darkness to a point where standing alongside a huge old boiler type steam receiver was the official - Mr Fred Earnshaw. He was the day deputy who examined each safety lamp and then instructed its owner as to his work station. My turn came.

'Stand there, lad - wait for Sammy Marsden!'

Who Sammy Marsden was I hadn't the slightest idea. Underground officials in those days were however important persons and were given a very high degree of respect much of which has been eroded over the years. Fifteen minutes passed before Sammy arrived. The official turned to Sammy.

'Take this lad and help him gear up a pony for 9s North Dip.'

Intrigued with mixed feelings, dutifully with Sammy at my side, I walked further along the road to some stables and met the horsekeeper.

'I'm taking Duke,' said Sammy. 'What have you got for my mate?'

The horse keeper looked at me and said, 'You're the new lad, aren't you? You'd better take Peter the donkey; show him what to do, Sammy.'

Sammy Marsden, having initiated me into the mysteries of equipping the animal with sling gear harness and a pulling chain for working duty, we set off walking to our place of work some one and a half miles into the mine, with my colleague chatting and explaining along the way. He was quite reassuring and we became instant mates.

After about twenty-five minutes walking, and sometimes near crouching, we arrived at 9s North, a level roadway into which rise-benks and dip-benks merged (each benk being a working place for two pick and shovel miners who broke down the undercut coal by hand and loaded it into 7 cwt. capacity tubs). Sammy worked the level, taking in empty tubs and pulling out full ones to and from a point at which they were hauled away to the pit bottom, for hoisting up the shaft by a direct rope system. Sammy handed me over to my colliers, father and son respectively, a common arrangement, the latter having a loaded tub at the bottom of the ninety yard 1 in 14 dipping roadway. He showed me how to haul it away to the level and bring back a replacement empty, how to handle my oil-lamp carefully so as not to lose the light: 'getting it in the dark', he called it. He then instructed me how to lift the tubs back onto the track after they had become derailed which was quite often since the track was of a temporary construction. Ten minutes or so covered my tuition and initiation. It is but little wonder that I had three lamps in the dark and four full derailments in the first two hours. The colliers' tempers began to rise, as they were losing potential wages. I had difficulty in avoiding a crying session. However, my colliers were tolerant and patient that first day and I managed to get through the shift without further incident. The following days were much the same. One soon learns to adapt after having one's ear cuffed by the collier a few times.

PHOTOGRAPH 4: of No.1 Pit, Wharncliffe Silkstone Colliery, features the actual stables from which I first took out Peter the donkey on the first morning of my working life in January 1928. These stables held about forty ponies and were quite warm and spacious with the ponies being well supervised by a head stableman on the day shift and an assistant stableman on the afternoon and night shifts.

The animals generally were well treated by drivers and miners alike. Each pony had its own individuality. Some would work giving everything they could; others could be quite stubborn and, in odd cases, vicious. One pony called Fly was a legend of the latter class and could only be handled by one particular driver. Together they worked in the Warrendyke Fenton, at a lower level than ours. On one occasion the managing director's son, F. Colin Swallow (a Brough Superior 500 cc motor cycle enthusiast at the time) a mining graduate on a visit underground, was rather badly roughed up by the pony Fly as he tried to pass it in the mine roadway. Some time later it was on its way to the knacker's yard, being considered a risk and danger to the workmen. From time to time a pony would get killed, usually by being struck by runaway tubs or trains. We always found such occasions to be particularly sad and disturbing.

PHOTOGRAPH 5 depicts a typical pony drivers' situation underground, showing part of the gate road leading into a benk, or working place. Ventilation was usually weak in these benk roads; it was therefore warm and we sweated very readily.

Note that the lad wears no safety helmet or modern safety boots. We usually had wooden clogs with the sole and heel reinforced for extended wear by special irons nailed on, replaced when badly worn.

PHOTOGRAPH 6 shows the very early type of machine then used for under-cutting the coal seam. This was a two-cylinder CA (compressed air) longwall coal-cutting machine designed by John Gillot & Son, Barnsley, in the 1850s. An electrical direct current (DC) version was built by Clarke & Stevenson's at its Hoyland Common factory adjacent to the old Pillbox Colliery and the old LMS Wentworth & Hoyland Common railway station.

My brother John Round operated one such machine for over five years and undertook his own electrical and mechanical maintenance before being appointed a junior official underground.

PHOTOGRAPH 7 shows an alternative type of chain machine. This machine was an improvement on the earlier 'disk type' coal cutters, and may be considered as of being at the early development stage of the modern machines. Here the machine is working through a hand worked stall. The steel screw-jack sprag, which anchors the haulage rope, is visible in the bottom LH Corner, the arrangements being such that the steel haulage rope is maintained deflected along the foot of the seam. The steel box situated on the front bedplate carries some 50 cutter picks which were sharpened daily.

PHOTOGRAPH 8 shows the site, location and present condition of the former Clarke & Stevenson's coalcutting machine manufacturing adjacent to the old LMS Wentworth & Hoyland Common railway station. These early machines were considered to be robust and well designed, and quite a number were built. However, for more than seventy years this factory has been turning out nuts and bolts of all types, with the newer buildings associated with that process. The works were formerly built on a flattened dirt tip subsequently extensively over-grown with gorse and rough grasses, but is now a heavily tree clad area and not so readily accessible as formerly.

FIGURE 3 shows the kind of slot cut in the foot of the seam. Here the slot is actually cut in the seam and gummings are fine coal, but in a great many situations the slot was made immediately below the floor of the seam and dirt gummings were disposed into the void (goaf) behind the face supports from which the coal was previously extracted. Cutting these slots in conjunction with the use of explosives made the coal much easier to work which contributed to increased production.

The working layout of the East Fenton Seam was somewhat similar to that shown in FIGURE 4, with the north and south side faces being some 1,500 yds in length. These long faces were split into 33-yard working places in which two miners carried out all operations such as getting the coal, building stone packs, constructing the benk roadways, laying tub track, etc. The speed of mining was comparatively slow in those early days. Within a couple of weeks I was working as well as the rest of the pony drivers, and in a very short time feeding my colliers with tubs equal to the best of them. I soon found there were no inhibitions of language and everyone was treated equally, although out of the mine the process reverted to normal. Things were hard for everyone in those early days of my career but one grows fast in such situations.

After a few weeks I was transferred to pony driving for three development headings opening up the coal reserves for future working. Joe Cooke of High Green and Harry Rogers of Pilley (he, like my former teacher, was a good violinist) were the two men I worked for. The process of driving the coal-heading was mechanised with a Jeffrey heading machine, which undercut the face of the heading to a depth of three feet. Following blasting the men would fill-out the coal and the pony's job was to haul the coal up to a rope haulage pick-up point about ninety yards maximum from the face of the heading. I remember one incident during the time I was working with these two miners which involved the primitive electrical arrangements. In those days, the electric switches were simple on/off provisions with electrical overloads being catered for by bell wire fuses all housed in a cast-iron box. Joe couldn't get the heading machine to start. He went to the switch box to examine the fuse, came back and tried to start up the machine again, but there was still no response. He then did a very foolish thing, pushing his finger into the power plug socket. He was suddenly ejected across the face of the heading, having been electrocuted. Harry Rogers immediately applied artificial respiration, and within a minute or so, to our great relief, he came round, his hand being badly burned. Having had his hand dressed he continued his work and never lost a shift as a result of the occurrence. When asked why he had acted the way he did, we were told that he thought the plug sockets were filled with dust preventing proper electrical contact. Since that time electrical equipment and the safety of such apparatus has made tremendous strides, so that it is impossible for such an incident to occur in the manner described.

After twelve months pony driving I was moved up to 'rope runner' feeding and clearing all the levels. This involved collecting a train of thirty empty tubs which ran down a gradient of about 1 in 12 dragging the rope by gravity, the full train of thirty tubs being hauled back at about 6 m.p.h. against gravity by a single drum mechanical haulage driven electrically.

Shortly after taking over my new haulage job the system of mining was changed; the benk system was discarded and a concentrated advanced longwall face with the machine cut coal handloaded onto face conveyors was introduced. In simple terms this meant that the same output from approximately 2,000 yards of benk coalface with an average advance of six inches per day could be produced from some 200 yards of machine mined coalface advancing at the rate of five feet per twenty-four hours. Manpower savings were substantial and efficiency increased with the costs of production being greatly reduced.

To feed this system required fifteen trips (3,000 yards per trip), each shift covering a total distance of upwards of twenty-six miles. Riding the trip in either direction was prohibited since the height of the roadway varied from between five and six feet and to do so would have been dangerous. Tub hook and links shackle connections occasionally broke with resultant train runaways. One could have one's head well above the top of the tubs and have it strike low roof areas, or one could slip whilst trying to jump on or off the train and be dragged underneath and run over. To work the arrangement as required one would need to have to run in front of the train at upwards of 8 m.p.h. and 6 m.p.h. on the inbye and outbye journeys respectively. Under the circumstances we rode the journey in both directions on the rope shackle one way and between the tub-shackles the other, without any thought of danger. On more than one occasion during my retirement I have vividly relived this situational experience in a nightmare and awakened in a dread sweat.

From a haulage hand I later graduated to the coal face, working as a coalcutting machine assistant clearing the gummings (fine coal or dirt) from the undercut (four to six inch slot cut to a depth of five feet at the bottom of the seam); conveyor shifter (advancing the face conveyors following coal clearance); and collier loading twelve to fifteen tons of coal onto the face conveyor per shift. For a period of some four months I was a member of a ripping (roadway construction) team, which was the heaviest of all coal face occupations.

In November 1936, for the purpose of gaining wider experience, I found employment at Grange Colliery (Newton Chambers Ltd) on the outskirts of Rotherham. This experience was unique to me, in that I became a member of a 'butty-team' undertaking preparation work on machine conveyor faces on the afternoon and night shift. The butty system involved a single contractor taking responsibility for completing a prescribed volume of mining work, getting a team together and paying his men wages at different wage levels according to his agreement with the individual. It had proved in the past a pernicious system and was supposedly banned. I never met the contractor but was a member of his sixteen man team from the rough areas of Rotherham. They were quite a wild bunch of men but I got on very well with them.

The surface drift entrance to the coal mine was situated within the Roughwood woodland area. On the Fridays (following the men receiving their wages) it was not uncommon for some of them to gamble away their week's earnings in coin tossing schools. Such things as radios and watches were easy to come by at very low rates providing one did not question the origin. I was quite happy being paid fifteen shillings per shift which was quite a fair sum in those days. I stuck with them for four months during which time I did a week-end ventilation survey for the management but finally had to leave because heavy weekend shifts seriously interfered with my ability to study. The undermanager during my employment at the colliery was Cyril Dickenson, a Hoyland Common native who was formerly a workman. The former manager of the colliery was Johnny Longden, probably one of the most unique men of his time within the industry. He is referred to later within this narrative.

Moving on I found work at Elsecar Main Colliery (Earl Fitzwilliam's Collieries), firstly as an underground corporal in charge of a number of haulage hands in the East Plane 9's District Parkgate Seam and later progressing to a backshift junior official (deputy or fireman). Here the mining system was exclusively pick and shovel hand-loading stalls, two colliers per stall and a trammer. The Silkstone Seam was being worked by mechanised mining of the type applied at both Wharncliffe Silkstone and Grange Collieries with compressed air used extensively for the machinery and pneumatic picks. At weekends the job I had was such as to enable me to join the fitting staff and ropemen on maintenance and other work. It gave me a great opportunity to get a much wider insight into the colliery operations and organisation. The standards at the colliery were most immaculate and rigidly maintained as such by the undermanager Albert Naylor, a strict disciplinarian but highly respected by all throughout the colliery.

Whilst employed at the colliery in 1937, I obtained my First Class Certificate, which in the early stages had its disadvantages. I had become the only person at the colliery other than the colliery manager to be so qualified. At the age of twenty-three I was somewhat naive, exposed and certainly lacking political acumen, being motivated at all times only with the practicalities of trying to do each job efficiently rather than with personal aggrandizement.

About November 1937 the R.G. Baker Colliery manager left me a written instruction to meet him in his colliery office. Accordingly I went. He opened up by saying,

'Round, I understand you have recently obtained your First Class Colliery Manager's Certificate. What are you going to do with it?'

'Hopefully sir, I'd like to be able to attain a position such as yours', I replied.

He smiled and commented, 'And how do you propose to achieve that?'

My response was, 'By trying to be good at everything I am asked to do.'

Again he smiled, remarking, 'Then we will have to see what can be done. I'll get in touch with you later,' and dismissed me.

Such was the beginning of what later worked out to be a fine future and an inspiring relationship. In later years, whilst working under Donald Severn, the Area General Manager No.4 (Huthwaite), East Midlands Area, my former manager met him travelling on a train to London. R.G. Baker told him of the incident in the following manner.

'How is Charles Round faring? I recall quite a long time ago being told he had obtained his Colliery Manager's Certificate at the age of twenty-three, at a time when a number of my university acquaintances had failed the examination. I thought I'd have a look at the lad and called him to my office. What a rough diamond he appeared at the time. He needed a start which I was glad to give him, following which he did very well for me and never let me down in any way.'

Within three months of the meeting with R.G. Baker my status was raised to that of a 'shotfirer' working under Charlie Swift, deputy in charge of the Silkstone Seam 14's unit preparation shift. The coal was undercut - slotted in the manner already referred to – and my job along with a colleague was to blast the coal by explosives so that the colliers could load it onto the face conveyors more easily. Usually it took about eighty shots for the whole face, the shot holes being loaded with about 6 oz. of ammonium nitrate-based explosives in order to prepare a coal face of some 180 yards in length with an advance of about 5 ft. 6 ins. per twenty-four hours. Some twelve months after the above appointment new workings were opened up in the Thorncliffe Seam which was about 3ft. 3ins. thickness, the coal being extracted by the above prevailing mining techniques of the day. I was later appointed preparation shift deputy working under the senior day deputy Frank Knowles, an excellent practical official with wide experience and an ability to apply it. He had ambition which didn't go down too well with his team-mates. Even seasoned colleagues found him difficult to work with. I was now to get the feel of real pressure and responsibility with little or no support.

At this point it is necessary to re-introduce Albert Naylor, the undermanager. Mr Naylor was born, I believe, in Flockton, West Yorkshire and at the time was in his mid fifties. He had lost an eye in an accident many years previously. Mrs Naylor had been permanently ill in a nursing home for very long time. He visited her every Tuesday afternoon for over thirty years and never missed an occasion. They had two sons, one of whom became the colliery surveyor and later died at a premature age; the other, Arthur Naylor, became a very highly qualified medical at the Bradford Royal Orthopaedic Hospital and produced a number of medical books and treatises. For a number of years Mr Naylor worked under R.E. Horrox, the colliery manager whose preoccupation with Renishaw Park Colliery (which he owned), near Eckington, Derbyshire, virtually left Albert in charge of the colliery for very long and continuous periods.

During his working life he had a routine which was strictly followed. At 5.20 a.m. he descended the mine. Slowly he walked through the pit bottom surveying the junior officials (deputies), checking the men's safety lamps and instructing them as to their working situations. The deputies positioned themselves at fixed stations on either side of the roadway leading out of the No.1 Pit bottom. Where he found need to comment or chastise the men or their officials he did so but usually everything was comparatively quiet and orderly. At about 5.50 a.m. he retired to the underground office (boxhole). At 6.00 a.m. precisely the No.1 Shaft would start winding coal continuously without hindrance until the end of the day shift at 1.30 p.m. and later the afternoon shift at 10.00 p.m.

The underground office (boxhole) had two levels: the upper level a small office in which time and piecework rates were worked out, and the large lower level one with two very large tables upon which the deputies from the previous shift made out their reports and requests for materials. Often one received cutting comment at the lower office level, usually with reference to shortfalls relative to the visit the undermanager or an overman had made the previous day to a working district. If one was summoned up into the top office, one was really in trouble.

After being dismissed to ascend the shaft at the end of one's shift one never felt at ease. On many occasions whilst under the bathroom showers a telephone message would come from below instructing the recipient to stay behind until Mr Naylor came out of the pit to go home for breakfast - he'd overlooked something invariably to the detriment of the official waiting to see him. Between 9.00 and 10.00 a.m. he would interview potential employees and deal with the NUM trade union officials. Following the morning session he would go back down the pit and make an inspection of one of the many producing districts in the Parkgate, Silkstone and Thorncliffe coal seams. He claimed, and we believed, that his examinations took in more with his one eye than any of us did with our two eyes!

Even following this we were still under stress. As preparation night shift deputy, on arrival home, after a meal, we would go straight to bed at about 8.00 a.m., get up about 2.00 p.m., go to a good vantage point where we could see the colliers returning from the day or filling shift and ask the first one to be seen:

'Have you filled off?'

If the answer was 'yes', immediate relief; if the answer was 'no', the next question was:

'What was the problem and how many tubs had been filled before you ran up against it?'

If this answer was over 100 tubs (usually a fill-off meant about 500 tubs or 450 tons), again relief, but if not, the rest of the time that day was anticipating Mr Naylor's reaction the following morning. On one occasion I was taken up into the top office having fallen out with Frank Knowles over the sort of situation described above - but one in which 187 tubs had been filled before the conveyor belt broke and the fill-off became late. He reported me to the undermanager on the grounds that I hadn't inspected the conveyor-belt joints. After a dressing down by Mr Naylor, I was sent to meet the colliery manager, who suspended me for two days.

Later Frank and I became quite good friends as for quite some time he was engaged in a correspondence course studying for his Second Class Certificate during which I helped him considerably. Unfortunately he did not complete the course but he would have made an excellent undermanager. He was a great help to me a few years later however.

I had been a deputy for about three years when R.G. Baker again interviewed me. He had a number of things he wanted undertaking, including:

a) introducing steel face props on a big scale in replacement of the wooden ones then being used;

b) organising the dust sampling procedures in accordance with Statutory Regulations;

c) setting up a training scheme for new entrants and faceworkers;

d) sorting out ventilation problems;

e) stock control;

f) developing mechanisation, new methods and recovering manpower;

g) safety inspections, accident investigations;

h) meeting special assignments and setting up schemes for machine and equipment maintenance.

He asked me to undertake these for him. Thus I became assistant to the manager - not assistant manager. There was no increase of my basic deputy's wages, but breaking into upper management was more important. At the time Mr Naylor didn't favour this move and on one occasion the senior overman, Joe Machin (with whom he was very close), told me there was no room for schoolboys at the colliery and that I ought to apply my talents somewhere else. It took about two years to win over the undermanager into acceptance of the situation. The first major contribution towards this resulted from a serious problem of ventilation which severely constrained the development of the new Thorncliffe Seam, ventilated at the time by about 20,000 cu.ft. of air per minute. One weekend, together with a couple of men, I spent thirty-six consecutive hours carrying out a detailed ventilation quantity and pressure survey, analysing and committing to record all the problems encountered which were contributory to the unsatisfactory situation. The colliery manager accepted many of the recommendations in my report, and it was agreed to drive 2,000 yards of coal headings (roadways) to provide parallel air intake roadways within the ventilation circuit. In a little over twelve months the ventilation was progressively increased to 49,000 cubic feet of air per minute, and on one occasion, in an unguarded moment, Mr Naylor called me aside and said in his West Yorkshire accent,

'That Thorncliffe ventilation job, lad - tha's done a grand job.'

That was really something for very rarely did he make such comment, no matter what the situation. Things eased from then on as I extended my duties by taking interest in and initiating action to save manpower. In doing so I received great encouragement and support from my colliery manager. By partially mechanising a number of loading points and haulage terminals, very shortly, with simple applications of existing equipment and modifications more than forty-five men were saved. On one occasion Mr Naylor called me aside, saying,

'Charlie - Jack Bedford's having a difficult time on Silkstone North 3; he's got a 2 ft. 6 ins. fault on the face and can't get his conveyor to run. We're losing tonnage. Go down to him, lad, and see what you can do to help him.'

I immediately changed into my pit clothes and went down onto the district. Jack, the day deputy, was really in trouble. It wasn't difficult to determine the nature of his problem, in that there was a need to have some form of belt-trapping arrangement built into the face-conveyor to prevent the top carrying belt rising out of the conveyor structure and tipping off all the coal that came from the face below the lower side of the geological fault. Staying over until nearly midnight, I designed a simple two roller mangle device which fitted into the conveyor structure in the vicinity of the geological fault. The fitting shop mechanics were able to start its construction before I went home. In a little over twenty-four hours it was sent into the district and I fitted it into the conveyor. Staying over to assess its working potential, I awaited the colliers having tested the installation satisfactorily with an empty belt. The first reaction from the collier in whose stint the device was installed:

'Chas – what's this organ? Where's the monkey?' he asked.

'It's just come,' I replied.

After a small amount of lateral (sideways) adjustment to get the belt running central on the rollers, the belt was adjusted and ran quite well throughout the shift and thereafter, with the face being filled off without undue trouble that day and those following. FIGURE 5 gives an outline of the device *in situ*, indicating how the coal was transferred from the lower to the upper side of the fault disturbance.

When any new item of equipment or machinery was installed at the colliery the men invariably applied a local reference. On one occasion we had imported a new stone loading mechanical shovel, manufactured in Salt Lake City, Utah, USA. Within twenty-four hours it was known throughout the pit as the 'Mormon shovel' and retained that title throughout the whole of its use.

Another request made of me by Mr Naylor was, because of the potential shortage of junior officials (deputies and shotfirers). Could I urgently train selected men to get the appropriate qualification certificates?

'How long will it take you?' he asked.

'Give me a fortnight!' I replied.

'You'll have some lads on Monday. I'll give you a list later,' was his comment, seeming quite pleased with my approach.

On the following Monday morning I had seventeen candidates. Basically the Deputies Examination called for a candidate to be able to read down to ½% gas caps on the top of a reduced flame of a spirit safety lamp; to measure the velocity of air with an anemometer or smoke injected into the air-current flowing through a mine roadway; and to determine the quantity of air flowing by calculating the area of the particular roadway shape. Good sight and hearing were important requirements.

The men he had chosen were amongst the best he had and were in their mid-thirties, having a varying background of mining skills. For two solid weeks, eight hours a day, I really drilled these men. I let them measure air underground, made them carry out checks for firedamp in the air currents and in cavities, and taught them simple mensuration and air-quantity calculations. They responded well. A fortnight later they took the examination at Mexborough Schofield Technical Institute and all of them passed at the first attempt. Within a very short time they were all appointed as junior officials. Every five years all such officials were subjected to a compulsory medical sight and hearing test or had to take a re-examination.

A development to this was when I ultimately inherited these men and others I had also trained at the Kirk Balk Hoyland Evening Institute. After misdemeanours or serious omissions, I would usually finish by saying, 'You are not wholly to blame. I trained you myself!'

I never had any rancour; moreover, they made good officials: quite a tribute to Albert Naylor's assessment.

At the time a surface drift (tunnel between the surface and coal seam) was being driven down to the Moor Haigh Seam which was about 4ft. 8ins. thick. R.G. Baker took me on one side and showed me his projected ideas as to how he proposed to work the seam by a mining system he termed 'retreat buttocks': essentially a continuous mining approach which embraced the technique of retreat mining.

FIGURE 6 represents the proposals which were put into effect. The basic concept behind them was to establish early continuous mining as opposed to the cyclic mining systems which had been developed on the longwall mechanised faces, in which coal-filling on one shift was followed by two shifts necessary preparation work, i.e. cutting the faces, moving over the conveyors, blasting the coal, constructing the roadways etc.

Two pieces of plant were required which we did not have: a shuttle car coal transport, and a simple belt drive for the thirty yard conveyor. R.G. Baker sent me along to Sanderson & Newboulds, gear manufacturers of Sheffield, to look at their heliocentric-drive in which the gears were installed internally in a drum of 30 inches diameter (very compact and ideal for what he had in mind). I designed a conveyor drive round this unit until I had it exactly as my colliery manager wanted. On being put to work it functioned perfectly and was quite satisfactory for a long time, but finally succumbed to the adverse working conditions.

PHOTOGRAPHS 9A AND 9B are views of the finished design which was put to use. The belt wrap which had something over 270 degrees of driving drum contact is indicated by the white arrowed line.

However, dust and the rough environment ultimately had their toll. We eventually had to replace the unit with the standard BJD R-9 drive, in use on the normal faces. It was like asking a precision Swiss watch to carry out the duties of Big Ben. The shuttle car was to run on tracks and carry a load of 30 cwts., with a belt discharge fitted and a chain driven back axle provided for transit between the coal heading face and the conveyor pick-up point, using the compressed-air provisions for the Siskol Heading machines as a source of power. Such a unit was designed and built, but proved to be underpowered for the duties required.

At the time Maurice V. Kelley was the Chief Engineer at the colliery, an excellent skilled and widely experienced person. He produced an alternative design which was trackless and incorporated a Jowett 8 motor car gear box. Whilst it worked a little better it was not the answer, so that we finished up with simple mine tub and assisted tipping arrangements.

With the ensuing appointment of R.G. Baker to assist and subsequently take over the duties of managing director of Earl Fitzwilliam's Collieries, A.G. Douthwaite was appointed manager of Elsecar Main Colliery and until 31 December 1946 I continued to carry out my duties as set out. Several months before I had been given the task of developing ideas for a number of reorganisation schemes for both of Earl Fitzwilliam's collieries, Elsecar Main and New Stubbin. Nationalisation of the collieries was pending and all the private companies were seeking to present their assets in the best possible light with the object of maximising their potential compensation. My application to this task was with the greatest of enthusiasm. I hadn't the slightest idea how I would be affected by the approaching change and wanted to show that I was fit to be considered for any opening that might arise. Mary, occupied fully with our two children John and Dorothy, aged four and two respectively, and her sister's young baby Derek, fully understood and encouraged me to augment my efforts by working long hours at home which I normally did in any case.

Having produced a number of alternative schemes for both collieries I submitted them to Mr R.G. Baker who considered them quite impressive. What was done with the schemes I do not know but when compensation was paid to the private owners I received a nice cheque with a letter of thanks. I had never even thought of reward, having spent some company time for which I was paid in the process. What was more important lay in the fact that my mentor came up to see me at the colliery in late November 1946. We spent about one hour together when he gave me a summary of what was to happen on 1 January 1947, nationalisation's Vesting Day. He was to become sub area manager in control of a number of pits including Elsecar Main, New Stubbin, Cortonwood, Aldwarke, Rotherham Main and Silverwood. Mr H.S Haslam, formerly General Manager of John Brown's (Silverwood Collieries) Ltd., was to be the Area General Manager for the Yorkshire Division's No.3 (Rotherham) Area; Mr G.C. Payne, former manager of Manvers Main Colliery, was to be sub Area Manager in control of Manvers Main, Barnborough, Kilnhurst, Wath Main, Denaby and Cadeby Collieries. Then he turned to me and said,

'Now, Charles, how and where are we going to fit you in?'

I was too bewildered to reply. He noticed and continued,

'How would you like to become manager of Rotherham Main Colliery? My uncle formerly managed it and it was my first managerial appointment.'

'Very much. I believe I'm dreaming, Sir,' I managed to respond.

He continued, 'Charles, I want you to go there now and to be appointed before Vesting Date. Go straight to Aldwarke Main Colliery Offices and ask for Mr Reg Lees. I've told him to expect you; he is fully aware of the situation.'

I hadn't a car at the time but I am sure that had there been no transport I could have flown the journey. Having thanked him the best way possible in the circumstances I left and caught the Yorkshire Traction bus to Rotherham where I changed to the trolley bus for Rawmarsh and walked the remaining distance to the offices. Duly I met Mr Reg Lees, a mining agent of the John Brown's Company. He told me the terms of my appointment and the nature of the job I was about to undertake. I signed the prepared contract without hesitation. It was thus that I was transferred with Rotherham Main Colliery into the new scheme of things on 1 January 1947.

PHOTOGRAPH 4. UNDERGROUND NO.1 PIT PONY STABLES, WHARNCLIFFE
SILKSTONE COLLIERY

These stables were in the East Fenton seam and were located about 400 yards from No.1
Shaft Inset. Fresh water was piped in from the surface both for the ponies' use and for
the daily cleaning down of the pony stalls and open spaces. Each day, as pony drivers at
the end of each working shift, we had to brush each pony and wash its legs. Pony feed
was sent in daily and the horse manure cleaned out continuously throughout each day,
being removed to the surface in mine tubs. Pony drivers were allotted an individual pony
with which close attachments developed. Each animal worked a single shift of 8 hours,
but occasionally they were called out to work overtime up to 8 hours duration. In later
years the ponies were taken out of the pits during strikes and holiday vacations.

No two animals were alike. Each pony had its own personality. In those early handcut
coal days they contributed greatly to the working of a mine. Following the nationalisation
of the mines the vast majority were withdrawn from the collieries although the last ones
to be withdrawn were at Ellington Colliery, Northumberland in the early 1990s. There
they were working alongside the most modern of American production machines, dragging
equipment and materials (such as roof supports and other supplies) into the working
places.

PHOTOGRAPH 5. PONY DRIVER, COLLIER AND WORKING STALL

This depicts a typical pony drivers' situation in the mine at the time I started working underground, and shows part of the gate road leading into a benk or stall. The coal tub probably has a capacity of 10 cwts and runs on a 2′ 0″ track. The pony is dressed exactly as on my first day below ground. Note that substantial pulling chain. Here the loaded tub is at a junction leading into an adjacent collier's gate roadway. The wooden sprag below the nearside tub wheel is holding the tub at rest. On gradients favouring either full or empty tubs it was customary to use lockers, 12 to 15 inch lengths of wood about 1.75 inches diameter. These were inserted between the spokes of the tub wheels, preventing rotation, causing the tub to slide thus increasing friction which controlled the rate of motion. Double lockering provided for locker insertion in both front and back wheels.

The oil lamps provided about 1 candlepower of illumination and were completely sealed. They were the forerunners of electric portable lamps which gave about 2.5 candlepower. The introduction of the cap lamp with its battery carried on the miner's belt was a great boon. It ensured both arms were free and the light was always directed onto the object upon which the eyes were focused. Note the absence of roadside support. Generally the benk roadways were supported by wooden bars (beams) carried at roof levels by slots cut into the roadway sides.

Although at the age of fourteen years we were not quite so physically well built as the youngster featured we quickly learned how to use wooden levers plus the pony's strength to deal with both full and empty tub derailments and only in the most dire situation would we seek the help of our colliers. Ventilation was usually weak in these benk roads so that it was warm and we sweated very easily. Thirst in those early days was quite a problem for many of us – this was overcome by the circulation of water tanks throughout the mine during each working shift.

PHOTOGRAPH 6. OLD TYPE DISK COAL CUTTING MACHINE:
JOHN GILLOT AND SON, BARNSLEY, YORKSHIRE

This machine is representative of the very early coal-cutting machines which were developed during the mid-1850s. In this case the motive power was compressed air although DC 440 volts electrical power was later used at Wharncliffe Silkstone Colliery. When run on light or no-load, motor armatures would build up speed and often burn out. These disk machines were invariably used on very long faces up to a mile in length and generally cut upwards from the lowest to the highest point in the face workings, taking about two to three weeks to cut the total length. The rate cutting averaged about 60 to 70 yards per shift. Having cut through the top of the face they were partially dismantled, loaded on bogeys and fitted down to the the bottom of the face where a stablehole had been prepared to accommodate the disk.

The machines ran on a short length of track which was frog-marched in stages, i.e. the length of track behind the machine was transferred to the front. Rope haulage involved running the free end of the wire rope from the haulage drum and anchoring or staking it along the face over a distance of about 30 to 40 yards. The machine was hauled along the rope whilst undercutting the seam. The simple 'ratchet' propulsion mechanisms are in evidence, together with the replaceable cutter picks sharpened after use in the blacksmith's shop. My brother John and his colleagues carried out their own maintenance, both electrically and mechanically, during those early years. Later John was appointed to the post of Deputy, a junior official underground.

PHOTOGRAPH 7. CHAIN TYPE COAL CUTTING MACHINES

This machine was an improvement on the earlier 'disk type' coal cutters, and may be considered as of being at the early development stage of the modern machines. The basic difference lies in the machine's cutting section in which the original disk has been replaced with a flat jib with a chain race within which an endless chain equipped with cutter picks is caused to travel. It had many advantages over its predecessor: it did not require stable holes, could undercut to depths of over 7 feet, required no tracks and was bidirectional. Classic developments included BJD's (British Jeffery Diamond's) 15 inch; 'Ace' Anderson Boyes' AB17; and the old Mavor and Coulson war horse, the 19 inch Samson. Although they were used on the original longwall handworked faces, it was with more concentrated 250 yard units that the machine came into prominence. Here the machine is working through a hand worked stall. The steel screw-jack sprag which anchors the haulage rope is visible in the bottom LH corner, the arrangements being such that the steel haulage rope is maintained deflected along the foot of the seam. The steel box situated on the front bedplate carries some 50 cutter picks which were sharpened daily. Note that the face support system consists of wooden props and bars which are in some cases ledged into the top of the coal seam. This support is reinforced by 'chock' hardwood blocks 4″ by 4″ or 6″ by 6″ erected in the form of columns and spaced at intervals of about 15 feet. It was behind one such BJD 'Ace' machine that as a youngster of 18 I used to crawl cleaning out the undercut.

EARLY SUPPORT SYSTEM WOODEN PROPS & BARS REINFORCED WITH STONE WASTE PACKS

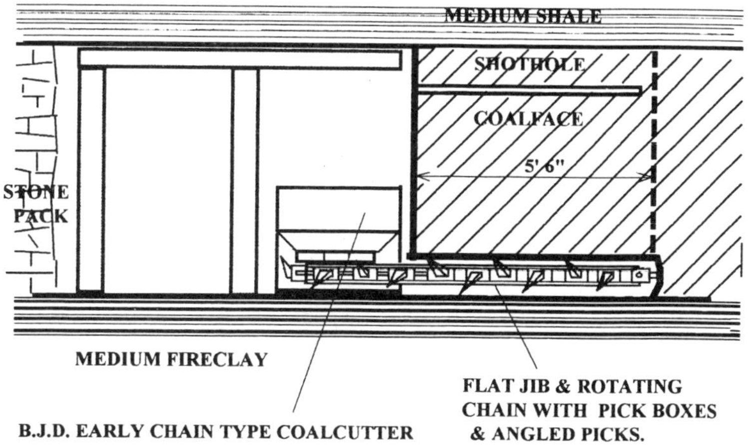

MEDIUM SHALE

SHOTHOLE

COALFACE

5' 6"

STONE PACK

MEDIUM FIRECLAY

FLAT JIB & ROTATING CHAIN WITH PICK BOXES & ANGLED PICKS.

B.J.D. EARLY CHAIN TYPE COALCUTTER

FIGURE. 3. EARLY TYPE OF FLAT JIB CHAIN COALCUTTER VERTICAL SECTION THROUGH COAL SEAM & STRATA

The sketch illustrates the 'Longwall Chain Type Coalcutter' which represented a very substantial improvement on the early disk machines of the mid-1850s. My first experience with the machine sketched was in about 1932 when it was introduced into the Fenton Seam, Wharncliffe Silkstone Colliery in connection with the major reorganisation of the mining system, as referred to in the text, where undercutting a 180 yard face was undertaken in two shifts with a total production of some 300 tons per day. For some twenty-five years or more in this format it represented the basic workhorse of longwall machine-cut handloaded systems. During its early inception it handled machine-cut longwall faces of upwards of a mile in length, embracing 'Stall Units' of some 33 yards or more long in which the coal produced was handloaded into tubs taken into the workings. The average rate of cutting performance was about 75 yards per machine shift in those very early days. In 1951 at Gelding Colliery the average performance was 200 to 240 yards per shift in the Low Hazel seam where the cut faces, concentrated to 400 to 480 yards in length, were handloaded onto face conveyors, maintaining a cyclic advance of 5' 6" per 24 hours. The production yield in these circumstances was about 600 to 750 tons per day.

With later developments of cyclic mining, blasting was introduced on a large scale with the object of increasing both production and productivity. The shotholes were placed somewhat as in the sketch at intervals of about 7 feet and usually took from 5 to 8 ounces of low density ammonium nitrate based explosives per blasting.

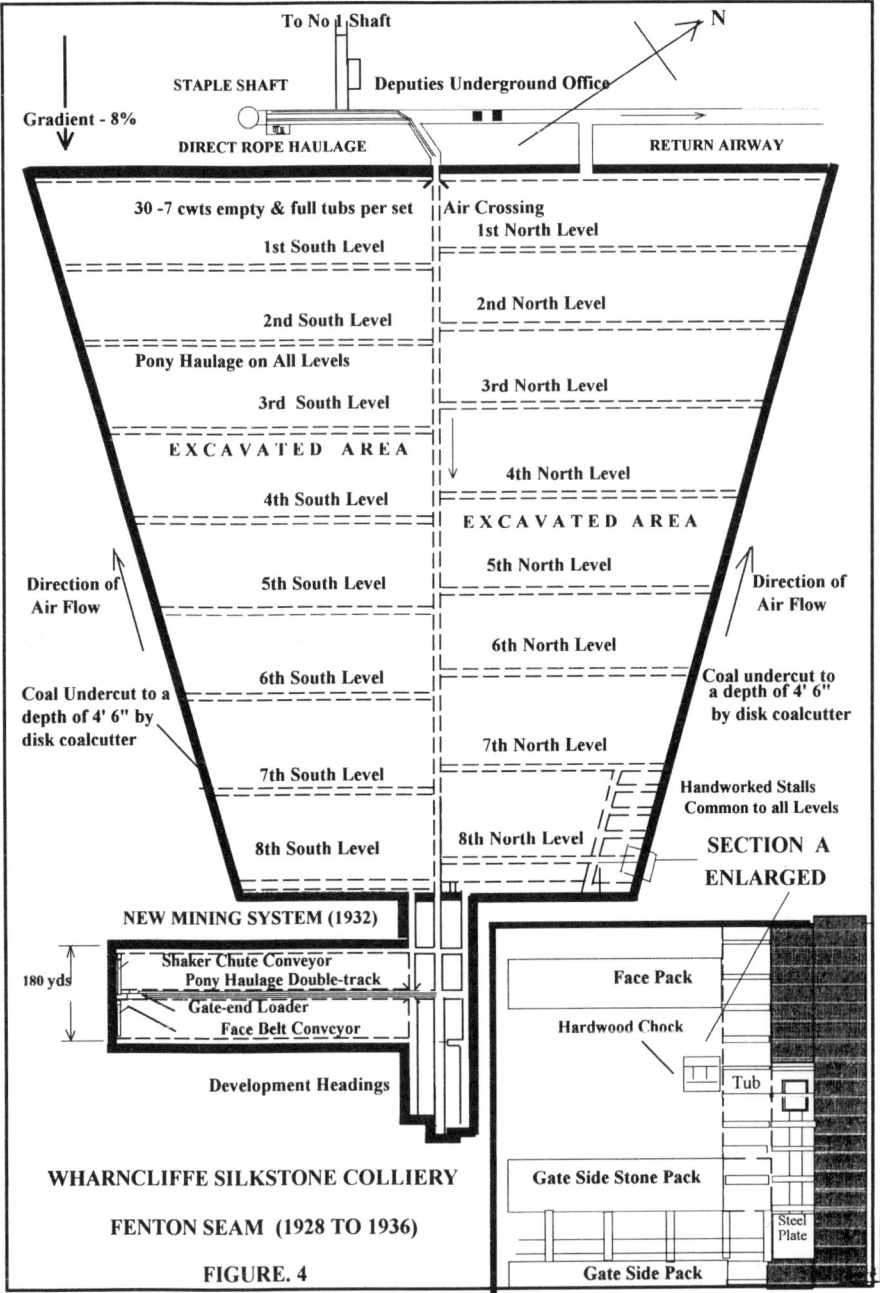

To No 1 Shaft

STAPLE SHAFT   Deputies Underground Office

N

Gradient - 8%

DIRECT ROPE HAULAGE   RETURN AIRWAY

30 -7 cwts empty & full tubs per set   Air Crossing
                                        1st North Level
1st South Level

2nd North Level

2nd South Level

Pony Haulage on All Levels   3rd North Level

3rd South Level

EXCAVATED AREA   4th North Level

4th South Level

                                        EXCAVATED AREA

Direction of   5th South Level   5th North Level   Direction of
Air Flow   Air Flow

                                        6th North Level

Coal Undercut to a   6th South Level   Coal undercut to
depth of 4' 6" by   a depth of 4' 6"
disk coalcutter   7th North Level   by disk coalcutter

7th South Level

                                                               Handworked Stalls
                                                               Common to all Levels

8th South Level   8th North Level

                                        SECTION A
                                        ENLARGED

NEW MINING SYSTEM (1932)

180 yds   Shaker Chute Conveyor
           Pony Haulage Double-track   Face Pack
           Gate-end Loader
           Face Belt Conveyor   Hardwood Chock

                                                        Tub

Development Headings

WHARNCLIFFE SILKSTONE COLLIERY   Gate Side Stone Pack

FENTON SEAM (1928 TO 1936)   Steel
                                                        Plate

FIGURE. 4   Gate Side Pack

The sketch plan roughly shows the layout of the East Fenton Seam and the system of mining with handworked stalls operating at the time I started working there in January 1928. Two 1,500 yard longwall faces north and south were in operation as shown. Each face was split into 45 stalls employing two miners who in addition to getting the coal performed stonework in making the gateroads and building stone-packs in the waste for the support of the face-workings. Usually the colliers were happy to be able to load out 20 tubs per shift which equated to 6 tons each during coal-getting operation. The coal was undercut to a depth of 4' 6" with the early-type disk coal cutting machines, which took from 14 to 18 days to complete the whole 1,500 yards north and south, following which they were then flitted on bogeys back to the original starting points, the lowest points in the workings.

Haulage of the coal consisted of tramming (man handling) along the level stall gateroads. Ponies were used where full tubs had to be pushed against the gradients, and also along the levels as numbered. Rope haulage transferred empty and full tubs into and from the various levels to the pit bottom via a staple shaft (a mine shaft sunk between two lowest points in the workings within the strata as opposed to being extended to the surface).

This system of mining was very slow, that of advancing a total length of 1,500 yards of face at an average rated of 6 inches per 24 hours, a situation which gave rise to many problems relative to the maintenance of both the underground roadways and the face workings. Ventilation of the workings was most inefficient due to ineffective air circulation through the workings, in that air leakage was excessive between the intake and return roadways and a large percentage of the total air circulated never reached the working faces.

In 1932 production was concentrated on a 180 yards machine-cut conveyor equipped coal face, the coal handloaded onto the conveyors being thereafter transferred via a gate-end loader into the mine tubs for haulage by rope to the surface. This unit advance of about 5' 6" per 24 hours on a cyclic basis was the first step towards the concentration of the area and increasing the speed of coal extraction rates.

From this beginning ultimately substantial improvements in ventilation and the haulage of coal from the mine evolved with the introduction of more efficient rope-systems and underground locomotives. Later belt conveyors emerged as the primary form of continuous coal transport, developing progressively to ever increasing capacities, currently in excess of 1,000 tons per hour.

From the original Davy Safety Lamp a range of oil and spirit lamps were developed extending their lighting power from about 1 to 3 candle power, the latter generating great heat and often inflicting burns on the chest, arms and legs of the carrier. With the evolution of electrical accumulators portable electric lamps replaced the oil-lamp for lighting purposes. The early electric lamps were heavy, inconvenient and restrictive. These later evolved into the present form of 'cap lamp' in which the lighting element was separated from the battery by a short length of cable which enabled the lamp accumulator to be secured at one's waist by a belt and the headpiece carried on the protective head helmet. This was a great step forward as hat illumination is automatically focused towards the object being looked at. With improvement of lighting provision the then prevalent eye disease nystagmus was eliminated from the mines.

The detail of a miner's stall is shown in the bottom RH corner of Figure 4. Both north and south faces fully manned would employ some 180 miners per working shift, two production shifts being worked: days, 6.00 a.m. to 2.00 p.m. and afternoon 2.00 p.m. to 10.00 p.m. with repair work and coal-cutting undertaken on the nightshift. The 'new mining system' is shown in the bottom LH corner of Figure 4. It represents a situation which was the fore-runner of current single unit mechanised faces of up to over 300 yards in length.

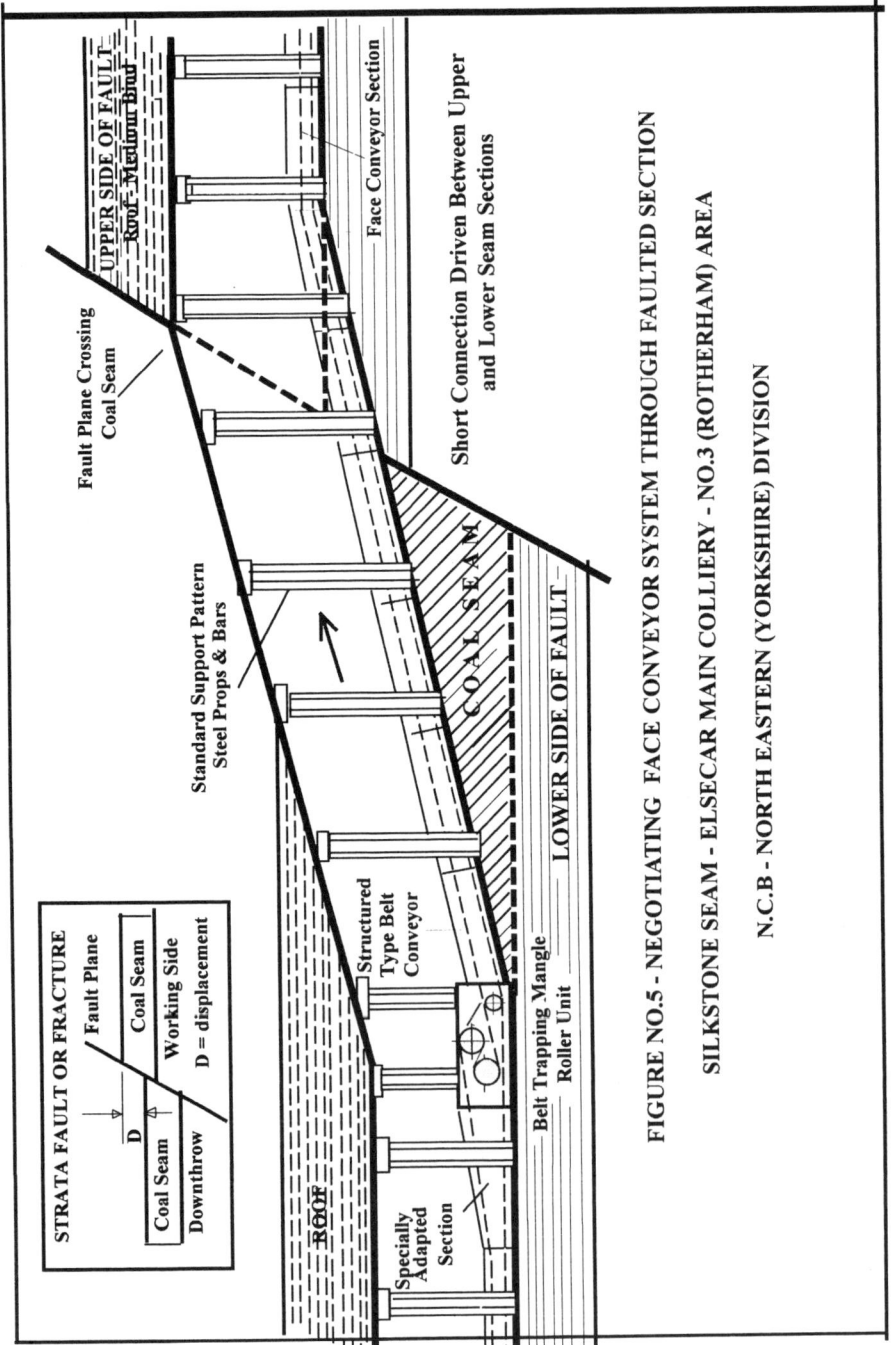

STRATA FAULT OR FRACTURE

Fault Plane

Coal Seam

Working Side

D = displacement

Coal Seam

Downthrow

UPPER SIDE OF FAULT
Roof - Medium Bind

Fault Plane Crossing
Coal Seam

Face Conveyor Section

Short Connection Driven Between Upper
and Lower Seam Sections

Standard Support Pattern
Steel Props & Bars

COAL SEAM

LOWER SIDE OF FAULT

Roof

Structured
Type Belt
Conveyor

Belt Trapping Mangle
Roller Unit

Specially
Adapted
Section

FIGURE NO.5 - NEGOTIATING FACE CONVEYOR SYSTEM THROUGH FAULTED SECTION

SILKSTONE SEAM - ELSECAR MAIN COLLIERY - NO.3 (ROTHERHAM) AREA

N.C.B - NORTH EASTERN (YORKSHIRE) DIVISION

(OPPOSITE) FIGURE 5. NEGOTIATING FACE CONVEYOR SYSTEM THROUGH
FAULTED SECTION: SILKSTONE SEAM, ELSECAR MAIN COLLIERY – NO.3
(ROTHERHAM AREA) NCB NORTH-EASTERN DIVISION

The top LH inset depicts a geological fault which has cut through a coal seam within the
strata, in this case the Silkstone Seam, 2ft. 10 ins. in thickness, of South Yorkshire. Such
situations create difficulties both with regard to undercutting the coal, and in conveying
the coal through the affected section. Each aspect calls for a degree of modification to
normal practice. The section of face in the vicinity of the faulted area has to be graded
through in the initial stage, following which the coalcutter can be made to prepare each
successive grade thereafter. The conveyor needs to have some form of 'belt trapping'
provisions such that both top and bottom belt can be made to run within the structure
or to proper loading horizons without the risk of coal spillage across the disturbed area.
Here the provisions took the form of a 'mangle roller device', which trains the belts to
run in horizons parallel to the floor along all sections of the face. Often in weak strata
situations special modifications to the support system have to be undertaken across the
broken face area.

FIGURE 6. HANDWORKED BUTTOCK OR "STALL" RETREAT SYSTEM

HAIGH MOOR SEAM – ELSECAR MAIN COLLIERY

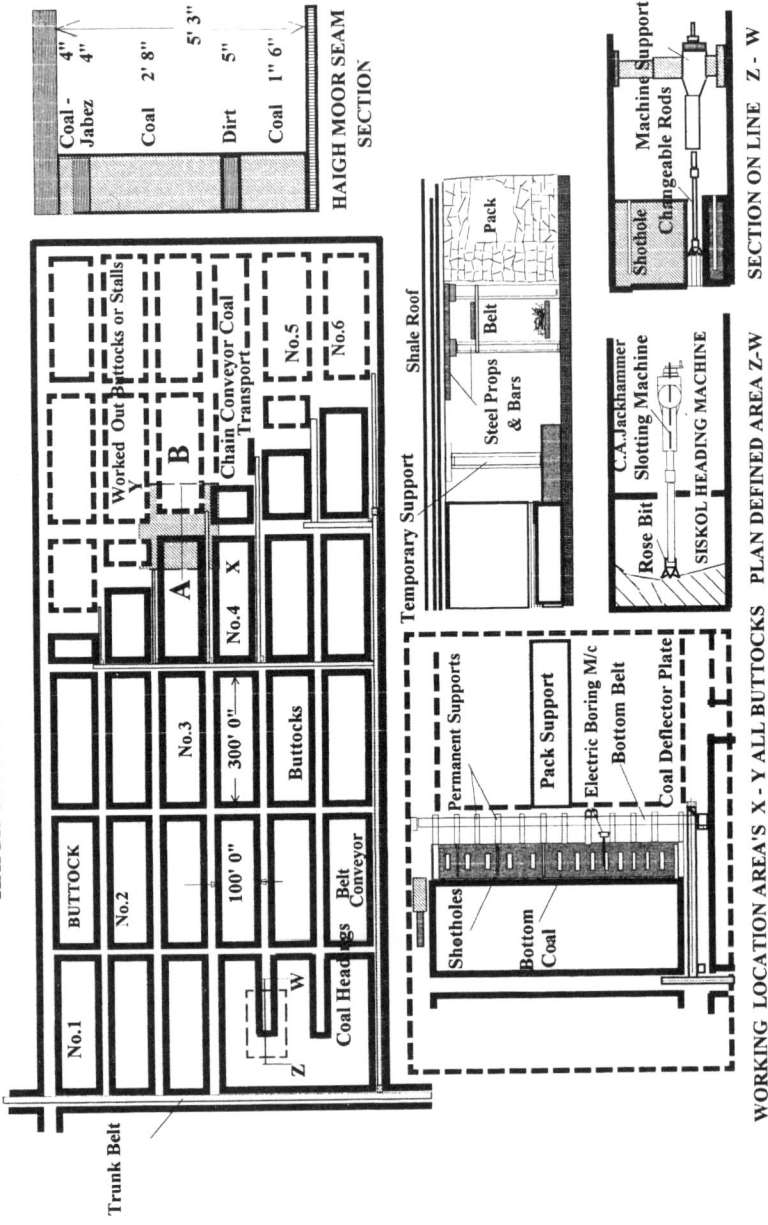

HAIGH MOOR SEAM
SECTION

Coal - Jabez	4"
	4"
Coal	2' 8"
	5' 3"
Dirt	5"
Coal	1" 6"

WORKING LOCATION AREA'S X - Y ALL BUTTOCKS   PLAN DEFINED AREA Z-W

SECTION ON LINE Z - W

FIGURE 6. HANDWORKED BUTTOCK OR 'STALL' RETREAT SYSTEM:
HAIGH MOOR SEAM, ELSECAR MAIN COLLIERY

The basic feature about this form of 'retreat mining' is one of 'in seam continuous mining', one in which the old hand-got systems of Figure 5 have been mechanised by the application of chain and belt conveyors, and coalcutters. In this instance Mavor & Coulson Samson coalcutters were modified with the back or cutting section inverted so permitting the machine to extract a 7 inch band of middle dirt.

The 'in seam' mining aspect provided for all operations to be undertaken within the confines of the seam's thickness thus avoiding the production of dirt on a large scale. The machine cut and blasted loose coal was loaded out onto 'bottom belt face conveyors' eliminating the use of pit tubs, the coal being transferred directly between the coal face and the surface by a gate and trunk belt conveyor system. Chain conveyors installed in the coal headings eliminated the hand tramming of pit tubs. Gate and trunk conveyor belts eliminated pony and rope haulage, the total effect of which was to provide for a continuous system of coal transport.

The sequencing of coal-heading drivages with the retreating buttocks ensured that temporary roadways had a short finite life in which road maintenance was either eliminated or considerably reduced. The plan and section in the bottom RH corner of Figure 6 show how the coal headings were driven with a simple jackhammer pneumatic machine. Two men were employed in driving the coal headings.

The mechanised 'stalls' in both plan and section show how the coal was mined and the workings were supported. Eight miners were allocated to a single mechanised stall; they undertook all the work associated with the extraction of coal in 5' 5" slices including building stone packs on the face. The panels shown in the layout plan were 1,800 feet long by 536 feet wide, the buttocks or stalls were 300 feet in length and 100 feet wide. Performance of the system was a considerable improvement on the old methods whilst production costs were a great deal lower.

These proposals were the brainchild of R. G. Baker, the Colliery Manager at the time (early 1940s), which I had the good fortune to inherit and implement – and which had a profound beneficial effect on my thinking at the time and indeed throughout my whole career.

PHOTOGRAPH 8. CLARKES & STEVENSONS:
FORMER MANUFACTURERS OF LONGWALL COAL-CUTTING MACHINES

This photograph shows the old workshops in which the early type Longwall Disk Coal-Cutting Machines were manufactured. These machines were electrically driven DC in the early stages by series wound motors. The voltage I believe was 440. Where DC series motors operated on light loads they tended to race, heat up and burn out.

At Wharncliffe Silkstone Colliery at the time I started underground the electrical supply system was AC 440 volts, 3hp 40 cycles generated at the colliery by gas engines fed from the coke oven plant. DC machines were however used underground.

During the past few decades the factory has been engaged in the manufacture of 'nuts and bolts' of all types. The large open area behind the factory was extensively worked by bell pits and during our childhood days was desolated and bare. As can now be seen it is rich in natural growth and less accessible than it formerly was.

(LEFT) PHOTOGRAPH 9A. BOTTOM
BELT CONVEYOR DRIVEHEAD

This view takes in the LH and back side of the belt-drive unit for the specially designed Elsecar Main Colliery 'Haigh Moor Seam – Buttock Retreat Units' shown in Figure 6. The degree of 'belt wrap' is traced by the chalk line.

The power input from the motor shaft is on the RH side of the unit. The vertical bar secured to the output shaft is locked to the base of the drivehead unit, a situation which causes the cylindrical outer internal gears container to revolve thus driving the belt running along the buttock to be loaded. The unit worked very well for several months. However, after a while the adverse dust and mining conditions took their toll such that an alternative had to be found.

PHOTOGRAPH 9B. HELIOCENTRIC DRIVE DRUM AND POWER INPUT

Front view of 'heliocentric gear drum' with vertical front cover removed. A standard 3ph 550v 15hp motor flameproof motor provided the input power to the drive.

The unit was quite light, speedily and easily manoeuvred by hand by a couple of men. Anchoring the drivehead in its operating position was simple, usually taking the form of one or two staker supports set in front of the leading channel base girder.

With an advance of 5ft. 6ins. the tonnage loaded out by hand was of the order of 120 long tons (2,240 lbs). This early approach towards 'in seam – retreat mining' really made an indelible impression at a very early stage in my career as to how mining should be approached in terms of cutting out wherever possible non-productive operations.

3

Colliery Manager Appointments

Rotherham Main Colliery

Vesting Date, 1 January, duly arrived with no special event of significance in so far as the miners at the colliery were concerned. Mr Reg Lees, the mining agent responsible for both Rotherham Main and Aldwarke Collieries, met me at about 9.15 a.m. at the colliery office. He introduced me to the colliery staff and trade union officials, following which we had a walk round the colliery surface and then to the adjacent coke oven plant which absorbed the total colliery output. Here, on being introduced to the coke oven manager, I took the opportunity to determine the restrictions and constraints involved in the use of the coal being taken.

The whole of the afternoon was allocated to discussions and an examination of the mining plans with Mr J.T. Templeman, a mine surveyor of no mean ability, well known and highly respected within the Institute of Mine Surveyors. He was much older than I was and had a pleasing personality. We got on extremely well together during my association with the colliery, and he helped me understand many things about the colliery's former history, operation and layout. In its early days coal was wound from two shafts at different levels, the undermanagers of each pit being fiercely independent so that the tubs and tracks were of different gauges and sizes, a very costly approach by today's standards. Subsequently, being short of the type of tub in use, I was unable to convert and employ any of the large number of non-standard tubs available, so necessary for the colliery to be converted into a good profit making entity. From our discussions, it was discovered that I had inherited a colliery which had a single market and was able to supply that mart by winding coal three days a week. The Barnsley Seam coal being produced was of excellent quality and primarily used to upgrade coal inputs to the coke-oven plant from other sources.

My first trip underground the following day was a very long and analytical one. The coalface workings were upwards of an hour's travelling distance from the surface and involved about 1,000 yards of manriding provision with the rest being walked. The Barnsley Seam (approx 5ft. 9ins. thick) was being worked by a single face of some 150 yards in length, equipped with face conveyors and coal cutters. Immediately above the seam was a blue bind shale type roof which at the time was badly broken and generally inhibiting efficient production. Mining standards were variable though good in parts, and the saleable output was of the order of 300 tons per day. The main haulage facilities consisted of an endless rope system which, by reason of heavy gradients favouring the full sets, needed strict supervisory control in order to run safely and effectively.

The undermanager, Sam Roberts, was older than myself and was of the Albert Naylor school. He was a practical type of official and during my tenure at the colliery gave me good support. My immediate impressions as discussed with him were firstly that basic roof control standards needed to be substantially improved and to this end the supplementary waste packing system called for both reorganisation and a substantial upgrading of the workmanship involved in the construction. Supervision by the firemen (colliery deputies) needed to be strengthened. Secondly, that basic service communications (ventilation and transport) should be greatly extended.

The first consideration needed immediate and interim improvement remedies; the second was more long term and involved re-planning of the mine workings. Sam and I together, over the next three months, concentrated on the former, steadily effecting improvements in performance. With regard to the latter, J.T. Templeman and I prepared new layouts designed to shorten communications and also provide for more efficient standards of roof control. These I passed on to Mr Reg Lees, the mining agent I was working under.

The first three months of 1947 brought severe weather conditions, frost, snow and ice, later followed by heavy flooding. In the early part of April 1947 I recall one Friday afternoon returning to the surface from an underground visit to find that the office bathing facilities had failed, so that I had to go across the colliery yard to the deputies' showers, located on one side of the River Rother along which was a flood prevention wall (the colliery yard being quite flat and vulnerable). Looking through the bathroom window I noticed that the River Rother had risen to some three inches from the top of the wall. I had visions of severe flooding. Fortunately it had peaked at that point. However, although the water level was well below entry into the open pit shafts, water had broken through the downcast shaft lining at six points with jets of varying intensity spraying into the shaft. Together with the old colliery engineer and the skilled men we were able to muster I spent a full weekend dealing with the situation before getting it under control by caulking the leaks.

On the Friday afternoon of my third week at the colliery I faced my first colliery strike, based on 'alleged lack of adequate canteen facilities'. It was an unnecessary and stupid situation designed more to test out the new manager than to put forward a legitimate complaint. It was quite easy to ride and I held my ground. The men returned to work early in the following week and I had no further trouble in that direction.

It was about this time that I discovered a set of unused American Lend-Lease 'duckbill' coal loading equipment at Aldwarke Main Colliery which was potentially suitable for driving coal headings at high levels of productivity. Persuading my agent to let me have the equipment, we set out to prepare for its installation in some old Parkgate coal-pillars near the pit bottom, as a means to boost output temporarily until more long-term provisions could be applied. Whilst working to this end H.S. Haslam and R.G. Baker, then Area General Manager and sub-Area Manager respectively, decided to visit the colliery. Travelling alongside

Mr Baker on the manriding train inbye and having heard that A.G. Douthwaite had been appointed Mining Agent for Elsecar Main and Cortonwood Collieries I asked him if he could confirm this.

'Yes, Charles, but why do you ask?' he replied.

'Because if the vacancy at Elsecar Main had occurred three months hence I believe I could have proved, by what we are doing here, to be worthy of consideration for the Elsecar Main post.'

'Couldn't you do it now, Charles?' was his response.

'I feel confident I could and would dearly love to try,' was my answer.

Nothing more was said. The inspection by the Board's senior officials passed off well apart from a few minor complaints. On returning to the office the Area General Manager, sub-Area Manager and mining agent retired to one of the upper offices. After about thirty minutes Reg Lees phoned me to say that Mr Haslam wanted to see me - 'Come straight away.' On entering the office I noticed that they had been poring over the new project layouts that Templeton and I had prepared. Mr Haslam turned to me and said,

'In the comparatively short time you've been here things have greatly improved, more particularly so since my last visit, although there is much yet to be accomplished.'

I was encouraged but more especially so by what followed:

'Mr Baker tells me he'd like you to go back and take charge of Elsecar Main. How do you feel?'

'Over the moon, Sir!' emotionally I replied.

'Then we will make arrangements for you to start in May - but think about this. Mr Baker has put a great deal into that colliery - you must not let him down,' finished Mr Haslam.

Thus I returned to my first love in terms of collieries.

Whilst I was winding up my situation at Rotherham Main, I had a telephone call to tell me that Mr Albert Naylor had in an emergency been taken into the Sheffield Royal Infirmary a few days previously. I rang Mary (we were then living in Rawmarsh, Aldwarke Cottages, way out in the beyond, with its three approaches each of twenty minutes walk), told her I would be home very late and why, and caught a bus to Sheffield. During visiting time I was able to go in and see him. He was alone and feeling a little better than he had been on admission.

His immediate greeting on seeing me was,

'I'm so glad you've come, Charlie - I've been wanting to see you.'

'Sorry I wasn't able to come earlier. It is but an hour or so I was told you had been brought here. How do you feel now?'

'Much better, lad. I've had Mr Danby [former managing director of Earl Fitzwilliam's Collieries] and Mr Baker in to see me. They have been looking at Robert Clive's [former secretary of the Yorkshire Coal Owners Association] son for the managership at Elsecar Main. Why, I asked them, when we've trained a lad who knows the pit inside out, one that will do a good job for you? What about Charlie?' Then he paused and continued, 'I understand you are going back. I'm so glad, but I'll be watching you with my one eye,' he finished with a broad smile.

He kept his word. As he lived over the fence from me at Armroyd House. Armroyd Lane, Elsecar I visited him a great deal after coming back home. I recall on one occasion when we were together he turned to me and said,

'Charlie, tha's paying Harry Lowe £12 a yard for that Silkstone scouring [driving a tunnel through a worked-out area], I nobbut paid him £10, tha's slipping!'

On another occasion he opened up and told me of the problems he had faced during his life: his wife's unfortunate health problems; the loss of his oldest son; how he lost his eye and kept it secret for more than five years; the difficulties he had faced getting started and during his career. When I asked him why he had kept the loss of his eye secret, he said, 'I had a young family. Being able to carry on in the job I was doing with one eye was unlikely and any thoughts of compensation would have been devastating in consequence with the risk of being dismissed very high.' With great pride he then showed me a medical treatise which was about bone fractures and treatment. I am fairly sure that in this book was a sketch of an adjustable clamp designed for use with spinal injuries. Both the design of the clamp and the authorship of the book was the work of his younger son Arthur Naylor, really a most beautiful piece of work in my opinion. At the time, Arthur Naylor was looking upwards and was in the process of applying for a higher medical post. Telling me this, my old friend then produced a copy of an application he had used for a particular appointment, refereed by eminent medical authorities and beautifully prepared. In all my subsequent experience within the National Coal Board I never saw its equal. The presentation backed by wide medical experience was superb. I felt more like a son of his than the raw school boy he'd hardened and licked into shape over the years. Fate, together with competent and kind people, can and does work wonders beyond comprehension. My regard for this fine man increases with every thought I have of him.

Elsecar Main Colliery

Returning back as Colliery Manager of my old colliery had its problems and required a great deal of thought. Having for a number of years been a workman alongside 1,800 or so colleagues in the same position, one experiences a form of respect between mates which is totally different to that which is expected to prevail between the manager and the managed. Providing one is capable and competent and does not act in a manner giving rise to ridicule, when one moves into a new environment, respect comes much more easily and tends to flow with the position. It still however needs to be maintained. Moving to Rotherham Main was a classic example.

In the Yorkshire coalfield to be addressed by one's Christian name by either men or junior officials was considered by members of higher management, to show a lack of respect, a mark of familiarity with a low standard of discipline. In the American coalfields I have been present when the President of a very large coal company producing upwards of 20 million tons per annum was addressed by his Christian name by workmen and officials alike, all of whom held him the greatest of respect and esteem. Nobody batted an eyelid. This to a much lesser extent was the situation in the old East Midlands Coalfield. Respect naturally finds its own level.

My problem was to adjust between the two naturally, without incurring ridicule from the managed. With regard to this I recall a true story heard countless times amongst the Elsecar miners. It was the case of a workman being appointed to become a deputy (underground junior official). His name was Joe Bunting. On making the first of his inspections through his district the first man (as was usual) greeted him with, 'Good morning, Joe.'

Joe reared up to his full height and said, 'You from now on address me as Mr.!!'

The man said nothing, but this occurred almost with every man he met. A group of men collected together during snaptime were discussing this new situation when one miner piped up with,

'They tell me Mrs Bunting is no longer able to use the *Green Un* [the local Sheffield football paper] for a table cloth, it has to be the *Yorkshire Post*.'

He got the nickname 'Tablecloth Joe'. I don't think he ever lived it down.

I decided to deal with the situation by being strict but fair in terms of mining and safety standards, to fulfil all promises, and to adopt a straightforward and honest approach in all my dealings with a determination to have things done as I wanted either as an individual or as agreed by consensus. Within six months I had the respect between manager and managed of my predecessors greatly cemented by solid technical and mining achievements.

In my early days as Manager of the colliery I had the impression that my late manager A.G. Douthwaite appeared to resent my appointment. I may be wrong in this for George, as we knew him, was a fine person and in later years whenever we met we had pleasant discussions. However at the colliery we had an enigmatic Trade Union president by the name of Councillor Albert Wilkinson who was employed by the miners as 'check weighman', his job being to check alongside the 'company-weighman' the loading of each collier's daily production as carried within the identified tubs they had loaded out. We were having a strong dispute over the tare weight of the empty tubs. Over the years the colliery had been introducing steel tubs which were much heavier than the wooden ones formerly in use. I had a strong feeling that re-taring the tubs would greatly favour the management with a loss of wages to the men and, having just returned to take charge of the colliery, I would be accused of trying to cut wages. The colliery was quite profitable and I could not at the time afford any disturbances for a number of reasons. George Douthwaite used my office for lunch on many occasions and during a visit I referred this difficulty to him but nothing was said nor discussions entered into. On leaving the office immediately on finishing his lunch I said, 'Have you any suggestions?' He turned as he passed through the office doorway and replied, 'You're the manager,' and left. I felt rebuffed but vowed within myself that I'd never seek his help or advice on any further issue. I never did, and George did me a great indirect service by his approach on that occasion.

The outcome of the problem was that, because of heavy pressure from the men, I had to yield and agree to the re-taring being demanded. The NUM branch union wanted taring to be exclusive to wooden tubs. We countered by requiring such to be confined to steel tubs as being the more numerous. In the end we agreed on proportionality; thus the tare took place. I believe we tared about two hundred tubs out of the several hundreds in use. This was done one weekend. A number of the men carefully selected each tub during the week preceding the event and with brushes thoroughly cleaned them, to the extent of wiping the oil from the tub axles. On behalf of the management Clem Hyde, the pit top foreman, selected with them, on the basis of tub for tub. The tare-weight came out well in our favour so the men were distressed and demanded another exercise which again we agreed to. This time I asked Clem to be a little easy with his choices. As such the tare weight was less in our favour but still useful. We compromised within the range which, following acceptance by the men, was put into practice and things settled down.

ADMONISHED BY THE AREA GENERAL MANAGER.

In the Silkstone Seam we had a 'state of the art 1947' hand-loaded coal face advancing down a gradient of about 1 in 12 producing about 300 tons per twenty-four hours. The coal was loaded out in a single shift onto the face conveyors and transferred into the tubs by what we termed a 'gate end loader' and this, being portable, was moved down with the advance of the coal face about 5' 6" per day. From the loader the loaded tubs were hauled away to the pit bottom by a system of endless rope haulage. This system had grave disadvantages which frequently disorganised systematic production, such that we were losing tonnage with further losses pending unless the prevailing difficulties were cleared. R.G. Baker had formerly designed a very useful flexible rope attachment and chain which was easily and quickly applied, but the system was costly and tied down too many men. Moreover the loading station was temporary and moved along with each advance of the coalface. I had designed and had built a transfer station at the same time, which made it both easier and safer to transfer the loaded tubs between two rope systems on an adverse gradient and which worked admirably. Studying the system carefully, it became obvious to me that, by extending the gate-end loader structure with the daily advance of the face and keeping the loading station in a semi-fixed position, it would be possible to have more permanent loading arrangements which could be moved down every two or three months, offering greater stability, manpower savings and other advantages.

Taking a spare standard gate-end loader we started to undertake modifications in the fitting shop and being restricted for room had to leave the entrance doors open. Out on a particular inspection the new sub-area manager John Howatt observed the work in progress.

'What are you doing there?' he demanded to know.

'Repairing a gate-end loader,' was my reply,

'Right, we will go and look,' he responded, striding off towards the entrance to the shop. Going into the shop, he turned and said quite angrily,

'You're doing no such thing - you're modifying it.'

I tried to explain the need and purpose but to no avail.

The following Saturday I was instructed to go down to see G.C. Payne, Area General Manager at Manvers Main Area Headquarters. He had succeeded R.G. Baker who had been appointed as Deputy Chairman to the Divisional Board. I'd never personally met the man before. Seated behind a large desk (which I inherited in later years) he glared and said,

'I understand that you have been altering the Board's machinery.'

I stood to attention, not unlike a schoolboy who had transgressed and was being severely admonished by his headmaster.

'What do you mean, Sir?' I asked.

'You know what I mean. Who gave you authority to start messing about with the Board's equipment?'

I tried to explain but again to no avail.

He continued, 'You know the Area rules. Did you consult Colin Rudge [Area Mechanisation Engineer]?' still in the process of dressing me down. Incidentally Colin Rudge had formerly worked closely with him at Manvers Main Colliery prior to Vesting Date for quite some time. I had no idea of the rules he was referring to, but nevertheless couldn't say so in such a situation. He curtly dismissed me with the instruction,

'You do not in any way undertake anything in future without first consulting and getting the approval of the people at Area.'

Returning home it crossed my mind - 'you won't last long under him!' Despite his instruction, we subsequently completed the modifications and installed the system. It worked admirably, enhancing both production and productivity performance.

When I recounted the incident to Mary, in her quiet voice she said, 'Look love, don't worry. We managed before you got the job, we will manage if we have to, should you lose it.'

Mary was the rock I could swim to in the stormiest situation of which the future held quite a number. Within a week I'd forgotten all about the incident and went about things as before.

Shortly after the incident we needed a double-drive gearhead belt conveyor unit. R.G. Baker's original Huwood design was not available but having two British Heffrey Diamond R.9 driveheads spare I sat down and designed an arrangement which combined them as shown. (See PHOTOGRAPH 10)

It had to be compact to fit into the standard roadway drivages of the size we employed.

Recognising that it was possible to remove the unit's delivery head and mount it at right-angles under the driving rollers with the motor turned upside down, the rest was easy. This concept is clearly seen on the RH unit which shows the thread of the belt under its normal 'top belt' operating condition using the face conveyor structure for which it had been designed. Mounted on goalposts at either end which accommodated the plough scraper, it made a very solid and reliable unit, fitted with wheels which ran in a short length of channel irons, making advance and retracting operations quick and easy to perform.

Each unit had two 15 h.p. motors which were capable of operation on face lengths up to about 1090 ft. or more where a good system of 'belt rollers' supporting the bottom loaded belt was provided.

I made no mention to either my Agent or my Area. A few weeks later Colin Rudge came to the colliery and saw the equipment working; he came to see me and told me he'd developed something similar using the same gearheads but it was larger. I pointed out to him it would be a very costly exercise for us to size roadways to fit large machinery. It had to be the other way round. Following his visit I expected Colin to report this latter 'misconduct' to the Area General Manager with subsequent serious repercussions to myself. Whether he did so or not I do not know but, thankfully, there was no backlash. Colin and I ultimately became good friends and remained so for very many years until his recent demise towards the end of 1994. Incidentally, subsequent to the tragic death of his first wife, he married one Margaret Duckworth, a beautiful and delightful lady with an exquisite contralto voice who came into great international prominence after winning the Kathleen Ferrier Memorial Prize Competition in Hilversum. Holland. For many years she was featured throughout Europe, over the Dutch radio and regularly appeared on UK concert hall platforms and in the leading musical festivals. She is currently one of the foremost voice trainers in the UK. Colin himself in his early days had a good tenor voice and formerly sang in oratorio throughout many parts of Yorkshire.

REORGANISATION OF NO.1 SHAFT COAL WINDING ARRANGEMENTS

During and after my first year as manager, the position at the colliery was that of assessing a situation of detrimental change which had unobtrusively taken place over the previous four or five years. Output from the Parkgate hand-loading stalls had decreased. Mechanised output from the Silkstone and Thorncliffe seams was slightly increasing and the Haigh Moor with its separate coal winding circuit was being developed and tooled up for production. The provision of additional manpower needed to be considered. I had however a strong desire to find them from within our existing labour force of 1,700 to 1,800 men. In order to do this a programme of labour-recovery, involving changing techniques and the application of mechanised systems in all fields, was at the forefront of my mind and as such was initiated and progressively followed through.

Jim Hirst (a former shotfirer colleague) was my undermanager and on the retirement of William Downs, the former senior day overman, I appointed Frank Knowles to his post. Studying the coal winding statistics of the No.1 Shaft it appeared quite obvious to me that, by reason of changes which had taken place, it now had a capacity well in excess of that required.

Frank Knowles was asked to carry out a comprehensive survey of the pit bottom manning and operational arrangements. Locomotive haulage had already been introduced in the main haulage section underground which offered in prospect greater and wider flexibility within the underground transport and winding systems.

In the meantime, directing detailed research into the No.1 Shaft coal-winding facilities (which embraced coal being loaded into both decks of the two-deck cage from two levels simultaneously, as opposed to coal being loaded into the two deck cage one deck at a time from the lower level), it wasn't difficult for me to deduce that by developing the latter arrangement quite a reservoir of manpower could be recovered. At the same time the current coal winding requirements could be met. Frank Knowles came up with a very simple low cost reorganisation in the pit bottom whereby we could extend the work of the locomotive system most effectively to fit in with what we had in mind.

I did however have quite a serious problem: that of obtaining simple control gear relative to the restrained movement of the tubs at both pit bottom and pit top levels. I hadn't referred my ideas to Area or to my immediate Agent, George Douthwaite. I disregarded it as an Area, Divisional or Nationally approved scheme in terms of the accepted procedural sense (bureaucracy had not as yet spread all its tentacles; this was pre Dr Fleck, Chairman of ICI, referred to later). I presumed that obviously my Agent would become aware of what was going on, but we never discussed it. He never raised the situation, nor did I. The matter of his former observation had bitten deep inside me. However, I formed an alliance with Leslie Rutherforth, the Group mechanical engineer. Leslie found me the pneumatic controls from Martin Air Products Ltd, which we ordered and charged to revenue in controlled amounts over a period of time. Other operating costs were intensely squeezed to permit purchase of this equipment.

I found the underground tub handling gear through Tom Adam, a Dowty representative, having previously seen the applicable equipment working at Holmewood Colliery in Derbyshire.

We worked quietly at the scheme, prodigiously extending our normal hours day after day, working out and undertaking the maximum amount of completed work prior to the proposed actual holiday changeover, thus reducing the actual work to be undertaken at that time to a minimum. A great deal of No.1 Mine shaft work had to be done at weekends by the fitters and shaftmen, who, to me at the time, embodied the souls of saints and had all the characteristics of true dedication. They were Walter and Sam Hunter, Horace Howes, George Simpson, Freddie Firth, Albert Earnshaw, Dick Haigh and other members of their staff. They worked weekends in abnormal conditions, some 250,000 cubic feet of cold air passing through the shaft per minute in a winterly situation. All the comforts I could procure in the form of protective clothes, hot beverages and transport home in the early hours were a priority.

Working at the colliery during weekends, we drew up schedules of the recovered labour potential, worked out the proposed coalface workers from amongst them and rang the changes over a wider number of men with regard to the others. The respective placing of both the identified potential coalface workers and other workmen occupied a different set of finalised schedules. Each man was interviewed by the undermanagers, so that when the revised system of winding came into operation, there was no confusion.

Occasionally I took my children John and Dorothy with me to the colliery at weekends. The men would make a fuss of them and give them sweets and short rides on the locomotives. John was not over thrilled but Dorothy would have gone with me every weekend.

During the annual holiday vacation the revised winding changeover was undertaken without any serious difficulty and all work was completed in time for the pit to start up thereafter. The Monday start-up following the changeover arrived. It was one of the most dreadful days of my whole career. While the winding enginemen formerly wound at the rate of 300 tons per hour, under the new system it wasn't possible to touch 50 tons per hour and output was being lost. Controllers were failing, air valves were sticking, everything that could go wrong did. I was there in attendance almost sixteen hours per day over several days. This situation prevailed for more than three weeks. I had visions of G.C. Payne, the Area General Manager, calling me to book. Had he done so I would have had no defence and what follows might never have been written. It was remarkable how sympathetic the men were towards me at this traumatic time. The onsetters loading the cages and the banksmen unloading them endeavoured, in every way, to ease my position and tried to encourage me by such statements as:

'Don't worry, boss, it's all new to us; we will sort it out for you.'

The fitters and others who carried out the reorganisation would quietly come along to see what was happening. After a while they would disappear and shortly afterwards reappear, along with a piece of lathe turned metal, start some activity in between tub movements, and behold, the equipment involved started working much better. How many times they did this I don't know but within about four weeks they had the surface and underground control systems working with great reliability.

The basic problem now was to improve the winding rate. Although it was a little better, it was insufficient to cover the current requirements. During a visit to the winding engine house, I recall talking to William Maleham (his son was, and I believe still is, a celebrated organist in a London Cathedral and in the City generally). Bill, as we knew him, was a really fine person with a gentle disposition. Turning to me, he said,

'Don't worry, Mr Round, we're breaking through; you know I have been winding here for over thirty years automatically. It's somewhat second nature. We have more movements with the controls under these new arrangements, but in a short while it will be second nature to us again.'

How right he was. Over the next three weeks things were back to normal; thus we were in a position to expand output in the Haigh Moor seam, and now my future career prospects if any within the Area were intact.

The work we did, in retrospect, really was tremendous, having effected a manpower saving of some 150 men (surface and underground), all on a low revenue cost basis. Capital expenditure to save a man at the time was about £3,000; for 125 this equated to £375,000. The actual revenue cost was less than 10% of this figure spread over several months. The circumstances then were favourable, but what was equally important, we had recognised that the situation at colliery level called for a major change at about the right time.

Through the preliminary manpower scheduling we had undertaken, it was possible to transfer production labour into the Haigh Moor Seam almost immediately. This increased the colliery output by over 3,000 tons per week, raising the average weekly output of the colliery to a little over 18,000 tons per week.

At this point I refer to the pioneering efforts of George Douthwaite and Merton Gullick of Gullick's Ltd, with respect to the advanced technique of power loading, i.e. the use of machines to cut and load out the coal. My first knowledge of the proposed scheme, which involved the installation of a chain conveyor, the Samson Stripper and Dowty Hydraulic props, was when the equipment duly arrived, although I had been told to get a small face South 3's in the Parkgate Seam ready. I wasn't 'in on' the preliminary discussions but had the active job of setting up the actual installation and operation of the equipment. I wasn't sore about this. Anything that came my way to improve myself technically and to increase coal production was to me a providential act.

The Samson Stripper was really a vertical jack, and horizontal jack combined in such a way that the vertical jack secured the machine between roof and floor whilst the horizontal jack, with steel wedges, pushed into the coal and sliced off about nine inches of the full thickness of the seam which fell onto the face conveyor. It was a very simple machine. The vertical jack exerted a pressure of 90 tons and the horizontal jack thrust the wedges into the coal under a pressure of 200 tons. The installation was undertaken under ideal conditions: a hard floor and a strong sandstone roof.

The machine worked quite well in the circumstances but it had a number of limitations (these basically were the small nine-inch depth of cut taken, and its slow rate of travel. One needed to be thinking in terms of a two-foot wedge cut depth, and lineal speeds of 30/50 ft.min. for mass production. It therefore never really broke through into extended development and use. During this development the Dowty hydraulic prop, a descendant of the undercarriage jacks of the aeroplane, made its appearance and was greatly superior to the German friction-type props which were making their appearance into the mines at the same time.

PHOTOGRAPH 11 'THE SAMSON STRIPPER' shows quite clearly its format and particular application in side view. A special chain single strand, flighted face, conveyor was designed in conjunction with the machine. This conveyor was pushed over bodily by a series of specially modified sylvester (see Photograph 33).

PHOTOGRAPH 12 is a second view, this time showing the machine slicing off a nine inch strip of coal from the front of the face. This also gives a good view of the Dowty Hydraulic Props which was probably the third installation of its kind. Power 550v 3ph AC cables are shown hanging from the props.

JOHN TODHUNTER SNR

Elsecar Main Colliery had many home and overseas visitors from time to time and one such visitor during my time as manager was Richard Todhunter, President of the Barnes & Tucker Coal Company, Barnsboro, Pennsylvania, USA. Accompanied by C. Payne, John Howatt, and George Douthwaite, he came along to visit the Haigh Moor seam. About that time we were either contemplating or had just replaced the Drift endless rope system by three tandem conveyors. The belts were delivered endless and had only one high quality vulcanised joint in their looped format. This was indistinguishable from any other point of the belt structure. During the installation the drive units were built into the belt which was quite a novel approach at the time. The situation called for, and received, a great deal of careful planning.

Richard Todhunter Snr was greatly interested in this and questioned me at length as they were considering a 54 inch conveyor belt installation into one of their USA Lancaster Collieries surface drift-mines. Throughout the trip he kept close to me and questioned me at length on many aspects of longwall mining. At the end of the trip he quietly drew me on one side.

'Charles, have you ever thought of working in the States?'

'Not as yet,' I answered

'When you do, let me know. By the way, I'm sending my two sons, Richard Jnr and John, over to see you when I get back.'

Within six months Richard Jnr came and we had a great time together. He was interested in beagles and together we visited the Ecclesfield Kennels which were in existence at the time. Many years of correspondence passed between us, during which time I went over to the States on four occasions to meet him. The last time I went over, Richard Jnr was Mayor of Barnsboro. His father, who was in his late eighties or early nineties, called to see me. Apparently he'd been in hospital for two weeks and following an operation had called a cab, discharged himself and gone home. He was somewhat incoherent but he did seem to recall his early trip over to Elsecar Main Colliery. The younger son, John, came over shortly afterwards and gave me the history of his father who was apparently a quite unique character, very highly respected on the American scene.

There are two sequels to these events. The first was that Richard Jnr installed a power-loaded mechanised unit (later described herein) in one of his Lancaster Mines using British equipment and methods which achieved fantastic results by our standards. During a subsequent visit over there by Lord Robens, accompanied by W.V. Shepherd, the Board's Director General of Mining, and others, at dinner His Lordship turned to Richard Todhunter Jnr and said,

'Dick, how is it you are able to get the sort of mass production and high level performance productivities you do with our equipment that we can't?'

'Well, Lord Robens, we took advice from one of your people.'

'Who?' pressed his lordship.

'My friend Charlie Round. I sent him a telegram to which he replied in great detail. We followed his advice.'

'I know Charlie very well,' said His Lordship. Dick himself told me of this last time I met him, since when, unfortunately, he passed away at a comparatively early age.

I did receive the telegram during a fortnight in hospital following a minor operation, to which I dictated a reply telegram at length, although I was never appraised of any action taken relative to it.

The second sequel is of recent origin. On 22 October 1993 my son John flew me out from our UK Birmingham International Airport to Fort Myers, Florida, to spend a superb and restful holiday with both him and Sue my daughter-in-law on Sandibel Island. Alongside me on the Boeing 747 was a lady and since we were to be partners on the flight to New York for more than seven hours, we started a series of discussions which lasted throughout the journey. The lady, Mrs Lynne Adams (nee Roberts), was born near Philadelphia, and had happily married an English mathematician. She was a teacher and had lived in Coventry for the past twenty-five years.

She asked me if I had ever been to Pennsylvania. I had spent time in Barnsboro during my American visits.

'Do you know of the Barnsboro Coal Operators - Barnes & Tuckers?' she asked.

'Yes, as formerly run by Dick Todhunter Snr & Jnr,' I replied.

'Did you know them?' asked Lynne.

'Yes, we were friends,' I answered.

With greatly registered surprise she informed me,

'My parents knew them well. In fact my father was Company Treasurer and every Christmas when we were little girls Richard Todhunter Snr sent my sister and me dolls and presents.'

We were also together on the plane coming back from New York. Lynne had been to see her mother and recounted the incident on her flight over, following which, her mother went to a drawer, drew out a sheaf of photographs and handed them to her. She showed me a series of photographs of Lynne's mother (Wheltha H. Roberts) and father making a visit to one of the Company's mines, featuring her mother, Richard Todhunter Snr and Richard Todhunter Jnr (see PHOTO-GRAPH 13).

By now I'd been manager of the colliery some four years. We had made tremendous strides and things were going extremely well. My appetite for progress was insatiable.

One Friday afternoon I received a call from Manvers Area Headquarters, instructing me to see the Area General Manager at 11.00 a.m. the following morning. Dutifully I went along on time and stood before him. He was no less curt than on the previous occasion:

'Round, on Monday morning I am sending you an assistant manager. His name is Raymond Gill. He has just obtained his degree in mining at university. You are to give him every assistance and help him to develop. That is all. Good morning.'

I went home, really perplexed. I neither required nor wanted an assistant manager. Reg Burton had been appointed undermanager for the Haigh Moor and was doing the job admirably. Jim Hirst, undermanager for the No.1 Shaft, Parkgate, Silkstone and Thorncliffe Seams, had worked under me for some four years. Not one of us had university degrees but we all had a solid practical mining background with the necessary qualifications to support our positions. What was equally important, the colliery was making substantial progress and making good profits. Discussing the situation with Mary, we decided we were at a cross roads. It seemed obvious to me that there was little respect for those who worked their way up into colliery management from the coalface and that we would ultimately be replaced.

'Mary, the time has come for me to look elsewhere and re-establish our situation. I will never get anywhere under this new man and his setup and will ultimately be pushed out or reduced. How would you feel if I contacted Dick Todhunter in the States and let us try to make a new start over there?'

'What about John and Dorothy [seven and four years respectively], their education? Please don't ask me to go abroad. I'll go anywhere with you in this country but I should be very unhappy out of it,' she replied, her family being a very strong unit. Rather than create a situation of that nature I would settle for the best position I could obtain.

'OK, love, let's wait and see what happens for the moment.'

Thus I dismissed the thought of immediate action from my mind.

On the following Monday after a few phone calls it was found that my situation was not an isolated one. Other young graduates had been selected and positioned at various pits in the Area including John Holtom who I believe was a Quaker and who later worked for me. (He was a fine lad but I am sure he would have found running a colliery quite an immense task, although he did very well within the planning side of the organisation.) Geoffrey Thorpe who subsequently took my position was amongst those placed.

As I had found, getting the required qualifications for the colliery undermanager's or manager's position at a very early age does not instantly or even within a short time signify that one is ready for carrying out the responsibilities of the posts. To assume it to be so gravely depreciates the importance of the respective positions. Despite this however a traditionally well organised colliery with excellent senior and junior staff can carry the relatively inexperienced young manager for a time providing he does not behave with stupidity.

Turning to my newly imposed support, Raymond Gill: my secretary Mark Hague (who had been of great assistance to me and to all my predecessors) ushered him into the office to see me. The lad was nervous and most uneasy. Mark brought in some coffee and biscuits to set him at ease, following which I gently questioned him at great length. It was obvious after a while that he hadn't been subjected to such a degree of questioning before. He was extremely raw, uncertain and needed lengthy basic mining training. At the end of the interview I turned to him and said,

'Now, Raymond, how do you think I can help you?'

I honestly felt sorry for him. He certainly was no threat to me or to anyone at that stage and for that matter the prospect of maturity appeared to be something which needed to be developed over a period of time.

'Can I leave it to you, Sir?' he replied, which in the circumstances was most sensible.

My best approach towards the lad's development was to subject him to the type of training Albert Naylor and R.G. Baker had given to me. Although he was a very nice person, however, unfortunately for him, he was unable to develop the respect of the undermanagers and the officials. After a little over twelve months or more he disappeared (where to I never knew, nor what became of him). I had tried to encourage the youngster in any way I could and involved him any activity I felt might be helpful, including social ventures.

During the early days of nationalisation, the Chairman of the Yorkshire Division was Sir Noel Holmes who, on one occasion, came along to the colliery with R.G. Baker (Deputy Chairman) together with top Area officials. We had a good trip. He was most interested and very pleasant towards me. In the course of the discussions he turned to me and said:

'I've heard a great deal about you, Charles,'(presumably from R.G. Baker).

'I hope it was not adverse, Sir.'

'No, quite the contrary - but you work too hard,' he remarked.

'I enjoy what I'm doing, Sir.'

'That's no excuse, Charles you must also learn to live,' he responded.

COLLIERY CONSULTATIVE COMMITTEES

With the introduction of nationalisation, consultation at Colliery, Area, Divisional and National Levels developed on an extensive scale throughout the Coal Industry. The Consultative Committees were fairly well balanced in their constitution with provision for NUM officials, tradesmen. NACODs (National Association of Colliery Overmen, Deputies & Shotfirers), workmen and management. Terms of reference covered: output and efficiency; attendance; safety and welfare; planning and organisation; and general matters of interest.

Wages and contract matters were handled separately under the industry's national conciliation machinery which ran somewhat parallel to the former. Local contracts were negotiated at colliery level and approved at Area and Divisional level, but national agreements for the industry were undertaken at National or Headquarters level.

During the early days of nationalisation strong efforts were made to expand the Industry's total output to meet a number of projected minimum levels of production. In the early 1950s absenteeism became quite a serious problem such that a national procedure through the Colliery Consultative Committees was established to deal with it, the worst cases being dismissed. Usually, interviewing persistent absentees wasn't the most rewarding of activities.

One particular situation stands out very strongly in my memory. It concerned a former miner, Ernest Brammah, who had left the coal industry to join the Navy in his late teens. He served throughout the Second World War and was involved in the Dunkirk retreat as a naval rating. After the war he joined the Merchant Navy for a number of years. For some reason he found his way back to the colliery but rarely worked three days in any one week. The Navy had built him in stature to over 6' 0" in height and some 240 lbs in weight. He was both tough and mean when the occasion suited him. I knew his family well, his older brother John being one of my earlier school pals.

We interviewed him along with a number of others having similar attendance records. When his turn came, everyone was apprehensive of approaching him, as he was noted for a violent streak. As he walked into my office, those within close contact slid in their chairs to the walls on either side of my office; it was not unlike the parting of the Red Sea. He was facing me directly across the large table.

'Ernest, you are not very happy working underground, are you?' was my opening question.

'What man in his right mind is?" he replied.

'You are aware of why you are here, but knowing you and your brother John as I do, there is something else wrong. Can you tell me what it is?'

'Give me a Merchant ship and let me get out of this god-damned place,' he volunteered.

'Ah, that's it - now I can help you,' was my offer.

'What can you do?' he queried.

'So long as you are on the pit books you'll never get a ship - you can have your employment cards now, then you will be free to join whatever ship you can find. With your experience at sea that shouldn't be difficult.'

'Give me my cards, then, and let me get away from this bloody place,' was his expression of acceptance. Mark Hague my Secretary had them ready, got him to sign a receipt, and out of the door he went, much to the relief of all present.

One Saturday night, about a month later, outside the local Ship Inn at Elsecar he badly mauled one of the colliery's best overmen (over some dispute or other they had had during the time he was working underground). I approached my agent for legal support for the overman but was told that, the affair being off the colliery premises at the time, nothing could be done.

Post Interviews - East Midlands Division

Following these encounters with the Area General Manager, for the first time in my career I started looking for suitable openings in the mining journals and NCB vacancy lists. In due course one came along: the No.5 (Eastwood) Area of the East Midlands Division wanted a mining development engineer. I became interested.

Having had no previous experience in applying for jobs, I recalled seeing the application Albert Naylor's son had submitted to his medical authorities. To the best of my ability and recollection I modelled my application on it and was subsequently called to Sherwood Lodge, Mansfield, Notts, for an interview. Never having been interviewed by a panel before I'd no idea what to expect. I prepared myself with plans, photographs, statistical detail and drawings of ideas which had been put into practice, together with a load of material I thought relevant and so equipped appeared before the panel.

The panel was W.H. Sansom (Divisional Production Director), Arden Bowker (Area General Manager), William Unsworth (Area Production Manager) and the Divisional Staff Director. They looked at me astounded when I placed the plans, etc. in front of them on the large table between us.

'Not for the moment, Mr Round, please sit down. Let me introduce you to the panel,' said Mr Sansom.

The panel then started a long and searching interview which I really enjoyed for they made me feel most welcome and comfortable. Nothing caused me any real embarrassment. After about twenty minutes Mr Sansom turned and said, 'Now, Mr Round, what is all this you want to show us?'

I opened up my plans briefly, explained the respective situations, talked about the other material and showed them a number of photographs for a further ten minutes, following which Arden Bowker and William H. Sansom started having fun. I'd open one plan up only to be questioned on another; they were doing this alternately but I managed to keep my head. They were all nice people, understanding and very generous to a stranger. There was one question from Mr Sansom which I shall always remember and that was:

'Mr Round, where do you think good ideas should come from, the top or at the bottom of an organisation?'

I noticed quite a twinkle in his eye. There was something most unusual in this question, at least to me. I recall that my answer was:

'At the bottom, Sir; there are more of us with good ideas down there than are available at the top. If you take them exclusively from the top you'll miss all the ones developed by those below.'

The whole panel roared with laughter. Maybe he wanted 'take your good ideas from wherever you can find them', which in later life is what I always did. However, some fifty years later, in his book *The World in 2020* Hamish Macrae includes the following quotation with reference to attitudes in Japanese industry, of Konosuke Matsushita, head of Japan's largest electronic group, in a speech to visiting foreign managers.

We are going to win and the industrial west is going to lose, there is nothing you can do about it because the reason for failure lies within yourselves. With bosses doing the thinking, while the workers wield the screwdrivers, you are convinced that this is the right way to run a business. For you the essence of management is getting ideas out of the bosses' heads into the hands of labour. For us, the core of management is [the] art of mobilising and putting together the intellectual resources of all employees in the service of the firm. Because we have measured better than you the scope of the new technological and economical challenges, we know that the intelligence of a handful of technocrats, however brilliant and smart they may be, is no longer enough for a real chance of success.

I did not get the job which had been reserved for Wilfred Crossland, another Yorkshireman I later got to know very well. However, as future events unfolded I did much better.

It so happened that the Divisional research and strata control engineer was one Alf Wright who had undertaken research work at Elsecar Main Colliery with my assistance. He was close to W.H. Sansom. About a week after the interview I had a private phone call at home: the caller was Alf.

'Charlie, would you like to work in the East Midlands? If so, would you be prepared to start out as a manager?'

'Yes, Alf, if I could get a bigger pit than mine,' I answered with enthusiasm.

That was it. Someone was interested in me. I felt good. A fortnight later Alf rang again.

'Charlie, could you go for an interview at Bestwood offices on Saturday at 11.00 a.m? Ask for Mr Richard Pogmore.'

There was no doubt I could and did. Two days later I got a further call, this time from the Eastwood No.5 Area offices. The caller I can't remember but the message was very clear:

'If you are not fixed up at Bestwood come straight on to Eastwood. Let us know and we will wait for you.'

Taking off with young John, then aged about nine, in our little Morris Eight quite early on the Saturday morning we arrived in very good time. As directed, Mr Richard Pogmore was located. He settled me in his office, brought me a drink and asked me to wait. A candidate was being interviewed by the selection panel. After ten minutes or so my turn came and I went before the panel in the Area General Manager's office. The interviewing panel were Messrs. Allan Hill (Area General Manager), Norman Siddall and Mike Young (Sub-area Managers) and Richard Pogmore (Area Administrative Officer). I had a fine interview which lasted thirty or forty minutes, following which I was asked to stand by in the adjacent office for a while, as there was still another candidate to be seen. After a short period I was called in to see the panel again. Mr Hill immediately put me at ease and said,

'We are prepared to offer you a colliery manager's position in this area. Would you be prepared to go anywhere within the Area? My answer was yes, but that I would like a pit bigger than the one I was running. Norman Siddall responded by saying,

'We have two pits: one is Hucknall Colliery doing about 15,000 tons per week, the other is Gedling Colliery raising 20,000 tons per week, but here there is a large reconstruction to be carried out.'

Elsecar Main was then doing about 18,000 tons per week.

'I'd love to have Gedling,' was my enthusiastic reply. The reconstruction was a real carrot.

It was agreed that I would be given Gedling Colliery. What a fine choice it proved to be from many aspects. Moreover, it opened up a friendly relationship with Norman Siddall which is still as strong some forty years later as it ever was.

Preparing to leave, I went into the outside corridor intending to find a phone and call Eastwood, but Norman Siddall caught me before I'd moved twenty steps saying,

'Come with me to my office, I want to talk to you.'

As we settled I told him of the Eastwood call and what had transpired, for I wanted to start clean and openly.

'Don't worry,' he said, 'I'll do it for you.'

I could never forget his opening remarks when he made contact with that 'someone' in Eastwood. 'Is that you, Arden. We've got him!'

Further conversation followed at the end of which he said, 'Come with me, we are going for a drink at the Horse and Groom in Linby.'

The public house was little over a mile away and on my way back home. On entering the Horse and Groom at Linby, Norman Siddall settled young John by having the landlord's wife take him tea and sandwiches. He got the beers and we sat down in a corner of the lounge talking generally. After a while Arden Bowker walked in and joined us. Their open and direct approach to me was out of this world. I was comfortable, happy and relaxed in the presence of an Area General Manager, something I hadn't been for a long time following the transfer of R.G. Baker to the Yorkshire Divisional Board. They both knew Gedling Colliery as well as I knew Elsecar Main Colliery and they explained the basics to me most helpfully.

Having had no previous experience of Area politics it was quite revealing to me to be advised whom I could trust and where help could be found. These people were so different in approach from those I was currently working with, that instinctively I felt I could be really happy working loyally for them and again future events proved my early assumptions to be correct.

Looking back, I have no doubt the East Midlands Staff assessed me at the time, better than I assessed myself, for in my wildest dreams I could not have imagined the developments in my career which followed. On our way home I explained to young John what had happened and that we would shortly be moving into Nottingham.

'Does that mean a new school, Dad?' he asked.

'Yes', I told him, 'but Mum and I will do the best we can for you both.'

Mary had never travelled far from home but, although a little apprehensive, she was delighted. I felt a great load had been taken off my mind. It was such a great thrill to have before me a new challenge in a more promising set of circumstances.

It was expected that I could take up my new appointment within about five weeks but it didn't turn out that way. The Area General Manager had appointed Geoffrey Thorpe (one of the university recruits I referred to) as Manager of Elsecar. He, not unlike Raymond Gill, was quite raw and greatly inexperienced. I was kept back for over twelve weeks to help bed him into his new post before he took over my responsibilities as manager. I did my best to help him for I wanted to move on. Had it not been for the intervention between Sir Hubert Holdsworth KC and Sir Noel Holmes, the two Divisional Chairman, it is possible my delay would have been much longer. On my last day at Elsecar Main Colliery, about lunch time, the Area General Manager called in at the manager's office, wished me well in my new appointment and left.

Incidentally, about a year after I took over Gedling Colliery, G.C. Payne accompanied by Norman Siddall visited Calverton Colliery, a new unit in the process of major development. During the visit he turned to Norman and enquired, 'How are you going on with Round? He should never have been appointed higher than an underground overman.'

Norman in his direct manner turned on him, replying, 'You must be joking. There is little wonder the Yorkshire Coalfield is failing to make real progress if you can allow such fine talent to migrate.'

He rang and told me this on his return to his office. We both laughed. I was so happy working with him I couldn't have cared less about my former Area General Manager's assessment. However, thinking back over the situation, I shudder to think of the blight and damage to my early career (indeed anyone's career in similar circumstances) that such a public expression of opinion by one in a high level of management and with great influence could have inflicted. (More so in that we were employed within an autonomous monopoly.) Not having previously worked for the man, how he had developed such views I don't know. Subsequent events, I believe, indicate his views were not corroborated. Indeed, they suggest a lack of his ability in the recognition of potential talent and, further, a lack of respect for the actions of his predecessors so gifted.

During a holiday with Mary and the children in Scarborough, whilst in the harbour I saw my former Area General Manager working on the boat he had recently bought from a Scarborough boatbuilder. I had a few words with him during which he said,

'You'll burn yourself out before you are forty.'

I replied, 'I hope not, for there is so much yet to do.'

I hadn't the heart to tell him that Mary's nephew John Mason had been engaged along with others in the building of his new boat. Although I met him on more than equal terms in later years, the atmosphere was invariably cool between us, and we never entered into any real purposeful discussions. Shortly after this meeting Mary's nephew John, aged sixteen years, received fatal injuries whilst working for the boat builder. He was struck by part of the rim of a wooden wheel which burst whilst it was being trued in a lathe. Gertie his mother never overcame this supplementary catastrophe from which she suffered greatly all her life. She received no compensation for the accident.

I suppose it is possible for any of us to have prejudice, to dislike other people for no logical reason, and then to try to justify our feelings within ourselves. The tragedy is, particularly in a situation of position and influence, that to broadcast such views unbeknown to the subject whose future potential could be then blighted or gravely impaired. One of the greatest difficulties I have had was making an early assessment on the true merits of individuals, although I don't think I made many mistakes in that direction. Surely the best approach in such circumstances is to help each person overcome any weaknesses he may have, so to maximise his future contributions to whatever enterprise is being served.

NEWCOMEN ATMOSPHERIC ENGINE AND NASMYTH STEAM HAMMER

Before taking my leave of Elsecar Main Colliery it would be remiss of me not to provide a short account of some very old features which, despite the fact that the colliery is no longer working and the surface has been fully restored to its original form, still exist. The first such feature is an old Newcomen engine and pump which worked continuously between 1787-1923 and which was brought into operation during an emergency in the mid 1930s. It was however last run for the purpose of demonstration in 1954. The engine is situated alongside an alternative back path to the former colliery locally known as the Stillery Side.

PHOTOGRAPH 14 takes in a view of the enginehouse, part of the rocker beam, counterweights and stroke restriction frame. The Stillery Side original cottages which housed the engine and boiler firemen are clearly evident. This form of steam/atmospheric engine was the forerunner of the James Watt steam engine and opened up technical advances in design and application, from the simple single cylinder, hand valve operated (Wharncliffe Silkstone No.1 vertical single cylinder winding engine) to the twin cylinder, compound and cross compound steam engines with sophisticated valve and control gear which emerged at later dates.

PHOTOGRAPH 15 is a view of one end of the rocker beam at third floor level. The heavy link media which connects the piston rod to the end of the rocker beam is clearly seen. The vertical wooden beams on either side of the beam limit the length of the downward stroke by arresting the movement of the horizontal steel bar attached to the beam. Further along the beam the linkage connecting the pump rods to it can be seen.

PHOTOGRAPH 16 inside the building at third floor level looking outwards to the opposite end of the rocker beam to which the balance weights are connected. The same form of horizontal rod beam stroke limitation arrangements can be seen quite clearly. The new shaft headframe for access to the water level is seen alongside the Stillery Side pathway bounded on one side by steel railings.

PHOTOGRAPH 17 was taken at ground level and shows the steam/atmospheric cylinder, steam and water inlets and control gear. The pump rod connection is visible just in front of the vertical wooden supports.

FIGURE 7 indicates the principle of operation of this forerunner of the modern steam winding engines, in which atmospheric pressure is adapted to move the rods which operate the drainage pumps. This engine was in operation before the James Watt steam engine. Its simplicity made it an ideal subject for use with reciprocating motions such as pumping equipment. It was extensively used for water drainage in the tin mines of Cornwall and the coal mines throughout the UK during the late eighteenth and early nineteenth centuries.

Basically, it is a simple fabricated beam centre mounted on massive bearing support. On the RH side an upper open-ended steam cylinder, complete with piston, is attached through linkage. Connected to the LH, again through simple linkage, is a counterweight assembly to balance the inertia in the system. Valve control and pump rods are connected on either side of the beam as shown. In view of the high mass within the system, stroke limiting provisions are necessary to limit the length of stroke and avoid damage during operation. Such provisions take the form of heavy timber restraining structures at either end of the beam.

The engine operates on the principle that the first stroke of the piston is obtained from the application of steam pressure to its lower side; this lifts the 'rocker beam and pump rods' and delivers the pump's water through the drainage system. At the top of the stroke, water is admitted into the steam cylinder, which causes a rapid condensation of the steam therein, thus creating a vacuum. Atmospheric pressure, acting on the upper open side of the piston, forces it down to the bottom of the cylinder and lifts the balance weights, thus operating the pump for the next water delivery stroke. The balance weights are there to counter-balance the weight of the pump rods and inertia within the system. Thus for all practical purposes, the steam pressure raises the piston and the pump rods and atmospheric pressure returns the system back to status quo, with water delivery every other stroke. A system of rods and controls are linked to the rocking beam such that the plant operates continuously until the steam is cut off. The single vertical cylinder steam engine (James Watt format), originally installed to wind coal at the No.1 Shaft Wharncliffe Silkstone Colliery, had hand operated valves which were later replaced by automatic slide valves initiated by suitable eccentric rods. It is intended that the installation, which has been well maintained, will be provided with adjacent steam raising plant to make it operative for demonstration purposes. Incidentally the engine drained water continuously from the Old Barnsley Water Level (see PLAN 2), for over 140 years. Henry Ford is reputed to have made an attempt to buy this old plant for his Detroit Museum but Lord Fitzwilliam at the time refused to sell.

The second feature takes the form of some very old workshops containing an original steam hammer, which were operated on the basis of 'There is a place for everything. Everything in its place,' as etched into the outside front wall. These shops contained a number of James-Nasmyth designed tools, including a drilling machine and steam hammer which he designed in 1840, and which later were both in use during my association with this fine colliery. In 1850 Earl Fitzwilliam purchased a steam driven 14 ft. diameter ventilating fan designed by him for one of his largest coal pits (probably the old Hemingfield Pit which closed in 1911 at about the same time Elsecar Main Colliery opened up production).

James Nasmyth was born in Edinburgh in August 1809 and died 7 May 1890 in London. He was a Scottish engineer, known primarily for his invention of the steam hammer, which overcame a major problem Isambard Kingdom Brunel had in building his SS *Great Britain*, that of forging its paddle shaft. He designed and built machine tools of all kinds along with a variety of steam powered machines. Beside the steam hammers, he built over a hundred steam locomotives, a variety of pumps, hydraulic presses and other hydraulic machines. At the age of forty-eight he retired and devoted himself to his hobby, astronomy, a subject upon which he wrote a number of treatises.

Gedling Colliery

Here was a quite a different form of challenge developed by people with a very wide and progressive outlook, including Colonel and Robert Lancaster, D.M. Rees, Arden Bowker and Norman Siddall. Its background was cosmopolitan. It had no village mining roots and the men were recruited from anywhere in and around Nottingham. The labour turnover per annum was extremely high, probably the highest anywhere in the UK.

Under my predecessor George Thompson, a fearless and very outspoken North Countryman, training systems were introduced and pioneered which ultimately became the standard for the coal mining industry. George was more than ably supported by a deputy manager, another very strong character known throughout the colliery and beyond as 'Gus Wilkinson'. Gus worshipped George in a manner I suppose similar to the way I held great esteem towards Albert Naylor and R.G. Baker. I could understand this, although in my early days at the colliery it was quite difficult.

George Thompson started out as a miner like myself, qualified through Technical College, and moved up into management in the North-East, spending some time in the Mines Inspectorate before finally arriving at Gedling as colliery manager. The stories about him were somewhat legendary and included the following:

Immediately upon being raised to the surface, the men on the afternoon shift would rush off to get a drink but the nearest pub was quite some distance away. To offset the difficulty, George built them a canteen and equipped it with a licensed bar, following which there was a long-running battle between the AGM and himself about who should pay for the furnishings. This, in my experience, was the only licensed bar I ever heard of located on colliery premises.

During the 1957 Suez crisis, petrol to him was no problem. With his car tank full of prohibited red petrol he was caught in a police trap on Mapperley Plains, Nottingham, near the colliery. Abandoning his car, he thumbed a lift to one of his colliery overmen's home from where he reported to the police that his car had been stolen. Four hours later the police rang back to say that they had located it. Could he have it picked up? Sending the car driver from the colliery, he had it returned.

During one Durham Miners Gala he arranged for all the officials of his NUM and NACOD branches to spend a weekend in Durham. He went with them and introduced them to local lads he knew. Apparently they all had a wild time. It didn't please Allan Hill, the Area General Manager, but it certainly helped him to effect some control over the respective branch leaders.

George Thompson and Arden Bowker each had an identical model Standard car coloured black. The former's had done over 100,000 miles and was ready for change. Arden's car was relatively new. One day he left it parked at the colliery over a long period. On the instructions of George, the fitters changed the engines over. Arden on his return was not amused.

Immediately after the Second World War when cars were very difficult to obtain, a group of officials (including Arden Bowker, Norman Siddall and others) approached one of the leading car dealers in Nottingham with a block request. When the first car was delivered to the dealers they rang George Thompson requesting to know to whom it was to be delivered. He told them that was a decision of the group and that he would ring them back after a decision had been taken. Without undertaking any consultation with the other members, several hours later he called the dealers and told them he had been chosen for the first delivery. When the matter became known he suffered a short term decline in popularity.

MY VOLUNTARY PERSONAL MINDER

The final story I recall related to a workmen and is one which had a sequel involving myself. We had at the colliery a young miner, Bill Marriot, who was exceptionally strong. The lad was an outstanding miner performing two stints per day regularly; his working place was immaculate at all times with all the supports correctly aligned and securely tightened between floor and roof. His problem was drink which caused him to be violent under its influence. One evening he turned up for the nightshift 'worse for wear' and during an argument struck an overman. George was rung at home and told that the man was at the shaftside berserk and uncontrollable. George instructed the caller to get some mates, overpower the man, fasten him on a stretcher and place him between the doors in the return shaft airlock. It took five of them but they ultimately managed to secure the lad as told and placed him in a location which is probably the coldest part of any colliery. The heavy leakage of intake air draughts really are cold; moreover the man was clad with only a thin vest covering for his chest. After a hour in the airlock they went to visit him only to be met with heavy verbal abuse; the second visit, an hour or so later, yielded the same result. After a little over two hours the lad had cooled down and yielded.

On being released from the stretcher he quietly went onto his work, did his stint and came out of the mine with rest of the men at the end of his shift. He had no acrimony following the incident.

The sequels which involved myself were twofold. The first was that the man was an active leader of the first strike Gedling Colliery had for many years; but the second was quite different. One weekend in Nottingham he stole a large sum of money. The notes he kept, the coinage he threw in the River Trent. He was arrested on the charges of stealing and defacing the coins of the realm. He was sent to Lincoln Jail for over two years. After he had served less than twelve months he wrote me a letter asking if I could testify to his character: quite a difficult thing to do in the circumstances.

Thinking through the situation, I felt that the best thing I could do was to highlight his attributes and skills as a workman which I did, adding:

'Tell the prison authorities I would be grateful if they could let you come back to work as soon as possible.'

Within a month he was back at work. I recall that after his first shift had finished he came into my office and thanked me for what I'd done. At the office door he turned with his fists clenched in a fighting posture, saying:

'Gaffer, if anybody says anything about you, or does anything to you, he's mine.'

Thus I had a self appointed 'minder' long before Arthur Daley (George Cole) on TV's 'Minder' series had his, although I never had to use him.

So much for George Thompson at the moment. Lest anyone think him a rogue, far from it. He was a very likeable person with tremendous wit. Like myself he had a great distaste for snobbishness, arrogance and bureaucracy.

My first day at Gedling Colliery convinced me that although my feelings for Elsecar Main Colliery were very deep, here was another colliery which could absorb me even more. Encouragement to do the right thing would never be stinted. Going round the surface with Gus Wilkinson I was given a very detailed account of everything he could think of: what George Thompson had achieved, why Reorganisation Project 201 was a waste of money and why its execution had been held up. During our inspection I recall him saying,

'You can leave the running of the pit to me.'

'Mr Wilkinson, I shall need to discharge my responsibilities as manager of the mine, I can't delegate those,' was my reply. Gus went strangely quiet, following which he said very little more; maybe he felt he had received a snub (which certainly wasn't intended). However, as I found later he wasn't a man to carry ill-feeling or bear a grudge.

The next two weeks were spent exclusively in travelling and inspecting the underground workings. These people were achieving things which made some of our efforts within South Yorkshire Coalfield puny by comparison. Conveyors were being used for coal transport on a greatly extended scale, and individual coalfaces were producing over 600/800 tons as opposed to 300/500 tons per twenty-four hours. Performance rates by the men exceeded those I had been used to, and trade union relations were much easier. However it was clear to me that there were quite a number of ways in which I could make useful contributions: the execution of Project 201 and the development of mechanised loading being two of the main avenues. I was really happy. Norman Siddall was proving to be the person I had anticipated him to be.

Mary, I and the children moved into Stoke Lodge, Stoke Lane, Gedling, a very large house in very nice surroundings about one mile from Lowdham, Burton Joyce and the River Trent. Through Arden Bowker, Mary was introduced to the Gedling Ladies Welfare Committee where, at the Welfare Hall, Mapperley Plains, she was able to pursue her passion for whist two or more times weekly. Here she got to know a lot of the officials and men's wives who greatly respected her quiet and sincere manner. She settled in far better than I had expected. John and Dorothy were placed in a small private school at Burton Joyce and they too were quite happy. We did however have a problem with John, who had to leave this private school after twelve months at the age of eleven.

We had a real problem finding a suitable school we could send him to, for I was desperate to give both children a better starting out chance in life than their mum and I had had. Mary during her whist drive activities along with Mrs Jennings (an old family friend of Arden Bowker's) was acquainted with a Fuller's chocolate travelling salesman and his wife, who had three of their children at the Beckett College, West Bridgeford, Nottingham. He interceded with the head of the college and John was duly called to take the entrance examination. He failed this, basically because in his last five years of schooling he'd been shunted around four different schools. The principal's note following the tests indicated he was not up to the entrance standard for the school. However, the final sentence of the letter stated,

'The lad showed such obvious distress we hadn't the heart to turn him away. Send him Monday next.'

In the examinations at the end of his first year he was placed 23rd in his form of twenty-five students. Within three years he was in the top section of his class and, on finishing his education at the college, had obtained three Advanced Levels and nine Ordinary Levels in the GCE examinations. My thanks are expressed to the Beckett College. They turned out a fine qualified and balanced person, one whose subsequent progress would no doubt be a source of great satisfaction to them if brought to their attention.

After about five or six weeks following my appointment, the colliers, supported by the rest of the colliery, came out on strike, claiming their wages had been reduced. It certainly wasn't due to anything I had initiated. Norman Siddall knew that and he immediately rallied behind me. The staff, Walter Thouless, Douglas Cotterill, Alf Kirkby and Gus, were all at pains to remind me that there had never been a strike at Gedling Colliery since 1926. The inferences weren't too difficult to decipher. The colliers' contract was very different from the standard piecework arrangements which existed on a national basis. This was something special in that these men were being paid a fixed day's wage on the basis of 80 cwts. (coal face productivity) per manshift and for every hundredweight above or below that they received or were deducted an extra shilling respectively. The face productivity over the first few weeks before the strike had fallen about three or four hundredweight, yet there wasn't any significant fall in the basic output to support this decline, though wages had actually fallen.

Light began to dawn when discussions with the Low Hazel undermanager Nelson Limb and his senior overman Arthur Shelton told me that George Thompson, who spent quite some time in the adjacent village of Lambley, had called in at Mapperley Plains colliery welfare hall around the time the men were meeting about to the strike, observing,

'This clever bugger from Yorkshire has come to cut your wages.'

To anyone who has cut his teeth in the South Yorkshire coalfield, particularly in and around Doncaster, strikes are not a new phenomenon and certainly do not initiate panic. Calmly discussing the situation with Norman I asked him to support me in getting rid of the wages arrangement, which he did in no small measure, helping me to work out piecework contract provisions which would meet the situation. He passed these over to Gerry Hollyoak of the Finance Department who found that although we had followed the proper approach some of our detailed figures were in error.

Shortly after Gerry had reviewed our detail, Norman took his family to Scarborough on holiday, from where he sent me by post a small 'abacus'. Embossed along one side was 'Charlie's Computer' together with a short note:

'Now that you are mechanised no more of those stupid mistakes you made in the Gedling Colliery "Price List" calculations.'

Thus the first NCB computer was installed in the Gedling Colliery manager's office late in 1952! It proved very useful in a number of ways and many were the times I was able to get complainants in a good mood by some humorous reference or other to it.

The strike was in its second day and at 5.00 p.m. Norman and I, together with my staff, met the Branch leaders of the NUM (they had no allegiance to me, in fact they hardly knew me) who launched in with a barrage of complaints and orchestrated assertions. We rode these out in the normal manner, following which Norman firmly but quietly said,

'It's obvious to me that there is something wrong here. I can't find what it is but it could go wrong again. We don't want any more of this nonsense. We ought to devise some other system of payment.'

The men were completely taken aback. Whatever had been fed into them in this situation was now irrelevant. He continued by saying,

'Look, go back and discuss it amongst yourselves and the men. In the meantime we will work out something to put before you.'

These people knew Norman extremely well, and such a statement, coming from him, had the desired effect. We met the Branch leaders a day or two later fully prepared. Norman opened up by saying,

'We have worked out an arrangement which will sort out your difficulties and offer better prospects for the men's future.'

The men's views were mixed as to whether to retain the current system or, as I expected, to adopt a new approach. However, we laid great emphasis on the change and avoided all reference to the prevailing arrangements, carefully steering them back to what we had in mind on every occasion they strayed. The Branch NUM agreed to the change and contracts were negotiated, taken back to, and accepted by the men. The whole situation was settled in less than five days. This initial strike action opened up a whole new vista from which all facemen in addition to the colliers derived higher wages including substantial increased coal production benefits to the National Coal Board which later accrued.

The collier's day wage and bonus system was quite a clever control from the management's point of view. Whatever the tonnage produced, manipulation of the allotted face shifts could be undertaken to produce whatever level of wages the management wanted to pay. If the tonnages produced were high the allotted number of face shifts (which did not necessarily coincide with the actual number of face shifts worked) could be increased. If the tonnage was low then the apportioned face shifts could be reduced to keep the wage levels acceptable to the men.

One doesn't determine issues of that nature within a few days of starting a new job unless some incident of that nature highlights the situation. There is no doubt in my mind that the fall in wages was deliberately manipulated to discredit the new manager. When I really got to know George Thompson I accused him of this and of inducing Alf Kirkby, the wages clerk, to manipulate the face shifts in order to cause embarrassment to me, and so to create an initial bad impression within men's minds, which is not so readily eradicated once planted. He didn't deny the act and said,

'Before you started work at Gedling I wanted to meet you, and I asked Gilbert [Winder Mavor Coulson Ltd. sales representative] to invite you down to meet me, but you ignored my request so I thought "we'll give the clever bugger something to think about".'

I was astounded.

'George, I never received any such invitation. Did you ever ask Gilbert Winder if he had conveyed the message?'

'No,' was his answer.

'However, George, as things have turned out it helped me to clear out a system I wouldn't have been happy with anyway and we're making excellent progress,' I told him.

'Now I know you, Chas, I'm glad for you.'

We became good friends after that first frank discussion right through his retirement until he died.

PHOTOGRAPH 18 was taken at a much later date when Sir Norman Siddall, then Board Chairman, was recuperating from a hip operation. George Thompson and I were in retirement. Stephen Siddall, his son, was following an Army career; he is now a Colonel. Robert, the other son, followed his father into coal-mining and rose high within the Board's management hierarchy. He has however since been displaced by privatisation. With the characteristics of his father I feel confident he will weather the situation and find an outlet for his experience and talents.

George Thompson was a boyhood friend of Owen Brannigan, the excellent operatic bass singer who took the initial role of Billy Budd in *Peter Grimes* by Benjamin Britten. Together George and I went to hear him sing the role of Leperello in Mozart's *Don Giovanni* which includes the Catalogue Song in which he really was something very special. We also heard him at the Sheffield Lyceum in Wagner's *Flying Duthman*.

Owen started out as a colliery carpenter. On one occasion, following a concert he had given at the little village of Hope, Derbyshire, he told me that to get to the top in his profession he trod as rough a road as any of us did in ours. Subsequently he was given a honorary degree by the Newcastle University, following which I wrote him the pseudo summons included here. This pleased Owen inordinately, for he wrote to tell me how well it had been received by many of the Savage Club members which included himself, enclosing a number of photographs of himself in his various operatic roles including that of Don Alhambra in the film of Sir Arthur Sullivan's *Gondoliers* (PHOTOGRAPH 19).

One of the Mozart arias that has given me great pleasure over the years is the bass aria *'Per questa bella mano'*, a recording of which I have, sung by the Finnish bass Kim Borg, another fine artist. I remember asking Owen why he didn't feature this fine solo. He told me that the problem was getting an accomplished 'string bass' player. Unfortunately Owen died at a comparatively early age quite suddenly following a very short illness; thus a former mine employee and subsequent great artist was lost to the operatic and musical profession.

Incidentally whilst on the subject, our nightshift telephone exchange operator was one George Extall who lost both his legs in the 1914-18 World War, during which he was batman to John Christie, founder of the Glyndebourne Opera. George stayed with him on a number of occasions after being demobbed and was fully aware of his early intentions in the development of Glyndebourne. Several times he asked me if I would be interested in visiting his former officer's place, but somehow, regretfully, the opportunity was missed.

Following the re-negotiation of the collier's contracts Nelson Limb, Arthur Shelton and William Nicholas started to encourage the remainder of the facemen to extend their tasks to the point where we were able to change their contracts onto straight piecework lines.

William Nicholas was night senior overman in charge of the Low Hazel. Although in his mid thirties, he offered great prospect at higher levels of management. Taking him on one side, I persuaded him to study for his Colliery Manager's Certificate, furnishing him with the books, materials and notes I had collected over the years. He did so and ultimately qualified; and was later appointed as colliery manager in one of Arden Bowker's No.3 (Nottingham) Area collieries.

With the surplus facemen we were now in position to open up an additional district in the C.3's Section of the pit, in connection with which Gus Wilkinson and I had our first and only dispute. The new district was to be served by a rope and locomotive haulage system. I wanted the rope system to be as short as possible, Gus wanted to load further in-bye on an adverse section of roadway grade which I knew would be unnecessarily difficult.

George Thompson's influence was still strong amongst some members of the colliery staff. Normally I didn't mind this, but one morning Walter Thouless and Gus were extolling the virtues of their late boss in my office to the point at which I'd had enough. I reacted very strongly by indicating a number of weaknesses he had left behind, particularly his approach to the wages system and to the execution of the Project 201 reconstruction. Then I indicated that no progress towards getting the new district C.3's started was being made. I recall with a fair degree of anger turning to Gus and saying,

'If you can't get it ready, I'll do it myself.'

Gus turned almost white, got up, walked out of my office, changed into his pit clothes and went through the pit like a hurricane off the coasts of Florida. Anyone in his way, any fault he found, resulted in the victim being flayed. I think that was when I really won his total respect. Moreover, it cut George's influence at the pit so far as I was concerned, for he was never mentioned to me afterwards. When Gus came out of the pit later that afternoon he called in my office and quietly said,

'Mr Round, I'm arranging C.3's loading station where you want it and the men are starting on the job tomorrow.'

No malice, no complaint: that I found to be Gus at his best. He really was a fine person, having formerly served in one of HM Guards Regiments.

The loading station he put in, and the standard of workmanship behind it, was a great credit to his immense ability in that field. It was impeccable. We had the new C.3's unit working within about six weeks, boosting the total output by about 600 saleable tons per day. Never, in all the time I was in charge of Gedling, did we ever have a similar incident. In fact both he and Mrs Wilkinson opened up the beauties of Derbyshire countryside around Buxton, Via Gelia, Matlock, Crich, and so on to Mary and me, and we often visited them together at weekends.

Calling Gus in shortly after this I said, 'Gus, let's get started on Project 201,' to which he replied,

'All right, I'll get Jimmy Gilks and Bill Shaylor together with some men. They will start in the upcast shaft within a day or so.'

True to his promise, he made the arrangements, following which he became fully absorbed in its execution. Project 201, at a capital cost of over £2.5 million (1952) in simple terms was the installation of 2.5 ton mine cars in both the upcast and the downcast winding systems and the construction of a locomotive circuit interlinked with both coal winding shafts, with modern signalling and control systems to match. In part, it was one of the most difficult and dangerous pieces of construction work I have ever been involved with. We had to work right up to the shaft sides under-pinning the shaft walls, substantially modify the headframe and deepen one of the shafts and at the same time maintain production (now well over 24/25,000 tons per week). I was now really in my element and Gus was fantastic. He had a tremendous control of detail and a memory to match.

Together we built up the labour force and started further works at other points within the two circuits. Real progress was now being made. Norman Siddall was delighted whenever his pressure and support in any matter, finance, materials and whatever was needed; he responded swiftly and with positive results. Arden Bowker, a co-author of the scheme who lived within a mile of the colliery, occasionally called to see me and review the progress being made in its execution.

It was about this time that Harry Wharton, the gardener I left behind at Elsecar, came down to see me at Stoke Lodge. When I arrived home he was sitting with Mary and the two children, who naturally knew him exceptionally well, having tea. He was very depressed and unhappy.

'What's the problem, Harry?' I asked.

'I want to leave where I am and would like a testimonial. Could you please give me one?'

'Of course, but first how would you like to come here and work for me again as gardener?' I asked, having difficulty in getting someone for the job. His face lit up with tremendous relief.

'Can I?' he almost choked.

'Yes, you can live with us,' I answered.

Stoke Lodge (subsequently converted into a number of flats) had many rooms and there was no problem; thus Harry became a member of the family for some twenty-five years or more, almost until his death at the age of seventy-three. He had been born in the little village of Milton about a mile from Hoyland, worked at Elsecar Main for a number of years, contracted lung-dust problems as a collier and had to leave the pit. He had always been a keen gardener and was an excellent workman. He was a great help to us all, including the children and Mary. He gave us great flexibility to attend functions and conventions as they occurred. The conventions Mary enjoyed were those of the then National Association of Colliery Managers where annually she met friends she had made from many parts of the country.

With Project 201 moving well, now was the time to turn to the second intended phase, mechanised mining. We were working two seams at the colliery, the Low Hazel and the Top Hard, approximately 3' 6" and 3' 9" in thickness respectively. In the former seam we had very serious but long term problems. If working under large areas of Burton-Joyce, Lowdham or under the River Trent, surface subsidence of three or four inches would undoubtedly result in extensive flooding of a wide area of the Trent Valley. The current system, with faces of 400 yards or more applied in that particular area, would give rise to catastrophic disaster. We had many years lead time with which to find an answer but it had to be always at the back of our minds. The answer, when found, would assuredly call for some form of intensive mechanisation for the coal to be mined profitably.

In the UK the situation transferring the coal along and from the coalface evolved in much the same way as with other facets of mining. In the very early stages face transport was by sledged boxes. Subsequently they were fitted with wheels and rail tracks provided to make small tubs, which were filled then pushed and pulled along the face into the mine roadways. Here however we are concerned with the development of mechanised face transport. From the 1930s and into the 1950s face conveying underwent great technical improvement. The early steel shaker troughs and structured top belt conveyors were superseded by the bottom-belt conveyor, which is probably the simplest form available. Real progress followed in the early or mid-1950s with the importation of the German *Panzerförderer* (armoured chain conveyor).

In the Silkstone seam at Elsecar Main colliery, compressed air driven reciprocating troughs (pans) were used on dipping gradients and 16-inch face belts applied on rising gradients for clearing the machine hand-got coal from the coalface. The use of both shakers and top belt conveyors involved a great deal of manual work in breaking down and rebuilding the relevant structures every twenty-four hours whilst transferring them over into a new face-operating position. The simple idea of loading the coal onto the bottom belt and using transfer scrapers at the point of discharge, i.e. the junction of the coalface and the loader gate or roadway, was pioneered at that time by Messrs Huwood's Ltd, Newcastle-on-Tyne. This arrangement evolved to become the general approach until the final application of the armoured face conveyor was introduced.

In PHOTOGRAPH 20 the open bottom belt is seen to be loaded and running on idler pulleys. The top belt runs over lengths of steel tube secured between the supports. The driveheads for these belt units were normally independently situated on either side of the gate (mine roadway) conveyor at the roadhead. In this situation there was a degree of flexibility which allowed greater tolerance for difficulties between the LH and RH face conveyors loading out the two sections of the whole coalface.

In conjunction with Messrs Huwood Ltd, R.G. Baker had already designed and built a double bottom belt-drive unit, which I introduced throughout all seams at Elsecar Main Colliery following its initial delivery and early success. Extension of the system was greatly facilitated following the modification shown in Photograph 10. During my first few months as manager of Gedling Colliery, on the basis of the original designs, I was able to convert two standard M & C face driveheads to the same purpose and extended them successfully throughout the Low Hazel Seam.

PHOTOGRAPH 21 features an *in situ* installation. The arrangement saved a great deal of time during preparation work prior to the coal-filling shift. All the redesigned belt driveheads worked admirably, the loss of flexibility in the former arrangement being offset by a natural discipline which ensured the whole face, RH and LH sides, had to be properly worked to ensure systematic twenty-four hour cyclic production.

Early Steps Towards Coalface Mechanisation

The problem of total coalface mechanisation was high on our agenda for the next stage of progressive development. My experience was limited to the Huwood loader which gave a reasonable amount of success as a loading machine in the Thorncliffe seam at Elsecar Main Colliery and also the Samson stripper in the Parkgate seam at the same colliery. I had visited a number of 'coal plough' installations in the Ruhr coalfield, Germany, but none of these systems approached what was being sought as a final solution. The American mining systems differed widely from our own in that their approach took the form of 'bord and pillar' layouts as shown in FIGURE 8. This depicts a 'five entry' system in which the main productive accent was based on the high production and productivity drivage of coal headings within the seam. Over many years, they had built up arrangements and modified systems, embracing advanced forms of applied mechanisation which was ahead of ours.

In this connection, early short-wall-cutters and duckbill-loaders were later superseded by advanced short-wall machines and 'Joy Loaders'; thereafter followed by continuous-miners, pickup loaders and shuttle-cars, producing coal at many times our productivity rates and at a fraction of our production costs.

The next two photographs, which date back to late September 1958, show the later type of equipment that was used with this type of mining. They were taken in one of the Clinchfield Mines of the Pittston Company, Dante, Virginia, USA in a seam up to 23 feet in thickness.

PHOTOGRAPH 22 shows a pickup loading machine filling a large shuttle-car, which transports its loaded coal to a disposal point for discharge onto a feed or trunk belt conveyor by which it is taken out of the mine. The machine is fitted with a pick-crushing device (the three star wheels) which reduces the excessive size massive coal pieces into more manageable form.

In PHOTOGRAPH 23 the loaded shuttlecar is shown in transit to its particular discharge point. I cannot remember the capacity of this particular vehicle, but its appearance would suggest some 15 tons or more, all of which would be completely emptied in a matter of a few seconds.

The American Mining Engineer is particularly adept in analysing and shortening the times associated with non-productive operations, modifying and adjusting his systems to best effect accordingly, no matter what type of process is being performed. The current stage in coalface transport development followed the introduction of the *Lobbehobel* or German panzer conveyor, developed over the years to rated capacities and reliability performance inconceivable at the time of its conception. This really has made for the current formerly unimaginable success of power loading.

Gus Wilkinson had formerly had experience with the Mecco-Moore cutter loader, the original of which was available on the Gedling Colliery surface. It was overhauled and put back into service. Whilst reasonable results were obtained, it was not, nor was likely to be able to provide, what we required.

One morning, Norman called in and picked me up for an inspection of the new Calverton colliery. The shafts were being sunk and surface structures for housing the various plant were being erected. Although I had been before there was still much to see. Whilst travelling through the temporary stock yard, I noticed a German panzer (steel chain) conveyor equipped with a coal-cutting machine fitted with a swan-necked cutting jib. I knew immediately what it was and for what purpose it had been obtained: that of providing a continuous coal-producing system in which coal was being loaded out on three shifts per twenty-four hours, as compared with the single producing shift per day in operation.

'I'll take that, Norman,' I said.

'You can't have it. It belongs to Les Boyfield, Manager of Bestwood Colliery,' he responded.

Bestwood was another of his colliery responsibilities.

'I'll install it in a weekend,' I rejoined.

'Don't be a stupid clot. It's not yours. You're not having it,' he said.

'I'll bet you thirty shillings I can install it in a weekend within the next month,' I pressed.

I could almost see wheels turning in his mind. He certainly wasn't the sort of engineer that wanted plant standing idle for lengthy periods if it could be put to work. With a dry humorous twinkle he said,

'Shut up. We have got to get coal from other pits besides yours.' I went back to Gedling and told the staff what I'd seen and that it offered some interim prospects. They were quite enthusiastic. Within a few days I had a phone call and with the directness of expression that exists only between two Yorkshireman on friendly terms, a voice said,

'You dozy clot. Have you installed that panzer yet?' It was Norman. 'Come and pick it up. The bet's on!'

Within a few minutes of the respective phone handsets being replaced on their bases the yard lorries were on their way to Calverton Colliery. Gus, Arthur Lomas (undermanager Top Hard Seam), Ross Bramley (electrical engineer), Jim Stone (underground overman) and myself worked out a strategy for the work in hand. I hedged my bet by splitting it amongst the team. Together with my clerk I drew up a humorous legal document in relation to the bet, which we all signed. Two copies were sent on to Norman, one of which he signed and returned. Gus and I signed ourselves as 'Gus (Slusher) Wilkinson' and 'Charles (Panzer) Round' respectively. We intended to have some fun out of the exercise and Norman fully responded. In fact he retained that document throughout the whole of his career and maybe still has it.

We organised the work down to the proverbial nuts and bolts and left nothing to chance, for we really had no idea of the size of the task before us, being in new fields of action. By the Friday night prior to the weekend chosen, we had all the equipment in place in the relevant Top Hard 7s District. The last coal-stripping shift of the old system was being worked on the Friday afternoon and, as is usual in such situations, things ran badly. It was not until 2.00 a.m. on the Saturday morning that we were able to start the installation: a delay of six hours in comparison to that projected. However, by 12.00 noon, Sunday lunch time, we had completed the installation, although on test we had problems running the panzer (not surisingly as it may seem to be in retrospect). The problem was that we had the conveyor-head and the tail-end motors in opposition; since this was our first experience of a conveyor being driven at both ends it was somewhat excusable. By 5.00 p.m. everything was ready for the Monday morning shift. We came out of the pit to beer and sandwiches. Moreover we'd won our bet. Norman paid up and later he said,

'You have saved a great deal of money and given me a yardstick for the future,' and knowing him there is little doubt but that he used it.

Although the system was an interim one, men still had to load the coal out by shovels. It worked well. We had equipped the face with Dowty hydraulic props and N.R. Smith, the Area Production Manager, made his contribution by modifications in the support system in the shape of a device (christened 'cabbage--heads') which fitted on top of the hydraulic props thus allowing the horizontal roof-bars to be advanced without having to release the vertical support loadings.

PHOTOGRAPHS 24, 25, 26, 27 & 28 were taken of the top hard coalface installation working at the time. They show quite clearly the sequence of cyclic event which is common to modern installations.

PHOTOGRAPH 24 shows the face filled off with the original German friction type props and crown steel corrugated bars.

PHOTOGRAPH 25 features the changed support system in which the Dowty hydraulic props and 4" by 4" rolled steel joists replaced the original system.

PHOTOGRAPH 26 indicates the machine in its starting position and ready to undercut some 180 yards of coal face. (This to be subsequently blasted and loaded out by hand.)

PHOTOGRAPH 27 shows the situation after the coal has been cleared away and the machine is being flitted down the face to prepare for the next 3' 4" slice to be taken off. The Dowty hydraulic prop and steel joist bar roof supports embracing the cabbage-head devices are shown clearly in this sequence of pictures.

PHOTOGRAPH 28 completes the whole picture of the events. It shows the waste collapse after the back row of supports have been withdrawn. Note how a couple of friction waste control props have been reeled out by the flushing rock. The situation indicated is one of a heavy waste collapse: a little rough, but it has not affected the security of the working area.

The sequence examined just about sets out the basic organisation of the modern power loaded face, other than the collier's shovel which in subsequent systems has been relegated to minor use. Jim Stone the overman did an excellent job for us and the results we obtained were very good in the circumstances, better than with the normal machine cut hand loaded arrangements. Incidentally, Jim Stone's son Jim Stone Jnr ultimately became Director of the Doncaster Area of the National Coal Board.

The same system was set up on T.12s in the Low Hazel Seam and under the direction of Arthur Shelton, senior overman, we obtained face advances of 10 ft. per twenty-four hours as compared with about 5 ft. on the main producing faces. We obtained a great deal of experience throughout the colliery by the application of these interim stages of coalface mechanisation. Moreover, it whetted our appetite to work towards the ultimate stage of total coalface mechanisation.

Shortly after getting the system installed Norman Siddall was promoted to the post of Area Production Manager for the East No.5 Eastwood Area where he worked under William Unsworth, his new Area General Manager. Despite this, we maintained contact. I recall a most embarrassing incident. He rang me up one night and said,

'Charlie, come over. I'm going to look round the surface at Moorgreen Colliery; we're doing some reconstruction work.'

I joined him and spent a couple of interesting hours looking at things. I recall on returning to his home saying, 'Make sure you have no oil on your shoes,' having made a cursory look at my own. Peggy, his good lady, had just fitted a brand new costly carpet in her lounge. Whilst we were having coffee I looked down at my feet and on the carpet was a nasty oil smear.

I said, 'Peggy, look, look!' in real panic.

Quickly she applied Brylcreme and reduced the mark considerably, whether or not completely I can't remember, but I do recall she was most kind in the circumstances. Norman turned to me and said,

'You lucky sod. Had that been me, she'd have flayed my skin off.'

Had I been responsible for such an accident at home, Mary would given me the same treatment. What had happened was that a small blob of oil had lodged in my shoe instep, although the heel and soles were quite clean. My foot, sinking into the pile of the carpet, did the rest. On all future occasions I used to think of taking my shoes off in the kitchen, with Peggy making appropriate fun.

Each time I moved up into a new post, without exception I would receive a most encouraging letter from him together with advice based upon his own experience. Working with Norman, as with R.G. Baker, was a very happy and rewarding experience. I was never constrained and, once I had proved my point, I was fully supported thereafter. Apart from taking advice from time to time, I avoided embarrassment either to him or myself by never requesting favours, or asking him to use his influence on my behalf in promotional endeavours. Support in any difficult situation was to me far more valuable. Any progress I made had to be my own responsibility in being able to provide at any time what was wanted, by who-so-ever required it: a philosophy I have adhered to throughout the whole of my career.

Shortly after Norman left us, Allan Hill, the Area General Manager, called in my office one morning.

'Come with me, Charlie, I want to go out to see the Burton Joyce borehole and to talk privately - I can't do that either in your office or mine,' (shades of MI5).

We reached the borehole, got out of his car and stood together looking over the fence at the brick cover which marked the location of the borehole.

'Charlie, I'm going to reorganise the Area by splitting it into three groups. You know Walter Sharpe (Norman Siddall's replacement) has taken the No.1 Group and Mike Young has the No.2 Group. I'd like you to take the new No.3 Group. How do you feel?'

'Very much as though I had won the football pools, Mr Hill,' I answered.

'The new Group will be Gedling, Clifton, Radford and Wollaton Collieries and later a new development, Cotgrave Colliery, a new pit to be created on the outskirts of the village of Cotgrave,' he said. 'Keep this strictly secret. Don't breathe it to a soul. Prepare and send in a short application note when the job is advertised,' he instructed.

We returned to the colliery and he went back to headquarters. I called Denny Knight, the mine surveyor, an excellent person who had unsurpassed detailed knowledge of the colliery.

'Denny, bring me the plan of the potential Low Hazel workings under Lowdham and Burton Joyce and showing the Burton Joyce borehole.'

Denny brought in the plans from his office next door.

'Denny, Mr Hill and I have just been down to the Burton Joyce borehole and looked round Lowdham. He's worried about future workings out there and has asked me to give thought as to what form they should take,' (which wasn't untrue in that we had discussed the possibilities of workings out there). Denny's reaction was quite direct.

'I don't know why he should worry. It will be years before you want that coal. In my opinion it will never be worked.'

Denny was truly prophetic as events have proved. However this would satisfy the enquiring minds at the colliery, of which there was no shortage, as to the purpose of the Area General Manager's rare visit.

Although we had achieved great progress with the new panzer system and the results were very good, we still needed to get rid of the collier's shovel. Moreover, we had to do something to reduce the heavy manual work in the roadhead construction operations at the coalface.

At Shirebrook Colliery in the adjacent No.3 (Edwinstowe) Area, the manager, Jimmy Thompson (no relation to George Thompson although he did have a brother called Jimmy and he too had mining qualifications), had adapted a simple hoist and bucket machine for use at the ripping lips (the roadhead at the coalface where the roadway is formed and enlarged). The basic system was very much like that I had previously tried out for R.G. Baker at Elsecar Main Colliery. After making sketches of the system I handed them to Gus. We had much of the basic equipment, i.e. a small hoist, bucket, pulleys and signalling devices, available. Within a week he had the first unit installed and working successfully, saving two men per unit and eliminating much of the heavy manual work associated with such operations.

There was one disadvantage: the men were exposed for a short period to working under unsupported ground. To overcome this, I designed a portable canopy and gave the drawings to a young fitter to work on and construct. He did a good job and it worked quite well, following which we installed quite a number. Submitting the idea to the Awards Panel I received £250, half of which I gave to the mechanic, as implementer of the idea. Some twelve months or so afterwards, one Sunday morning, his body was found gravely dismembered and broken at the bottom of the No.1 coal winding shaft in mysterious circumstances. There was no work taking place at the shaft and he was dressed in his civilian clothes. I cannot recall what the coroner's verdict was, but the men at the colliery felt it to have been suicide.

PHOTOGRAPH 29 shows one of these developments which were installed at a number of roadheads throughout the mine. Here the safety canopy is in its closed position when a standard depth of strata advance has been blasted down, the sides and roof are trimmed and the top sliding section of the canopy is pushed into the unsupported exposed area to give temporary protection. The fragmented stone produced by blasting is cleared away by a rope hauled mechanical skip into R and L side roadhead packs and when totally cleared the permanent supports are erected.

Note the foot of the LH arch - it has, welded on either side, a short length of identical arch section into which a block of wedge shaped hardwood is inserted, for the purpose of accommodating early road subsidence and avoiding the initial damage to the roadway supports which had previously prevailed. This simple device earned me a small award from the Divisional Awards Panel.

In connection with our mechanisation aspirations, I persuaded Gus to join me on a trip to a huge mining exhibition in Düsseldorf, Germany in 1952. Kreuger Park I believe was the location. The range of equipment covered every aspect of the coal mining engineer's needs from boring, sinking shafts, transport systems, coal face mechanisation, ventilation, to coal cleaning and beyond. Together we spent hours absorbing all we could. Although unable to understand the German language spoken exclusively at the various lectures we attended, we did follow, with the projection of drawings, sketches and photographs, the gist of what was being considered. We really had a rewarding time and picked up a fair amount of helpful experience, our main interest being the 'coal plough' which could bridge the gap preventing a full realisation of totally mechanised production. Underground inspections were fixed up for us, and we visited working coal plough installations, the faces being supported by steel friction type props which functioned quite well in the soft coal of the Ruhr coalfield.

Shortly after returning home, Gus told me of a coal plough installation operating quite successfully at Pleasley Colliery, Nottingham. After coming out of Gedling Colliery following an inspection I rang up the Manager and asked if we could visit the installation.

'We're inundated with visitors on the dayshift. Your best bet is to come in the afternoon, can you do that?'

The outcome was not only that we could, but that we went that same afternoon. The trip was most illuminating since the equipment was working successfully in harder coal than was the case in Germany, something akin to our own situation. Within a short time we had a complete coal plough installation, with which the mechanised handloaded panzer operation on the T.12's Low Hazel district was upgraded. We had one hell of a time over a very long period, trying to make it work. Compressed air-jacks which advanced the equipment gave immense trouble (hydraulics at the time had not been introduced). The plough could not be controlled. Roof control also became a very serious problem.

PHOTOGRAPH 30 shows a view of the actual plough in operation. Coal hardness and soft floor conditions created such mining difficulties that the installation had to be withdrawn. It was obvious this was not to be the answer and a search for an acceptable solution needed to be found, with other mining systems yet to be developed and proved. Maybe we (all mining engineers) were somewhat like the old Knights of the Round Table, searching for the 'holy grail' in all directions.

Towards the end of my first twelve months in charge of Gedling Colliery we had an overwind at the Main No.1 Winding Shaft. The top cage was raised to a point in the shaft headframe at which the winding rope detaches from the cage which is then left hanging from the safety bell attachment in the headgear structure (otherwise it would have a catastrophic fall down the shaft). The bottom cage bounces with tremendous force on the pit bottom sump baulks, the impact being damaging to both the cage and the structure. In this case, the cage was destroyed. Thankfully overwinds are a very rare event. (This was my first and only one, within a period covering some sixty years association with the industry.) They are extremely dangerous, especially if men are riding in the cages. The winding engineman was a brother of the Group mechanical engineer and had been winding without incident for more than twenty-five years.

The colliery engineer and I investigated the occurrence and subsequently, with the mines inspectorate, were able to prove without a shadow of doubt that the cause was an instantaneous lapse on his part arising from defective eyesight. He protested his innocence for some considerable time but finally admitted that for more than two years he'd been having trouble with his sight. After having given excellent service for a long time it saddens one to have to remove a person from his post, but such a job calls for the maximum rating of physical, mental and moral capabilities. We were able to place the engineman in suitable employment which covered him until his retirement. The resultant surface and underground damage took three days to restore before we could get the shaft and winding system back into commission.

To offset tonnage and wage losses, we extended the winding capacity of the No.2 shaft to a maximum by a manipulation of the face men within the Colliery.

For quite some time the Gedling management took a great interest in its old retired miners. Once or twice a year they were taken by bus to the beauty spots in Derbyshire, Lathkill Dale, Bakewell, Buxton and other places for a drink, a meal and a day's outing, I believe George Thompson initiated the event on an annual basis. The funds were voluntarily given and augmented by the fines Gus levied on the men for breaches of work standards or contraventions of the Mines Act. Gus was so enthusiastic in his supplementary funding of the required resources that it virtually assumed the importance of the colliery's weekly output.

Each Friday afternoon, quite a large number of men would be standing in the office corridors waiting to see Gus and to sign a note permitting recovery and the transfer of the fine. He would call in to see me.

'Eighty-five pounds for the funds today, Mr Round.'

'Be careful, Gus, don't push too hard,' I replied.

In the early days the men didn't mind so much, but this had by then been going on for a long time and a reaction was beginning to emerge. Over a period of time I called in each member of the NUM branch and got his feelings and support towards a weekly stoppage of 1s per man and 6d per boy, the funds being donated to the retired pensioners' activities. They all agreed, so that finally it was put to the men and they accepted. Thus the fund was boosted by over £100 per week and the office corridors were much less crowded on a Friday afternoon. I felt quite guilty about this, as Gus got so much pleasure out of doing something for the old timers. However, he did not relax his efforts. Notwithstanding that a permanent fund is much better than a temporary variable one, he took it in his stride and was no less dedicated towards them, still augmenting the foundation funds in his own way.

Earlier in this narrative I mentioned the circumstances of splitting the No.6 Area into three groups. The advertisement was published and I applied as requested. The interview was a straightforward affair. The members of the interview panel were W.H. Sansom, Allan Hill, and N.R Smith who all knew me very well. I was given the job and with it potential experience of management at Group level.

My successor at Gedling Colliery was a youngster, Robert Scott, who had a good practical basic background and had received a fair degree of hardening under Wilfred Crossland, the then manager of Hucknall Colliery, following which he had been appointed assistant manager of Linby Colliery under its manager William Bowen. I felt him to have greater potential than the university graduates I was acquainted with and whom I had left behind at Elsecar Colliery in the Yorkshire Area. He was raw for a position the size of Gedling, particularly as it was his first managerial appointment. However, sandwiched as he was between Gus Wilkinson and myself, he was under great pressure to try to satisfy his superiors with an insatiable appetite for progress and achievement, in a situation which would have been a test for an older, more experienced manager. Although we treated him kindly, knowing that he would develop and emerge successfully in his new post (as might be expected in such circumstances) stress got to him, although he may not have thought so at the time. Gus gave him fantastic support throughout his career at Gedling Colliery as he had done with all his predecessors, including myself.

The situation broke one Saturday night. Mary and I were watching television when the phone rang and a very angry voice shouted down the phone,

'My name is Wilfred Scott. I'm not having you treating my lad the way you are and I can do something about it.'

It was Robert's father who was at the time Divisional Welfare Officer. He had been for a while, before nationalisation, Agent of Clifton Collieries. My reaction was quiet but direct.

' Mr Scott, if you want to discuss your boy's treatment you can meet me in the manager's office at Gedling tomorrow morning, at 10.00 a.m. I'll be there,' and I put the phone down.

'What's all that about?' Mary asked, to which I replied,

'Nothing much really. Bobby Scott's father wants to meet me at Gedling at ten o'clock tomorrow morning.'

'That's a bit late for you, isn't it?' was her response. She could be mischievously and pointedly dry with her humour on such occasions.

On Sunday morning at 10.00 a.m. I walked into the Gedling colliery manager's office. Wilfred and Robert, father and son, were waiting. Now speaking angrily over the phone is quite different from meeting face to face. The atmosphere was very pleasant: no mention of the phone call or of Robert's treatment. Turning to the son, I said,

'Robert, I understand I have been treating you badly. Now tell me the circumstances.'

Robert was distinctly embarrassed and couldn't reply, so that father Wilfred protested saying,

'You're pushing him too hard.'

'Is that what you feel? Let me tell you, Mr Scott, it is not me that's pushing him too hard, it's the job we have to do at this colliery, and if you are now saying he's not up to it, where do we go from here?'

Wilf, as I later used to address him, almost rocked and managed to say, 'It's not that. I am saying here and now that one day you will both be Area General Managers.'

Later this prophesy almost materialised. Concluding the meeting I said,

'You'd better leave the lad to me. He won't get hurt under my direction so long as he does his best; I have a loyalty to those who work for me.'

From then on Robert stood on his own two feet, obtaining sympathy as appropriate from Gus who sustained him like a father. I couldn't have wished for anything better for him than to be closely developed by Gus Wilkinson. I got very fond of him and I still am. He developed a delightful sense of humour and like both Gus and myself was industrious and open to progress.

On one occasion (his confidence having developed in the job) I called in to his office early one morning to find him sitting at his desk strewn with cost sheets and performance detail. He looked up and said,

'Boss, on Monday whilst I was driving over Mapperley Plains to work I thought, I'm sure the old sod will hammer me on the fall in saleable percentage, but you didn't, it was lack development progress in the Low Hazel Seam,' continuing, 'On Tuesday on coming to work my mind was turning over. It's bound to be material costs this morning - he hasn't raised that for some time. Oh no it wasn't, it was overtime costs,' still continuing, 'Again on Wednesday, it's sure to be Dowty Prop losses - no, I never get it right - it was tonnage loss through breakdowns. Please, boss, can I have written notice of your questions, like they have in Parliament?'

Now who but an arrogant self-opinionated snob could but not be humoured by such an approach? Robbie, so far as I was concerned, had matured and made excellent progress in highly creditable stages. Following the National Coal Board's 1967 administrative organisation reconstruction he was made Area chief mining engineer for the North Nottingham Area, quite an important post. Later he was promised the Deputy Area Director (Mining) post for the same Area; but when the position was made vacant the promise was reneged upon and another person was appointed. He subsequently, however, received recognition in the form of an important Queen's Award in one of the HM Honours lists.

Whenever I met Robbie in other areas, which I often did, he would take great delight in reciting these matters and indeed many others which time has erased from my mind, although he naturally tended to be a little prone to embellishment. At one time, he had a great interest in obtaining old motor cycles and undertaking their renovation. Wilfred, his father, and I became quite good friends. Following the incident we met quite often, although we never discussed the work we were doing at the time.

PHOTOGRAPH 10.
DOUBLE BOTTOM BELT DRIVEHEAD: ELSECAR MAIN COLLIERY

This arrangement, based on R. G. Baker's original Huwood Ltd design, I have always felt proud of in that it was compact, and was relatively easy to construct. Recognising that it was possible to remove the unit's delivery head and mount it at right-angles under the driving rollers and motor turned upside down, the rest was easy. This concept is clearly seen on the RH unit which shows the thread of the belt under its normal 'op belt' operating condition using face conveyor structure for which it had been designed. Mounted on goalposts at either end, which accommodated the plough scraper, it made a very solid and reliable unit which, fitted with wheels, ran in a short length of channel irons.

Each had two 15hp motors which were capable of operation or face lengths up to about 100 yards or more where a good system of 'belt rollers' supporting the bottom loaded belt was provided.

PHOTOGRAPH II. MAVOR & COULSON SAMSON STRIPPER:
PARKGATE SEAM, ELSECAR MAIN COLLIERY

This was quite a unique development somewhat on the principle of the German Coal Plough which was developed in the Ruhr during the Second World War.

Hydraulic power activated the machine across the face, eliminating the plough haulage chain, and powered the 'steel wedges' splitting off the front coal of the seam whilst in transit across the coalface, by hydraulic rams. Taking a 9 inch depth of slice of the full seam thickness it more than doubled that of the plough unit.

In simple terms, the machine consisted of two very powerful hydraulic jacks. The horizontal jack motivated the forward thrust of the wedges whilst the vertical jack anchored the machine between roof and floor during that horizontal movement. The horizontal thrust equated to 200 tons, the anchoring thrust being 90 tons. The forward movement of the shearing wedges was 2'6". At the termination of the forward thrust the vertical jack was retracted and the anchoring unit was closed up to the leading wedges, being reset in that position ready for the next cyclic stroke. In this manner it was advanced bidirectionally across the coalface. The mining conditions were excellent: a sandstone rock roof and a hard floor in a de-pillaring area in which the coal had been subject to heavy pressure making it easily worked. Although it worked quite well it had the disadvantage of a low depth of bite and slow transit across the coalface. It was not, nor could be, a high tonnage producing machine.

PHOTOGRAPH 12.
MAVOR & COULSON SAMSON STRIPPER: ELSECAR MAIN COLLIERY

This is a view showing the machine in operation and advancing towards the observer. The coal wedged off the front of the exposed coalface falls directly onto the face chain conveyor which was specially designed as an integrated part of the whole system. It was a centre strand chain type unit 18 inches in width and running at about 150 feet per minute. The conveyor was advanced immediately behind the Samson Stripper in both directions. Sylvester type jacks were used (see Photograph 33 for detail of format). Because of the shallow 9 inches of advance per strip this was no difficulty but it was a timely operation.

Another new feature of the installation was the Downty Hydraulic Prop, a spin-off from the Dowty Aircraft Undercarriage. These, being a huge success, swept through the industry, later to be developed as a hydraulic mechanised support. The German mechanical type 'friction prop' – developed alongside the coal-plough and armoured face conveyor – was being introduced within the different coalfields at the time. However the Dowty Prop was more easily handlable and, set with a predetermined load, was much safer to withdraw and recover. The 550 Volt 3hp 50 cycles power supply cables are to be seen suspended from the supports.

PHOTOGRAPH 13. RICHARD TODHUNTER SNR., PRESIDENT, BARNES &
TUCKER, PA, USA

Richard Todhunter Snr. was the first American mining engineer I met. Although quite advanced in years his youthful and enthusiastic approach I found to be most stimulating. Moreover he was most encouraging to me at the time I was beginning to emerge in my career.

His two sons Richard Jr. and John became quite good friends. I met them on a number of occasions during visits to the American coalfields. Here, together with Mrs Weltha Roberts and Richard Todhunter Jnr., they are about to make a trip underground into one of the Barnes & Tucker Mines, Barnsboro, Pennsylvania. The lady, Mrs Weltha Roberts, now well into her 80s living in Philadelphia, was the wife of the Company's Treasurer and the mother of Mrs Lynne Adams referred to in the narrative in a most amazing set of coincidences during a joint flight between Birmingham (UK) and New York, during which we shared our experiences over some seven hours discussion.

Lynne, an English teacher married an English mathematician, has lived in Coventry for more than twenty-five years. She was the stimulus which encouraged the writing of these memoirs, a creative task which has given me great pleasure.

PHOTOGRAPH 14. NEWCOMEN
'ATMOSPHERIC' ENGINE I: ELSECAR MAIN COLLIERY

This view takes in the Old Newcomen pumping plant, the Stillery Side cottages and, in the foreground, the more modern pumping arrangements which replaced the former plant in the mid 1920s after almost 140 years continuous working. Water was raised from an old Barnsley water level giving protection against inundating the deeper seam which followed, and in more recent times the workings of the Haigh Moor Seam.

The Newcomen Engine is situated within the enginehouse and at the time I last visited it during the early 1970s was in excellent condition. The workmanship associated with the plant needs to be seen to be really appreciated particularly in view of the primitive tools which were used by the craftsmen in those early days. In 1957 a steam range was taken from the colliery boiler plant and fed into a 'steam receiver' on the outside of the building. As such, the plant was commissioned for demonstration purposes over a short period.

The three storey enginehouse is quite unique in that a framework of specially sized and selected stones forms a structured framework as with modern steel framed buildings. This can clearly be seen in the picture. The Stillery Side cottages, it is understood, were built for the pump attendants and boiler firemen. In the foreground is the more recent access shaft and pumping plant on the left-hand side of the picture. This was fully automated covering starting, running period and closing down. Under a projected Heritage Project it is understood the plant is to be placed in an operating condition.

The Old Barnsley Water Level from which the Newcomen pump installation drained water for some 140 years can be seen on Plan No.2.

PHOTOGRAPH 15. NEWCOMEN PUMPING INSTALLATION II:
ROCKER BEAM, PART ASSEMBLY. ELSECAR MAIN COLLIERY

This view, taken within the enginehouse at 3rd storey level, covers the opposite end of the rocking beam outside the building. At this end, through a linkage of the type referred to, the balance weights are attached to the beam which serves to balance the pump road column and minimise inertia within the system. The 'Stroke Limiting' arrangements incorporate a stabilising frame extended upwards from ground level.

The massive beam supporting journals can clearly be seen at the bottom centre of the photograph. Construction of these bearings at the time of manufacture undoubtedly involved a great deal of heavy laborious hand work by highly skilled artisans, and are a fine tribute to their ability and enterprise.

The replacement pumping plant operated automatically on a time scheduled basis since its inception in 1926. Formerly similar installations were started by a telephone call. A whole network of pumping plants was established throughout the Barnsley district, being administered under the direction of Small Mines. Most of the comparatively shallow depth collieries were covered with provision for controlled drainage, although relevant provisions were made for water control in the deeper mines and linked in with the system. The Stillery Side footpath was a second back access to the Colliery and provided a connection with the old (New Yard) Workshops.

PHOTOGRAPH 16. NEWCOMEN PUMPING INSTALLATION III: ROCKER BEAM
PART ASSEMBLY: ELSECAR MAIN COLLIERY

This view, taken from within the enginehouse at third storey level, covers the opposite end of the rocking beam on the outside of the building. At this end, through linkage, the balance weights are attached which serves to balance the pump rod column and minimise inertia within the system. Again note the identical stroke limiting arrangements which incorporate a stabilising frame extending right down to ground level.

The massive support journals can clearly be seen at the bottom centre of the photograph. Construction of these bearings undoubtedly involved heavy handwork by highly skilled artisans of the time and are a great tribute to their ability and enterprise.

The headframe in the distance provided shaft access into the Old Barnsley Level, and replaced this old system by more modern automatic electrical pumping provisions of the mid 1920s started by a telephone call. This new system operated continuously up to the time Elsecar Main Colliery closed in the 1980s.

The Stillery Side footpath in the far background connects the New Yard Workshops with the colliery.

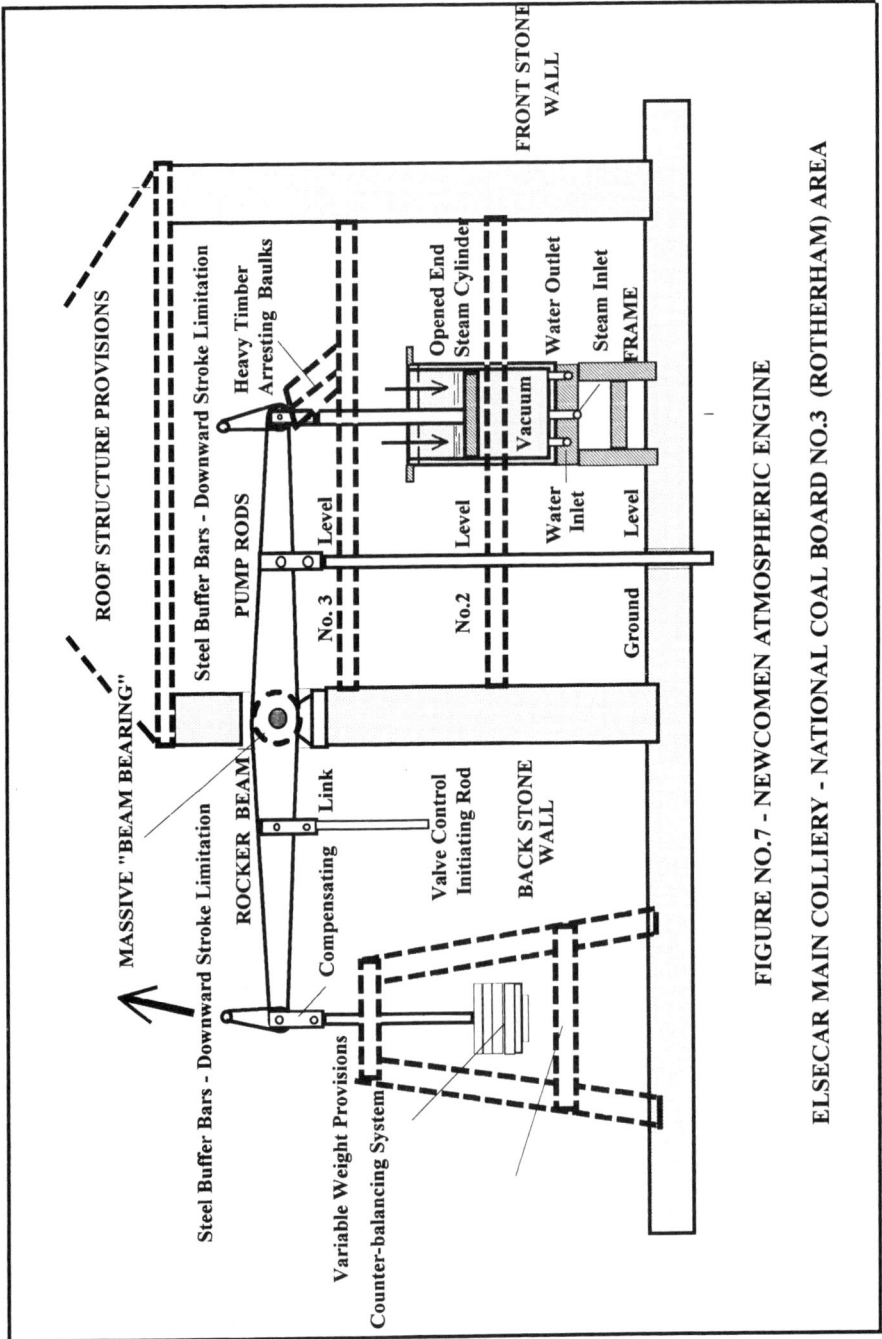

FIGURE NO.7 - NEWCOMEN ATMOSPHERIC ENGINE

ELSECAR MAIN COLLIERY - NATIONAL COAL BOARD NO.3 (ROTHERHAM) AREA

(OPPOSITE) FIGURE 7. THE NEWCOMEN
ATMOSPHERIC ENGINE: ELSECAR MAIN COLLIERY

This engine was in operation before the James Watt steam engines. Its simplicity made it an ideal subject for use with reciprocating motions such as pumping equipment. It was extensively used for water drainage in the tin mines of Cornwall and the coal mines throughout the UK during the late eighteenth and early nineteenth centuries.

Figure 7 has been produced with reference to the photographs as an aid to understanding the principle upon which it works. Basically it is a simple fabricated beam centre mounted on massive bearing supports. On the RH side an upper open ended steam cylinder complete with piston is attached through linkage. Connected to the LH again through simple linkage is a counterweight assembly to balance the inertia in the system. Valve control and pump rods are connected on either side of the beam as shown. In view of the high mass within the system, stroke limiting provisions are necessary to limit the length of stroke and avoid damage during operation. Such provisions take the form of heavy timber restraining structures at either end of the beam.

The engine operates on the principle that the first stroke of the piston is obtained from the application of steam pressure to its lower side. This lifts the rocker beam and pump rods and delivers the pump's water through the drainage system. At the top of the stroke water is admitted into the steam cylinder which causes a rapid condensation of the steam therein thus creating a vacuum. Atmospheric pressure acting on the upper open side of the piston forces it down to the bottom of the cylinder and lifts the balance weights, thus priming the pump for the next water delivery stroke. The balance weights are there to counter-balance the weight of the pump rods and inertia within the system, thus for all practical purposes, the steam pressure really lifts the water and atmospheric pressure returns the system back to status quo. A system of rods and controls is linked to the rocking beam such that the plant operates continuously until the steam is cut off. The single vertical cylinder steam engine (James Watt format) installed to wind coal at the No.1 Shaft Wharncliffe Silkstone Colliery originally had hand operated valves.

PHOTOGRAPH 17. NEWCOMEN PUMPING STATION, IV: STEAM CYLINDER AND
CONTROL GEAR: ELSECAR MAIN COLLIERY

The steam/air cylinder is shown mounted on a steel frame with the steam inlet valve and
input pipe visible in the bottom LH corner of the photograph. The pump rod linkage is
to be found to the right of the picture with provision lever control for hand operation
in emergencies. Projections are in being to rehabilitate the whole plant for demonstrational
purposes as part of a wider Heritage Scheme currently in progress. In those early days
the number of strokes made by the piston and pump rods during operational periods was
registered by a mechanical counter.

During a lecture trip to Delft University, Holland, I recall examining an old pumping
station (pumping system) museum protected installation built in the late 1700s in which
a central steam cylinder operated six satellite pumps lifting water about 10 feet. This was
of early Cornish construction and continuously operated reliably over a very lengthy
period. That too had the excellent artisan workmanship of the engine described. One
feels good to see early examples of such fine and industrious effort.

PHOTOGRAPH 18. SIR NORMAN AND LADY PEGGY SIDDALL

This photograph was taken outside Sir Norman and Lady Peggy Siddall's home in Mansfield, Notts.

Back Row L-R – Charles Round; Stephen Siddall; Sir Norman Siddall.

Front Row R-L – Lady Peggy Siddall; George Thompson.

Sir Norman was a former Chairman of the National Coal Board subsequent to Lord Ezra's retirement. He graduated as a brilliant student in Mining with a Honours Degree at Sheffield University following which he was employed by Bestwood Collieries Ltd., rapidly making progress to the post of Colliery Manager. On 1 January 1947 the coal mines were nationalised. It was in the early 1950s I first met him. At the time he was Sub-Area Manager in charge of Bestwood, Gedling and Linby Collieries and the Calverton Colliery Development. Following this he became Production Manager, No. 5 (Eastwood) Area; Area General Manager, No. 1 (Bolsover) Area, East Midlands Division. Transferred to Headquarters as Director-General for Mining in the mid 1960s, he later became Board Member for Production and finally National Chairman.

He has been for quite some time Chairman of the former National Coal Board's Pension Fund which has flourished greatly under his leadership.

George Thompson was my predecessor as Manager of Gedling Colliery. He is a North Countryman and formerly a Mines Inspector. Leaving Gedling he became a Sub-Area Manager in the No. 1 (Bolsover) Area of the East Midland Division.

Emperor's Palace,
Vienna
Austria

Herr Owen Franz Gerl Brannigan
(Metastasio Ascendi)

Her Gracious and Imperial Majesty Empress Marie Theresa has received knowledge by special courier of the esteemed award and degree conferred upon you by the world renowned Emporium of Learning and Art of Newcastle University in the Country of England.

Her Gracious and Imperial Majesty commands me to inform you of Her sincere pleasure in that knowledge, and also to express her congratulations and appreciation for the fame you have brought to your native province.

Her Gracious and Imperial Majesty recalls the delight and pleasure she has many times received from your roles as "Sarastro" and "Leporello" in Herr Mozart's operas Die Zauberflöte and Don Giovani performed so often with such great distinction.

Her Gracious and Imperial Majesty kindly commands you to appear at the Emperor's Palace on your return to Vienna, whereby she will be pleased to express her pleasure in person and once again listen to your expressive and delightfull renderings of the Folk Songs of your native province, together with a selection of Arias from Herr Mozart's works.

By Her Gracious and Imperial Majesty's Command

I remain,
Your Servant,
Carl Amadeau Rotundi
16th May 1791

PHOTOGRAPH 19. OWEN BRANNIGAN, OPERATIC BASS SINGER

This is a portrayal of Owen Brannigan, the operatic and concert vocalist, in the role of Don Alhambra during the shooting of the film *The Gondoliers* with libretto by Gilbert and Sullivan. Owen started out as a colliery joiner but, having a fine bass voice, subjected himself to intense hard vocal training, later emerging as one of the best operatic singers of his time.

One of the Mozart bass arias that has given me great pleasure over the years is 'Per Questa Bella Mono'. The recording I have is sung by the Finnish bass Kim Borg, another fine artist. I remember asking Owen why he didn't feature this fine solo. He told me that the problem was getting an accomplished double bass player who was capable of playing the obligato which required the difficult finger work of double-stopping. I mentioned Kim Borg to him. He told me that on one occasion, travelling down from Edinburgh to London by train late one night, he had a fellow passenger with whom he entered into discussion, later to find it was Kim Borg whom he knew of but had never met.

PHOTOGRAPH 20. BOTTOM BELT FACE CONVEYOR: ELSECAR MAIN COLLIERY

A mid-coalface loaded section of the belt conveyor is shown, together with the earlier primitive support system. The belt is running from left to right in the vicinity of a stone waste pack built to augment the use of steel crown bars and rigid steel props which support the face workings.

The bottom idler pulleys shown, positioned at intervals of about 12 to 15 feet, were for the purpose of reducing drag and to ease starting up after stoppages. The top belt was carried by a 1 inch diameter steel pipe clamped to and across the belt track supports placed about 30 feet apart. The section is part of a 200 yard conveyor unit running at about 180 feet per minute driven by a 25hp flameproof electric motor, and loading out some 300 tons per shift.

When stoppages occurred it was difficult to prevent the colliers loading up the stationary belt so starting up under abnormal inertia was often very difficult. Twenty-five men handloading onto these conveyors could really test the system and peaks of 200 to 300 tons per hour were common over short periods. In this case the independent drive units were located in a centre gate roadway which collected the coal from the left and right hard units feeding onto a top belt loaded gate and network conveyor system.

PHOTOGRAPH 21. DOUBLE BOTTOM BELT FACE CONVEYOR DRIVE:
LOW HAZEL SEAM, GEDLING COLLIERY

This is a copy of the original Huwood and adapted BJD drivehead used at Elsecar Main and Gedling Collieries. Built as a combination of two single M & C driveheads, it proved quite a useful modification and was extended throughout the colliery. When not in operation during roadhead construction it was moved back into the roadway by about 3 yards.

One of its main advantages was that it reduced the time required to set up the conveyors in working operation. Another was the reduced congestion at the roadhead whilst the rippers were constructing the head of the gate roadway.

After the work at the roadhead was completed the unit was moved forward, positioned and coupled into the RH and LH face conveyor belt system. The spring loaded 'scraper bars' maintained good contact with the belt to give efficient clearance of the coal from the face to the gate conveyor.

This unit regularly and reliably transferred over 600 tons during each coal loading shift onto the gate conveyors as shown.

FIVE HEADING SYSTEM

Note:- Similarly with this situation, with exclusive "bord & pillar" mining embracing high capacity Continuous Miners Shuttlecars and Ancillary equipment inclusive of the "Conveyor Belt Transport" provisions, great care is taken relative to heading widths, pillar& panel sizes and reduction of non productive time spent on and within all operations and activities so to concentrate in maximising performance.

Face Lengths Up To 375 Yds Circumstantial

Potential Distance up to 375 Yds

LONGWALL MECHANISED MINING PRACTICE

Main Trunk Belt Facilities

Manriding & Materials Shaft

Surface to Seam Access Drift

N

UNDERGROUND LAYOUT EMBRACING A " FIVE HEADING" SYSTEM

ROOF BOLTING IN COAL HEADINGS

Pickup Loader Filling Shuttlecar

4 x 4 ins Steel Plate

Continuous Miner Loading to Floor

Controlled Rate Discharge Bunker 15 tons.

Panel Belt

Boring & Bolting Machine

15 x 5 ft

PLAN & ELEVATION OF A COAL HEAD "A"

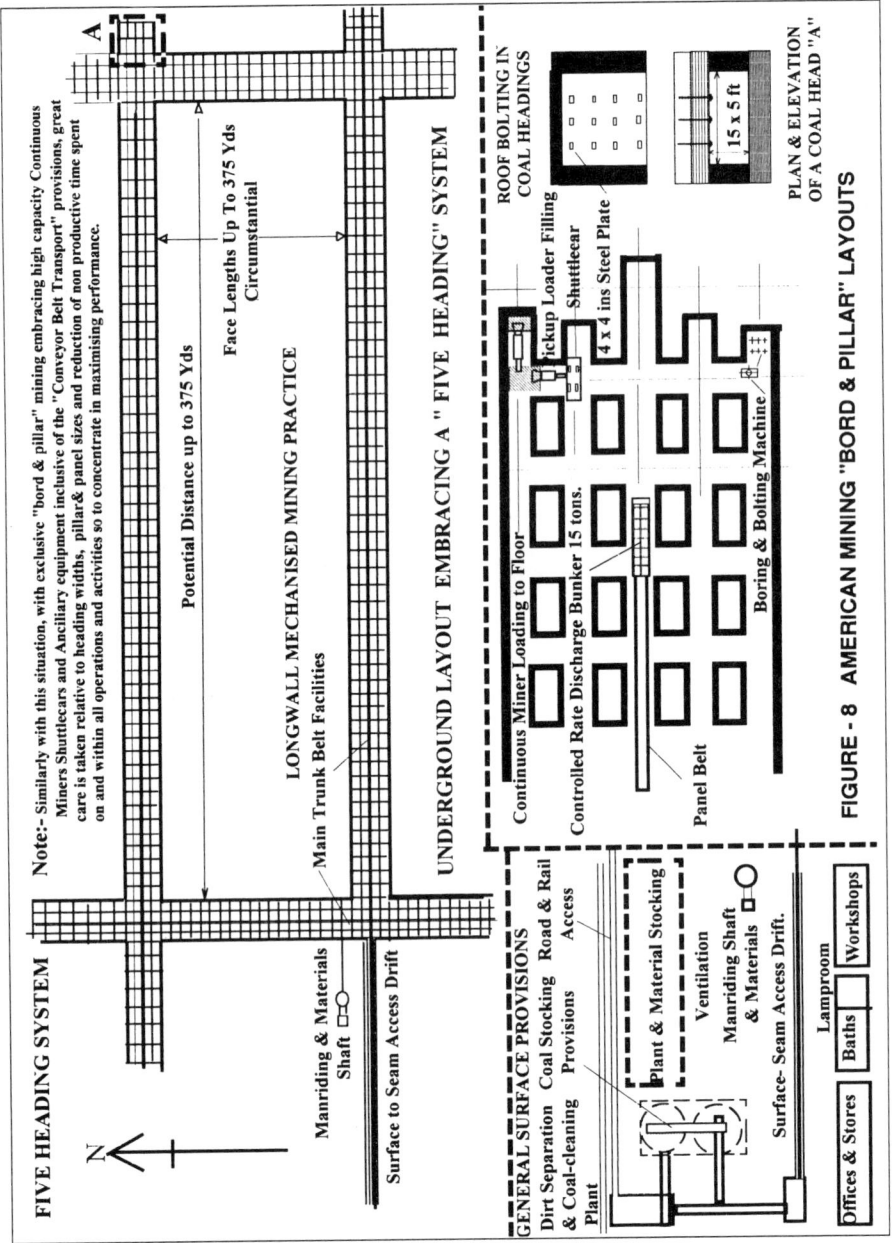

FIGURE - 8   AMERICAN MINING "BORD & PILLAR" LAYOUTS

GENERAL SURFACE PROVISIONS

Dirt Separation Coal Stocking Provisions
& Coal-cleaning Plant

Road & Rail Access

Plant & Material Stocking

Ventilation

Manriding Shaft & Materials

Surface- Seam Access Drift.

Lamproom

Offices & Stores   Baths   Workshops

(OPPOSITE) FIGURE 8. AMERICAN MINING 'BORD AND PILLAR' LAYOUTS

An example showing the nature of the American Bord and Pillar mining system layout in which the coal is extracted without recourse to cutting the roof or floor to construct roadways, i.e. mined exclusively within the confines of the seam. This is a '5-Entry' system, but the American mining engineer determines the actual number of headings and the logistics associated therewith with great care so as to maximise the production effect of the power loading and other equipment he installs.

Their early mining systems involved the men loading out the coal by hand, this being followed up later by the development of cutting and loading machines. These were subsequently replaced by continuous miners and shuttlecars currently used as the American basic mining system.

Formerly electric locomotives and large capacity mine cars constituted the underground haulage system. In later years the trend has been towards high capacity trunk and secondary belt conveyor arrangements fed by large shuttle-cars and auxiliary equipment.

The mining layout section shown is more in keeping with the current approach to mechanised longwall mining which seeks to integrate the best of both mining systems and which opens up even greater levels of achievement than the current best, rather than their conventional exclusive Bord and Pillar practice.

PHOTOGRAPH 22. AMERICAN 'BORD AND PILLAR MINING' PRACTICE – I:
CLINCHFIELD COAL COMPANY, DANTE, VA, USA (1958)

This photograph was taken at one of the Pittston Company's collieries, the seam being over 15′ 0″ in thickness and having a strong roof and fairly hard floor. Support to the 'entries and crosscuts' was by a gridded pattern of roof bolts.

The picture shows a pick-up loader transferring coal deposited on the floor by a continuous miner (power loading machine cutting and loading coal simultaneously) loading into a shuttlecar. The pick-up loader is fitted with a coal breaker which reduces excessively large blocks to a suitable size which the belt conveyors can readily transport out of the mine.

The shuttle car transfers its load to a conveyor discharge point for controlled loading onto the underground transport conveyor system. Sights such as this in which coal is flowing freely, continuously and without interruption, never fails to gladden the hearts and minds of all enthusiastic mining engineers anywhere.

Two of the great skills both of the American mining engineers and of their machine manufacturers I have admired over many years is their ability to pack tremendous power in the most restricted situations, and to give manoeuvrability, the two basic problems of coal mining.

PHOTOGRAPH 23. AMERICAN 'BORD AND PILLAR' MINING PRACTICE: II

A further part of the installation featured in Photograph 22. This picture shows a shuttlecar with a capacity of about 15 tons or so moving towards its discharge point where the coal is generally rapidly loaded into a feed control bunker type loader which feeds the coal onto the trunk conveyor at its rated capacity and allows the shuttle-car to move back to its point of loading with the minimum of delay.

In the top centre the support pattern of roof bolting can be made out: those little square plates define the placing and location of each bolt, which are machine installed and tightened in position.

It is under conditions such as these in both the UK and the USA that longwall mechanised coalfaces are producing coal at the rate of over 20,000 tons per day – which is way out beyond the capacity of the arrangements shown. Currently (1993) the USA is producing 100,000,000 tons of coal per annum by UK longwall methods.

(OPPOSITE, TOP) PHOTOGRAPH 24. ADVANCED CONTINUOUS MINING: I
GEDLING COLLIERY, EAST MIDLANDS DIVISION

This photograph was taken in the Top Hard Seam at the above colliery and is your first view of the German *Panzerforderer* chain conveyor, and friction type steel props as applied to a more advanced coalface installation. Here the coal has been filled out and the next operation to follow is that of moving over the conveyor into the cleared track.

The panzer conveyor was and is a formidable mining tool and to me represents the fundamental base upon which the current fantastic production achievements have evolved both in America and the UK. The original model was a two strand flighted chain assembly as depicted and driven at both ends through a gearbox by a 50 h.p. electrical motor. The steel structure accommodated both horizontal and vertical misalignment which enabled it to be slewed and snaked up to and within 1 metre over a distance of some 20 or 30 yards. This could be done without impairing the machine's running or conveying potential.

Over the past 40 years UK mining engineers, manufacturers and others have transformed its potential, capacity, reliability and application characteristics to a degree unimaginable in the mid or late 1950s. To transport some 800 tons per day in those early years gave us a great thrill and was most satisfying. Over many months, both here and in America, modernised panzer installations have conveyed some 25,000 tons per day and over 120,000 tons per week.

The German friction props were in a particular way quite something compared with the rigid supports we were using prior to their inception. They could be pre-loaded at the time of being erected; moreover, they could be more easily withdrawn and recovered whilst carrying intensive loads. In this photograph we see the shape of things to come. The characteristic of being able to sustain and support very heavy loads was useful in that the waste sheared and broke down along the last line of supports.

The corrugated bars that were formerly used with handloading methods lacked strength and were easily deformed under strata loads.

(OPPOSITE, BOTTOM) PHOTOGRAPH 25. ADVANCED CONTINUOUS MINING – II

The same section of the face as is shown in PHOTOGRAPH 24, but with a more advanced face support system. The German friction props have been replaced by Dowty hydraulic props which were easier to use, more safe to recover and capable of supporting heavy strata loads. The crown steel bars were also removed in favour of 4″ × 4 ″ section rolled steel joists.

The system of support was unique at the time in that it was possible to advance the steel joists without releasing the strata loads carried by the vertical supports. This was achieved by fitting the device (locally christened 'cabbage heads'), designed by N. R. Smith the Area Production Manager, on top of the hydraulic supports. It was far more flexible and secure than the former one (which can be seen in the background). Along the face side the cabbage heads were inverted and as the panzer conveyor was pushed forward they were transferred to the waste side in a conventional matter. Note that the leading Dowty support on the extreme right of the picture has its 'pump handle' or setting key in position. The coalcutter rope can be seen running along and over the top of the LH side of the panzer spill plates. The machine power cable runs along at the foot of the coal seam.

PHOTOGRAPH 26. ADVANCED CONTINUOUS MINING – III

Here the coalcutter is positioned in the RH stable ready to start cutting the face in preparation for the coal to be loaded out, the face conveyor having been moved over into the face track. All the supports have been advanced forward to support the working area. Shotholes about 1 metre in depth have been bored at intervals of about 9 feet some 12 inches from the top of the seam.

On starting, the coal-cutting machine hauls forward along the anchored rope, the picks carried by an endless chain in its race cutting a slot a metre in depth at floor level with the gummings being loaded onto the conveyor and carried away. At safe intervals behind the machine the coal is blasted and colliers start loading out the broken coal.

As the coal is cleared each workman advances his roof bars, withdraws his waste side supports and sets them along and adjacent to the coalface as shown. The whole system is continuous with the men being fully flexible undertaking their varied duties relevant to the different parts of the working cycle.

This photograph always reflects the joy and beauty of coal mining, capturing for me the delight and tranquillity of swans gliding along a river bank, together with the ingenuity and ability of our designers in the process of converting ideas into solid practical human aids.

PHOTOGRAPH 27. ADVANCED CONTINUOUS MINING – IV

This photograph shows the coal-cutting machine moving back to its starting point at a lineal speed of about 30 ft. per minute with its 'swan-necked' jib collapsed locked and swung over the panzer conveyor.

As the machine moves down the face the men fall in behind and push the panzer over to the coalface with the supports from the face side removed to the waste side line of the conveyor and reset, thus completing the total quantum of work involved in extracting a web of coal to depth of 3'4".

Reference was made much earlier in the narrative to jibs, chains, pickboxes and cutting bits. What you see in Photographs 19 and 20 depicts what such terms mean, even though they formerly referred to flat jibs. The picks are removed from the pickboxes and replaced by newly sharpened ones as and when required. The recovered ones are then re-processed by the surface blacksmiths.

The importance of this stage in mechanised production was concerned with 'continuous' as opposed to 'cyclic' mining. With the former coal is produced on each shift of operation; with the latter coal is produced only on one shift per 24 hours, the other two shifts being expended in integral preparation.

PHOTOGRAPH 28. ADVANCED CONTINUOUS MINING – V

This is quite an interesting and revealing photograph, which has captured a heavy 'waste break' situation, being one that also pays tribute to the efficacy of the support system employed with this continuous mining application. It is a view of the last row of supports along the waste edge, the boundary of the face working area. Here there has been a massive 'waste break', i.e. collapse of the unsupported void left behind after the supporters are withdrawn and the coal extracted. The friction props have been reeled forward having collapsed in the event. Four can be seen lying on the floor.

The Downty hydraulic props and roof joists withstood the impact and have maintained the security of the working area. Although the layman might feel apprehensive in such a situation, providing there is no displacement of the last row of supports, the release of such intense strata pressures is of benefit to the maintenance of a safe working area. Where there is systematic waste collapse without undue displacement of supports the mining engineer usually feels quite confident his support system is effective and secure. At this stage two basic aspects of operation still need to be mechanised: the first is hand filling and the second roof support mobility and setting. There is however at this stage a further vital element of operation yet to be resolved, i.e. the coal-cutting machine's ability to cut bidirectionally. Flitting is a non productive operation.

PHOTOGRAPH 29. MECHANISED ROADHEAD CONSTRUCTION: ROADHEAD
SAFETY CANOPY

This is a view of a tailgate roadhead in the Low Hazel Seam, Gelding Colliery. The safety canopy is in its closed position. When a standard 5ft. 6ins. advance of strata depth has been blasted down, the sides and roof are trimmed and the top sliding section of the canopy is pushed forward into the unsupported exposed area to give temporary protection.

The fragmented stone produced by blasting is cleared away by a rope-hauled mechanical skip into right and left side roadhead packs and when totally cleared the permanent road supports are erected.

Note the foot of the LH arch. It has welded on either side a short length of identical arch section into which a block of wedged shaped hardwood is inserted for the purpose of accommodating early road subsidence and avoiding the initial damage to the roadway supports which had previously obtained. This simple device earned me a small award from the Divisional Awards Panel.

PHOTOGRAPH 30. COAL PLOUGH INSTALLATION: GEDLING COLLIERY (1956)

These coal-plough developments were undertaken in the T's section of the Low Hazel Seam. The mining conditions deteriorated with the introduction of the system. The plough was too crude to control with sufficient precision to avoid cutting deep into the soft floor; moreover it failed to penetrate the hard coal sections which it encountered. Where it was possible for the plough to cut the seam the hard 'top coal' failed to collapse; strong evidence of this can be seen on the RH side of the picture.

The situation gave rise to face support difficulties which had to be reinforced by added timber and steel. Despite the fact that we were able to draw on the experience of German engineers they were unable to offer us any positive help in our situation.

The early compressed-air jacks for pushing the panzer forward with the face advance were bulky, inefficient and unable to cope with circumstances which obtained. Maintenance was extensive and expensive. The installation failed and had to be withdrawn. Not unlike the Samson Stripper discussed earlier, the prospect of bulk tonnage production was restricted by its limited depth of bite and crudeness of control. However, the experience we gained in view of the problems associated with the panzer face conveyor was more than valuable as our efforts were extended with and into later developments.

4

Management at Group Level

A s I had been previously informed I was given four working collieries and at a later date was to have Cotgrave colliery, a new development.

Gedling Colliery

At the time I took over the No.3 Group Manager's post, the execution of the colliery's reconstruction Project 201 was building up in tempo, with probably about 50% completed. Jimmy Gilks and Bill Shaylor were the two leaders of the Project 201 Reconstruction teams, both of whom were highly skilled in the heavy work and difficulties associated with such large projects. In the process of dealing with the difficulties of maintaining the production facilities at the No.1 and No.2 winding shafts without detriment to the execution of the project, both men were given freedom to organise their teams so as to accommodate each situation as it developed.

Weekends were used exclusively for shaftworks, so that deepening and plant installations could be undertaken safely, without risks of dangerous exposure to those men so engaged. In connection with the scheme, Harry Saviour (Area Chief Planner), Jack Wadsworth (Senior Planner) and myself undertook technical investigations with Rhymney Engineering, South Wales, into the provision of the control plant, signalling systems and service arrangements associated with the locomotive trains of full and empty mine cars operating in both winding circuits. Within the course of a further six months the project was completed without incident. In the early stages of commissioning the scheme we really had plenty of difficulties, but once the 'bugs' had been cleared, the system worked admirably.

With regard to the other collieries, my immediate need was to assess the situation within the collieries taken over regarding both potential and reconstruction needs.

Clifton Colliery

The colliery was situated alongside the River Trent, adjacent to which is the Clifton Power Station which took the total colliery output supplemented by coal from other sources. The management I had consisted of Russell Bracegirdle, colliery manager and Allan Walmsley, undermanager whom we appointed shortly after I took over the colliery. In later years Allan became Managing Director of Fletcher, Sutcliffe, Wild Ltd, a subsidiary of Bookers Group in the City, with Russell as one of his directors, subsequent to which he became a Board member of the 600 Group (T.W. Wards). Both were highly competent and industrious persons of real stature and certainly had the most difficult colliery assignment within the group. The services, communications, power, coal haulage, ventilation and material supplies system were greatly extended (being over three miles from the shaft side) absorbing a large amount of non-productive labour and restricting an expansion of output. Working time at the coal face was low compared with the other collieries.

The mining conditions were quite difficult. Heavy repairs were involved in keeping the inbye underground roadways open. Things would never be easy at this colliery and intense application was essential just to hold things steady, irrespective of effecting great improvement.

MAJOR BREAKDOWN - No.1 WINDING ENGINE

I would like to record a most significant event at Clifton Colliery which occurred in 1953. One Thursday about 11.00 a.m., Russell Bracegirdle informed me whilst at Gedling Colliery that his No.1 winding engine at Clifton Colliery was out of action owing to a breakdown. Dropping everything, I went across to the colliery in record time, as such a situation can be regarded as a major emergency, particularly as the Coal Mines Act 1911 required that all underground workings must have two means of exit to the surface at all times. Immediately the winder was out of commission we were in breach of statute.

Russell and I went up into the winding engine house. My heart froze immediately I saw the damage. One of the two winding cylinders was cracked along its entire length and totally beyond any emergency colliery maintenance possibilities. The engine was a very old engine and to the best of my knowledge there were no drawings available.

We arranged to withdraw all the men from underground by the return shaft forthwith and gave instructions accordingly. Alan Walmsley the undermanager was underground some three miles inbye and on being informed of the situation marshalled the men and supervised their return to the surface. Russell informed the Mines Inspectorate whilst I phoned Harry Hardwick, the Area mechanical engineer who on being told the nature of things said,

'The pit will be off for weeks, if not months.'

'Inform N.R. Smith, the Area Production Manager, Harry; I've got urgent things to do here. It will help me,' I requested.

The loss of potential coal production I found to be unpalatable. The situation had to be contained. Turning to my colleague I said,

'Russ, we've got to temporarily transfer the Clifton workmen throughout my Group. We can't have them unemployed for weeks and they are not going to any pits other than mine.' This was basically sensible, for it avoided a great deal of unnecessary surface travelling by the Clifton men.

'Can you do that?' he enquired.

'We'll soon find out. Get me Bob Scott on the telephone,' which he did.

Explaining the problem and what I wanted to do about it, I said to Bob, 'How many men can you take at Gedling?'

Bob thought for a while and said, 'Send all you wish, Boss, I'll fit them in. By the way, will the extra tonnage be mine?' (I told you he'd matured.)

'We will deal with that later, Bob. You will probably get two-thirds of the men, probably eight or nine hundred. Work on a ratio of 40% facemen. You'll get a scheduled list of producers and non-facemen within the next twenty-four hours.'

Getting hold of Alun Jones next I went through a similar procedure.

'Can you help, Alun?'

'Of course I can. I'll get Derek Francis [undermanager] and Tommy Mac-Donald to work things out, and then go down to Radford to get things moving there. How many men do you think I shall be getting?' was his immediate and helpful response.

Whilst I was dealing with both managers, Russell had initiated moves to get his NUM committee together for a meeting late in the afternoon. At about 2.45 p.m. N.R. Smith came to the pit and satisfied himself as to the situation. Realising the pit wasn't likely to work for some time he commented on the potential loss of production, to which I replied, 'It shouldn't be too bad; I'm proposing to move all the men round within the Group.'

'You can't do that. What about the problems of re-signing at the collieries and the attitude to all the Branch Unions? It's too great a risk.'

I felt as shattered as I had on seeing the damaged winding engine cylinder. Fortunately Alan Hill, the Area General Manager, called in to see us at about 4.00 p.m. His concern was,

'Can anything be done to ease the situation we are in?'

I told him what had been initiated and the response of the two managers. He was delighted.

'Go ahead, lads,' he said on leaving us.

Robbie Scott and Alun Jones had cleared the position with their own NUM branches on their own initiative; they rang in to tell me. We didn't feel at all worried about the NUM at Clifton as we all had a good relationship with the President, Jack Wiles, and his committee.

The previous year I had taken the trouble to take Jack with a couple of his Branch members to see the Durham Miners Gala as I'd heard so much about it at Gedling. In any case, Jack was a north countryman from Dawdon Colliery village, having worked previously at that pit for a number of years. On the pre-Durham Miners Gala Friday night, we reached the village. Jack guided us to some miners' cottages and on getting out of my car we were chaperoned to one of them and virtually lined up. He knocked on the door which was imme-diately opened by a middle aged lady with her husband standing in the back-ground. On seeing Jack she beamed.

'You've managed it, Jack,' but in Geordie accent, then she turned to us. 'Come in, lads, you must be famished. Come and get something to eat.'

A great deal is lost in the translation. The Durham or Geordie dialect embraces a warmth of humanity which is forfeited when expressed in cold English. Awaiting each of us was a large plate of ham and eggs together with goodies of all kinds, and with a gentle but continuous pressure from husband and wife, we had to eat our fill.

The cottage undoubtedly was the pride of the lady housewife. It was simply but beautifully appointed and everything shone with absolute cleanliness. Her husband had performed a labouring job underground at Dawdon Colliery and had been prematurely withdrawn from the coalface with a degree of dust congested lung problem. People in the mining villages of my own county are much the same in terms of hospitality but generally not to the extent we were experiencing on that occasion. At 7.00 a.m. breakfast the next morning we had the same treatment before setting off to experience at first hand the Durham Miners Gala.

Within a three mile radius of the centre of activities the roads leading to Town Moor were closed with all vehicular traffic grounded. The procession, led by a brass band, was followed by the first of the NUM Colliery Branches with its individual banner carried by two miners stabilised by four other miners (two back and front) holding cords. In windy weather, marching with these banners is very difficult to accomplish. An ordered sequence of brass bands, NUM colliery banners and their associated miners followed on behind. At its peak over 250 Bands and NUM Branches were associated with the procession which often took two to three hours or more to pass the point at which the main speaker of the event stood to attention with his supporters watching the proceedings.

The whole procession feeds into the Town Moor field, an open area within which the speakers' platform is located. On this particular occasion Hugh Gaitskell, the Socialist, was the main speaker with the procession having but fifty bands. However it was most impressive particularly as almost everyone treated the occasion with a great holiday spirit. On the Sunday morning, I requested and was given a tour of Dawdon Colliery surface before adjourning to the Dawdon Colliery Welfare Centre. When we walked into the bar-room at precisely 12.00 noon, placed on top of a very long wide bar were some two hundred pints of bitter beer, drawn ready for immediate consumption. Immediate was the operative term. In less than five minutes they had all disappeared. Prior to returning home on the Sunday afternoon I asked Jack how I could express my appreciation for the generosity and hospitality of this kind lady and her husband.

'For God's sake don't offer money; she will be most unhappy.'

I thanked the lady the best I could in the circumstances. On the way back Jack was given the wherewithall and instructed to find and send her the best bouquet of flowers he could lay his hands on, and judging by the beautiful letter I subsequently received from the lady I couldn't have addressed the situation in any better way.

Returning from the above diversion we met the Clifton Branch. They were most helpful and appreciated what we were attempting to do. Russell and Alan were fantastic in the circumstances. I stayed with them into the late hours before leaving them embroiled in the task of drawing up schedules of labour allocation at Gedling, Wollaton and Radford Collieries and sorting out the bus services to effect the relevant transport to each colliery. It took a further twenty-four hours to ensure that the men were informed which colliery they had to report to and what transport arrangements had been laid on for them. Other than the safetymen and those involved in the winder repairs, every man was placed.

After a loss of one working day (three shifts) only, all the Clifton Colliery men started in their new locations on the following Monday morning and continued thereat for just over two weeks before returning to their normal jobs. In the event, we actually increased the Group tonnage over the period involved. The managers at the receiving collieries had embodied a sizeable temporary labour force within their own with tremendous skill and application and with little notice. I was, and still am, grateful to each one of them.

With the help of the Divisional and Area mechanical engineers and Plowrights Ltd of Chesterfield a new cylinder was cast (a tricky situation in view of the winding engine's age), machined and fitted in just two weeks. Plowrights Ltd did a first-rate job on to this old engine in a remarkably short time. It was a most remarkable exercise which fortunately I never had to repeat. The only minor problem we had was that quite a number of the men didn't want to go back to their own colliery. They had experienced better working conditions and had earned higher wages where they had been placed. Co-operative acts such as these rarely get the same treatment in the media as do strikes and picket violence

Cotgrave Colliery

This was a new colliery development situated to the south-east of Gedling Colliery near the Nottinghamshire country village of Cotgrave. It was designed to develop and mine coal reserves stretching eastward. A special problem existed with sinking the pit shafts, as both sinkings had to pass through water bearing strata containing elements destructive to the concrete shaft linings.

Miners needed to be recruited into the area and this necessitated the planning of a sizeable housing estate. The colliery was to be designed for an annual output of over 1,000,000 saleable tons covering a wide range of markets.

Radford and Wollaton Collieries

These two collieries were jointly run by a single colliery manager, Alun Jones, together with an undermanager appointed for each colliery. Rope haulage existed at both collieries and basic communications were quite good. Prior to national-isation both mines had not been heavily exploited. Alun, I understood, followed his father before him and had been manager of the colliery for quite a number of years before the pits were taken over by the National Coal Board. He lived alongside the entrance gates to the Wollaton colliery yard and, at precisely 4.30 p.m. each working day, would lock these gates which remained closed until the pit started again at 6.00 a.m. next morning.

At Vesting Date, the time the collieries were nationalised, both pits were apparently outside the prevailing levels of mining technique and although Alun was quite an influential figure within the professional institutions - the Midland Branch of the Institute of Mining Engineers and the National Association of Colliery Managers - he was regularly passed over time and time again during the selection of incumbents required to fill various vacant mining posts arising within the No.6 (Bestwood) Area. My appointment as somewhat of a foreigner must have been quite a blow to his pride. I certainly felt that to be the case when I first met him. He very rarely went underground, only when the most senior people were carrying out an inspection. In his early days he didn't have to, as the mining pace at the time was apparently exceptionally slow. I didn't mind as he was moving up in age and keeping the pace with the younger man would be somewhat taxing.

Lord Mayors of Nottingham

Radford Colliery was sited within the City of Nottingham. In fact it was adjacent to John Player's cigarette factory. It was a neat little colliery and like Wollaton Colliery had been brought up to very good standards. I recall the underground wages clerk at Radford Colliery, quite a bright youngster, meeting me for the first time and attempting to create an impression with his knowledge of the colliery's history and general performance. Some time later at a function, now of more mature age, he sought me out from the guest list and said,

'Mr Round, I used to work for you; I was underground wages clerk at Radford Colliery.' I remembered meeting him.

'What do you do now?' I asked him.

'I'm the Lord Mayor of Nottingham,' he answered, following which he invited me to have lunch with him and visit the Council Offices. I regret I was never able to take him up on the offer, as Nottingham is such a fine city and is steeped in history. He was, however, not the first Lord Mayor of Nottingham with whom I had had contact. I had met one in 1957 (much to Mary's dismay).

We were attending a National Colliery Managers Association (Midland Branch) dinner dance at Arnold, on the outskirts of Nottingham, during the mid 1950s, at which time both Norman Siddall and myself were tied up in the development of two power loading machines, the 'Trepanner' and the 'Shearer' (referred to later). Immediately following dinner, whilst the room was being prepared for dancing, a group of ladies including Mary were talking to a tall impressive figure bedecked in medals and a spectacular chain of office. During a lull in the conversation I turned to the gentleman and asked, 'Do you have any "shearers" in your Area?'

Disdainfully he looked down at me and said,

'I don't know what you mean.' He was much taller than I.

'Power loaders - do you have any trepanners?'

Firmly he turned to me and responded haughtily, 'I am the Lord Mayor of Nottingham. We do not use the sort of things you are referring to. I cannot place them.'

I apologised, having thought him to be the president or some other dignitary of a visiting Branch of the National Association of Colliery Managers or Institute of Mining Engineers. Mary never let me live this down, even when she saw the funny side of the situation.

Some three weeks after this incident I received a late night telephone call during which an attempted disguised voice said, 'Is that you, Mr Round? This is the Lord Mayor of Nottingham. Have you managed to get all your power loaders working properly? Mine are working most satisfactorily.'

It was Alun Jones, the Manager of Radford and Wollaton. He had apparently heard the story from one of his acquaintances in Nottingham, or perhaps from his Radford wages clerk. Recognising the voice, I said,

'All right Alun - get ready; you're next on the list.'

In the early assessment of my new collieries one thing stood out loud and clear: the quest for the 'Holy Grail' (total and complete coalface mechanisation) was now more vital than ever, and the National Coal Board's knights throughout the UK in their respective roles were deep in mission. Within a short time leads were developed for appropriate investigation and action at each colliery.

Clifton Colliery: sorting out the services, manpower and coalface mechanisation.

Gedling Colliery: completion of the reconstruction scheme, development and extension of coalface mechanisation.

Radford Colliery: underground haulage transport improvements, manpower recovery, coalface mechanisation (although there were problems arising in that the average seam thickness was apparently below that which would accommodate the prevailing coalface mechanisation equipment available at the time).

Wollaton Colliery: reorganisation of haulage arrangements, manpower recovery surveys, application of coalface mechanisation.

N.R. Smith, the Area Production Manager, was my immediate superior. In his early days he had been associated with Manchester Collieries Ltd at the time their Sir Humphrey Browne was Managing Director of the company. Later he moved up to the North-East Division, becoming involved with Vane Tempest and other collieries within that Area. He started out as Area Production Manager a few weeks before I commenced my duties as manager of Gedling Colliery. He was quite a fair and decent person to work with but he hadn't for me the inspiring qualities Norman Siddall possessed and tended to keep one at a distance, although he generally treated his staff quite well.

During the first few weeks of my appointment as No.3 Group Manager we were engaged in a series of coalface contract negotiations at Clifton, Radford and Wollaton Collieries with reference to coalface handloading systems. N.R. Smith led the Board's representatives and Albert Martin, the General Secretary of the Notts NUM, led the local Branches. The negotiations undertaken took about two months to complete. With these out of the way I first turned to Clifton Colliery with the object of improving the belt conveyor transport systems. The conveyor belts were 30 inches wide and ran at speeds of a little over 200 ft. per minute and frequently broke down with a loss of production on each occasion. Capital wasn't readily available to replace the whole of the system upwards of three miles in length. It was at this point that I made contact with Ron Swindall, the Area accountant, one of those helpful staff associates Arden Bowker and Norman Siddall told me I could rely on during my Horse and Groom initiation referred to much earlier.

Ron, I understood, was formerly accountant with Bestwood Collieries Ltd prior to nationalisation. He was a lithe 6′ 3″ and an excellent local cricketer who played regularly every weekend with one of the Yorkshire cricket clubs in the Bradford League. He was a basically a fast bowler and quite a useful bat. He used to work late, often till after 7.00 p.m. each night. His office was directly adjacent to the main office entrance through which I had to pass in order to collect my car from the parking lot at the front. Having prepared in great detail what I had in mind for Clifton, I called in to see him on my way home. Although he knew me, we hadn't made real contact. He was most generous and helpful; we both sat together at his table and went through the detail submitted. 'No go' areas he identified and then helped me find alternative approaches. Where I had to employ revenue means to get what I wanted, he told me how to effect control which would not be too damaging to the colliery's monthly balance sheet. That particular night we both left the offices at 9.00 p.m. In less than an hour I'd made a close friendship which lasted a lifetime, right up to his death on 19 July 1993 when he was in his late eighties.

Thereafter, during the years I was employed in the No.6 (Bestwood) Area, many were the nights we left together around 8.00 or 8.30 p.m. Ron taught me how to understand balance sheets and accounting practices, and he always claimed I did the same for him in mining. Often, when he faced mining situations involving finance, he would come to my office and we would work things out together.

Ron and I together worked out the logistics of changing the Clifton underground conveyor belt transport system over to one of greater capacity, following which I passed our agreed arrangements back to the colliery. Russell Bracegirdle and Alan Walmsley applied themselves to the task with ingenuity and enterprise. They had to maintain current production whilst doing so, so it was by no means easy. The recovered material was used to set up a man riding system in the return airway thus enabling us to effect an increase in the effective working time at the coal face. In addition they were able to make great improvements in the materials supply system. Following completion of the project, the colliery results started to improve.

Our next approach within the group was at Wollaton Colliery with the object of sorting out the haulage arrangements and the heavy use of non productive labour involved within underground service operations, particularly the inbye loading station and coal winding shaft pit bottom. Having spent upwards of six hours thoroughly analysing the haulage and winding operations and their associated manpower allocations, late the same afternoon we had a full meeting to discuss the situation and to determine improvisations or mechanical devices which would release individuals.

The meeting was long and tiring and tempers were getting frayed when little Tommy MacDonald, the senior overman, piped up during the heated discussions,

'Mr Round, can I ask you a personal question?'

I was somewhat tired and impatient and generally not in a good a mood. I turned on him saying, 'Yes, what is it?'

He looked straight at me saying quite seriously, 'How many children have you got, Mr Round?'

I immediately said, 'Two, what's that got to do with what we are dealing with?'

He smiled sweetly and disarmingly before answering, 'I thought you might have managed with one.'

We all burst out laughing. All stress flew out of the windows and the meeting was adjourned for another day. We did, however, make very good inroads into the problem later.

Development of Mechanised Coal Loading Facilities

The next problem which needed to be resolved was at Wollaton Colliery. The loading arrangements at the bottom of the dipping haulage plane were inefficient and tied up many men to keep it operative. It was similar to the one I had cleared in the Silkstone Seam at Elsecar Main Colliery. It was absorbing twenty manshifts per day and coal spillage was excessive and gave rise to difficult manipulation of the tubs. The rate was probably little more than 150 tons per hour. I worked on the problem at home over about six weeks and in the event determined a very neat solution.

The system developed embodied moving the conveyor discharge point inbye along a level road and using the space created for the storage of about two hundred empty tubs which could be fed into the loading station by simple semi-automated push-button control. Once I had sketched it out in detail, Jim Hall, the colliery engineer for both pits, and his son were brought into the Colliery Manager's office where I explained in detail the principles involved and asked them to build it. Not only did they construct it but they jointly added some very useful features following discussions with myself.

FIGURE 9 WOLLATON LOADING STATION, PRINCIPLE OF OPERATION, is a simple sketch plan of the arrangements by which the capacity of the loading point was substantially increased by the development of a partially mechanised marshalling system and the provision of automatic spillage clearance facilities.

PHOTOGRAPHS 31 AND 32 show the empties side rope propelled travelling catch which marshals empty tubs forward, and the full stop controls, respectively.

The first shows the outbye end of the rope-propelled initiator jack catch and channel run, the fixed 'empty jack' catches which propel a continuous train of empty tubs over a standard tub track to feed the loading station. The rope and the initiator jack catch were quite ingeniously connected by linkage, so that on the return run the catch initiator retracted clear of the tub axles under which it passed, so cutting out distracting noise.

The loading point feel arrangements were such that eight empty tubs, disconnected from the main body of tubs by the control attendant, were in front of the rams. On initiation the initiators pushed the eight forward, dragging at the same time the column of tubs behind. At the end of the run the leading eight tubs gravitated backwards over rail crossings onto the full track down to the point of loading.

As further sets of empty tubs were fed into the roadway, these were coupled together and added to the column placed for initiation forward. Automation was provided in that, at the end of the forward run, the initiators were momentarily stopped, the electric motor reversed, and the catches returned to their origin and stropped to await the next feed sequence.

The second photograph shows how the spring buffered tub-stop is securely mounted on a heavy steel girder frame and is located to place the leading tub at its exact position for being loaded by the gate belt conveyor. The foot lever, when depressed, unlocks the tub stop and allows two tub axles to pass over and then closes, having placed the next tub to be filled. Even with simultaneous handfilling from two or three faces, the rate of coal flow was very high at certain periods of the shift and tubs might well be filled in 10 to 15 seconds, during which spillage, being very high formerly, blocked the system.

In the meantime, the necessary excavational work at the point of installation was undertaken, carried out and completed. This included provision for a small chain conveyor which automatically returned all spillage which had fallen during loading, back onto the gate conveyor. Jim Hall, his son and I spent some considerable time underground supervising the installation. Alun Jones kept us supplied with food and beverages. By this time he had grown quite dependent upon me and we were very friendly. Once we got the bugs out of the system it was fantastic. We were able to load out at peak rates of more than 700 tons per hour. Moreover we had opened up facilities which would meet the requirements of coalface mechanisation which was ultimately to follow.

I put the system up for consideration by the Divisional Awards Panel. W.H. Sansom, its Chairman, came down to view the installation. His remark was, 'Charles, you've used a lot of steel, but the system really is ingenious.' He made me an award of £450 (£300 of which I gave to Jim Hall and his son). Incidentally, the system was run by three men per shift, thus effecting a saving of fourteen men achieved exclusively on the basis of revenue cost.

It has always been difficult for me to contrast the encouraging treatment I got from W.H. Sansom with that formerly received at Elsecar Main Colliery but it did represent my experience of the difference in outlook between the East Midlands and Yorkshire Divisions which prevailed at that time. Here one received encouragement from Area General Managers right up to the Divisional Board Members Sir Hubert Holdsworth, Wilford Miron, W.H. Sansom, and others. Shortly after taking up my first appointment in the East Midlands, whilst Mary and I were attending an afternoon function in Bullwell Hall, Nottingham, the three above gentlemen went out of their way to locate us amongst the crowd, and then to settle us in with very kind observations and wishes. It certainly helped Mary who settled in more quickly and far better than I had expected.

Shortly after W.H. Sansom's visit to Wollaton, Alan Hill, the Area General Manager, decided to inspect the arrangements. Along with N.R. Smith, Alun Jones and myself the visit duly took place. Both N.R. Smith and Alan Hill were greatly impressed (personally I believed it to have been the best of the many technical ideas I ever put into practice). After watching the arrangements working smoothly and efficiently for quite some time, Alan decided to make an inspection of the working faces, but for some reason or other we had lost Alun Jones and this upset the Area General Manager.

'Where is Alun Jones?' he demanded to know of me.

'I've sent him back to the pit bottom; they're having a little, trouble.'

'You've done no such thing. I know Alun Jones much better than you do, where is he?' he asked, with rising anger.

'Look, Mr Hill, Alun's moving up in years and is not in too good a shape,' I told him, but he wouldn't let go.

'He's no older than me and he should be able to do what I can do,' he replied,

'There is one great difference,' I countered.

'And what's that?' he challenged.

'If Alun was as fit as you and was able to do the things you can and have done he could well have been Area General Manager in your place, and we both know that isn't so.'

He looked me and smiled (not generally a frequent event). 'All right, Charlie, the point is taken, but tell him I'm annoyed.'

I never did; I was quite happy with Alun Jones as he was.

In the meantime, with Jim Hall and his son, together we mechanised the coal winding shaft loading arrangements at both Radford and Wollaton Collieries, absorbing the costs on revenue in a manner which Ron Swindall, my finance mentor and friend, usually advised.

Gedling Colliery in the meantime was not being neglected, excellent progress having being made with the commissioning of the main reconstruction project which by now was operating very effectively, but although face mechanisation was being widely extended we still hadn't eliminated the collier's shovel.

In my early days as No.3 Group Manager I would call in the Gedling Manager's office around 8.00 a.m. almost every morning. Gus also started early. When Robert Scott started out as a colliery manager, he arrived more in line with office hours, around 9.00 a.m. It annoyed me greatly, as I felt it indicated an apparent lack of enthusiasm and a poor example of support to the undermanagers, junior officials and indeed the men who were called upon to start their shifts at 5.30 to 6.00 a.m. daily. I wasn't going to suggest or instruct that he started earlier. That had to come from a realisation of desirability, particularly as I didn't mind him finishing earlier than office workers, if he so desired. Arrangements at the time were such that I was fed substantively with the colliery's activities during the previous twenty-four hours and told about all emergencies. Being acquainted with the pit's total activities over the proceeding afternoon and night shifts, during my first week, immediately Robert walked into his office, before he'd had time to settle, I'd fire the most embarrassing question I could about a particular incident which had occurred within the past twelve hours, one of which I knew he had no knowledge. Embarrassed, he would say, 'But I've only just arrived,' which cut no ice. Within eight or ten days he was there before me each morning and he rapidly became more and more adept and versed in his knowledge of events. Maybe I did push the lad a little hard in those early days, though I strongly doubt, in retrospect, that he would have wanted things to have been different.

Subsequent experience I had on the American mining scene indicated that the mining engineers, vice presidents, and indeed presidents of coalmining operations often started as early as 6.00 a.m. or 7.00 a.m. at the latest, or where they had to visit a distant operation even earlier. What is more, they worked much later than their UK counterparts. Irvin C. Spotte, President at the Pittston Coal Company, the third largest of the deep mined coal operations within the United States, controlled collieries producing in excess of 20,000,000 tons per annum and started out each morning at 5.00 to 5.30 a.m. His day invariably took in an underground visit to one or other of his many coal operations, often at distances of more than two hours helicopter flight. Rarely did he return home before 7.00 p.m. In some instances, to utilise his time more effectively, he would stay overnight at a convenient hotel. He certainly got things done without fuss or rancour and is today probably one of the most respected mining engineers in the United States of America.

No.6 (BESTWOOD) AREA – EAST MIDLANDS DIVISION
WEEK ENDING 18 DECEMBER 1954

COLLIERY	PITHEAD TONS	SALEABLE TONS	SALEABLE OUTPUT PER MANSHIFT				
			FACE CWTS	U/G CWTS	SURFACE CWTS	TOTAL CWTS	ABSENCE %
Bestwood	25,641	24,938	92.9	54.0	243.4	44.2	9.19
Calverton	13,856	13,146	79.3	44.7	208.7	36.8	8.92
No.1 Sub-Area	39,297	38,084	87.7	50.4	230.2	41.3	9.09
Babbington	17,153	16,657	80.1	40.4	184.9	33.2	11.15
Hucknall	16,852	15,736	96.3	49.2	154.8	37.3	10.02
Linby	14,036	12,641	92.0	43.9	162.6	34.6	10.39
No. 2 Sub-Area	48,036	45,054	88.5	44.2	167.1	34.9	10.56
Clifton	11,348	11,135	94.2	47.3	209.7	38.6	9.24
Gedling	27,295	26,764	100.5	56.8	198.3	44.9	15.04
Radford	7,079	6,600	101.2	54.7	238.3	44.5	9.86
Wollaton	9,191	8,933	91.4	51.4	277.8	43.4	8.60
No.3 Sub-Area	54,913	53,340	97.6	53.4	215.6	42.8	10.40
AREA TOTALS	142,246	136,458	91.6	49.2	200.0	39.9	10.13

COMPARISONS WITH THE AVERAGE NATIONAL PERFORMANCES
WEEK ENDING 18 DECEMBER 1954

SUB AREA	PITHEAD	SALEABLE	SALEABLE OUTPUT PER MANSHIFT				
	TONS	TONS	FACE	U/G	SURFACE	TOTAL	ABSENCE
			CWTS	CWTS	CWTS	CWTS	%
No.1 Sub-Area	39,297	38,084	87.7	50.4	230.2	41.3	9.09
No.2 Sub-Area	48,036	45,054	88.5	44.2	167.1	34.9	10.56
No.3 Sub-Area	54,913	53,340	97.6	53.4	215.6	42.8	10.40
AREA TOTALS	142,246	136,458	91.6	49.2	200.0	39.9	10.13
NATIONAL SITUATION	4,738,000	65.14	——	——	——	24.62	12.21

The annual saleable output for 1954 was 211,812,706 tons of which 16,000,000 saleable tons were produced by a consortium of early machines such as the Meco-Moore; 127 of them produced over 8,000,000 tons. In 1958 out of the total of 897 machines the Anderton shearer loader and trepanner accounted for no fewer than 294 and 72 installations respectively, with the former settling down to become the universal single choice. The Anderton shearer loader, the embodiment of the modern longwall ultra high performance system, has, together with the panzer conveyors and powered support ancillaries, undergone some forty or fifty years of radical development, details of which, together with the associated current phenomenal performance achievements, will be referred to later.

MACHINE UTILISATION

During the days in which I was tied up with work at the Bestwood Headquarters I occasionally had lunch with the senior Headquarters staff in their special canteen. At lunch I often found the conversations quite interesting and revealing. Often the Area Production Manager would try to impress the diners by referring to snippets of news gleaned from the daily press, or by showing his smartness with the crossword or some special problem he had come across and solved. Other contributors would include the Area Administrative Officer who would tend to try and outsmart the former, or casually refer to purchases such as the early refrigerators or washing machines with learned discussions as to the merits and demerits of such household appliances. Over a period of time, domestic equipment gave them quite a run of conversation, particularly as most of the persons present had bought their wives one or more items of such equipment. I was being held responsible for denying Mary the advantages of mechanised equipment in my home and having no thought of reducing her domestic chores. They implied I had Scottish ancestry. Actually Mary wasn't interested in those early days. She claimed those early washing machines couldn't get anything like the degree of product cleanliness she could. On one particular dining occasion one of my colleagues had just plumbed a Bendix washing machine into his kitchen, one which had at the time made quite a substantial leap forward in performance. Thus I once again became the subject of enquiry and comment. On this occasion I mischievously turned to the Area Production Manager and enquired,

'Have any of you people worked out what degree of equipment utilisation you are getting from your washing machine investments? Mary and I have. We consider it to be about five hours a week, which is quite a reasonable estimate. With a 168 hour working week this equates to about 3%. If all the machines in this Area were operating with such low rates of performance none of us here would now be sitting round this dinner table.'

These figures were obviously distorted but, coming more or less out of the blue, the diners were taken aback, following which they started to look for justifiable explanations which I found quite amusing.

Several years later, on being introduced to a Board member, he turned to me and amusingly said,

'So you are the fellow who wouldn't buy his wife a washing machine by reason of its low utilisation factor.'

However, with the subsequent introduction of extended coalface mechanisation throughout the Coal Industry, machine utilisation became an important accountability element.

PHOTOGRAPH 31. MECHANISED LOADING STATION. VIEW OF TRAVELLING
PIT-TUB INITIATORS: WOLLATON COLLIERY (JUNE 1953)

This view shows the outbye end of the rope-propelled 'Initiator Jack Catch', its channel run, the fixed 'empty jack' catches which retain the long coupled train of empty tubs, and the standard rail track over which they run. The rope and the initiator jack catch were quite ingeniously connected by linkage such that on the return run the catch initiators retracted clear of the tub axles under which they passed, so cutting out distracting noise.

The loading point feed arrangements were such that eight empty tubs disconnected from the main body of tubs by the 'control attendant' were in front of the rams. On initiation the initiators pushed the eight forward, dragging the main column of tubs on behind. At the end of the run the leading eight tubs gravitated backwards over rail crossings onto the full track down to the point of loading.

As further sets of empty tubs were fed into the roadway they were coupled together and to the column placed for initiation forward. Automation was provided so that at the end of the forward run the initiators were momentarily stopped, the electric motor reversed and the catches returned to their origin to lie idle to await the next feed sequence.

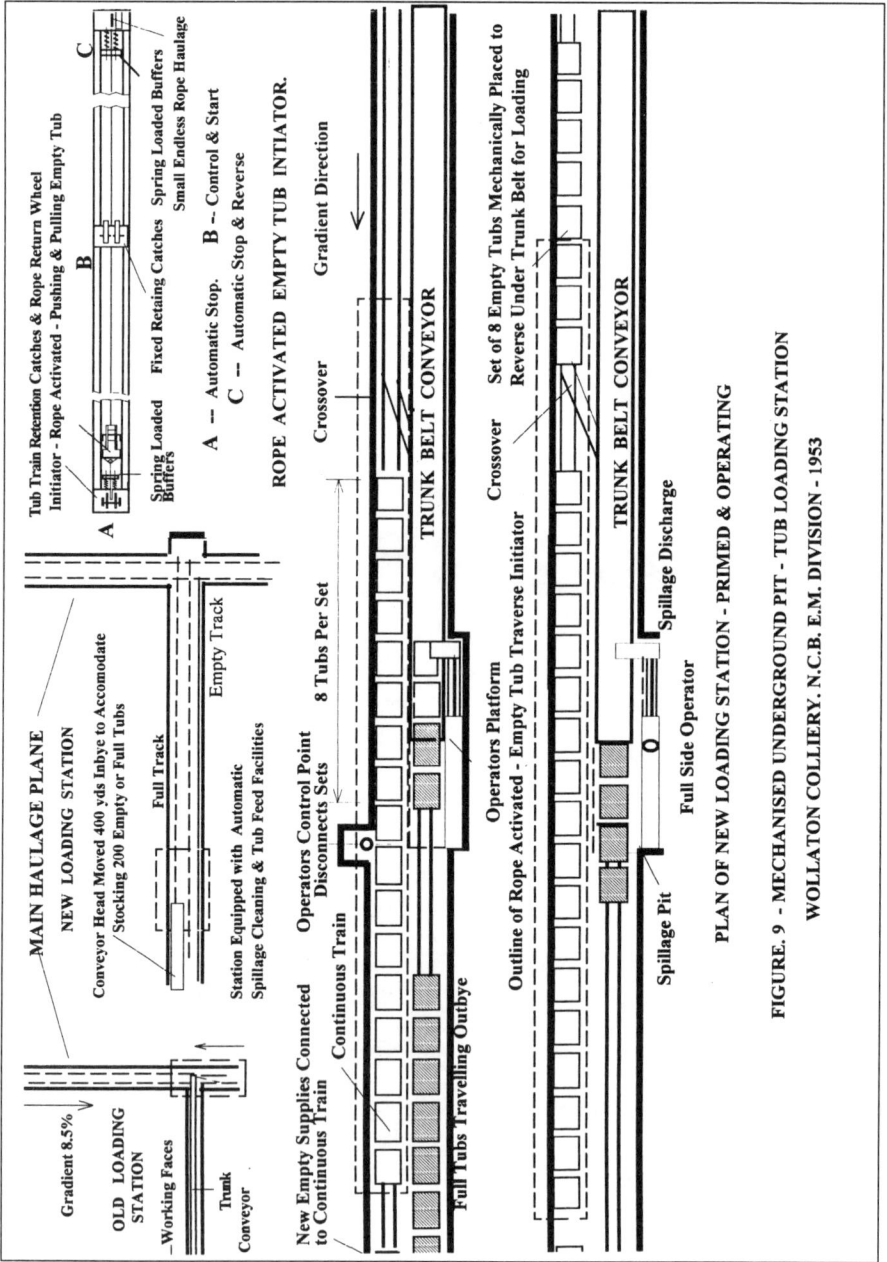

FIGURE. 9 - MECHANISED UNDERGROUND PIT - TUB LOADING STATION
PLAN OF NEW LOADING STATION - PRIMED & OPERATING
WOLLATON COLLIERY. N.C.B. E.M. DIVISION - 1953

(OPPOSITE) FIGURE 9. MECHANISED UNDERGROUND PIT-TUB LOADING
STATION: WOLLATON COLLIERY, EAST MIDLANDS DIVISION

The small inset in the top LH corner of Figure 9 shows the location and operation of the original loading point diagrammatically. Working on a gradient of 1 in 12 involved a contraflow system in which the loaded tubs had to pass under the loading belt as they were being hauled outbye to the coal winding shaft. Coal spillage was a heavy, often a disruptive, delaying problem. The former arrangements needed upwards of 8 or 10 men per shift tied up unproductively some 20 hours per day.

Moving the loading point of the trunk belt conveying system inbye some 400 yards enabled more efficient overall loading facilities to be created together with an increase of coal-stocking facilities in the event of meeting emergency situations.

The main sketches depict the operation and sequence provisions which incorporated a degree of mechanisation not readily possible in its former location. The upper sketch shows the system primed; the pit tubs E are ready to run back under the conveyor belt for loading in the situation indicted. Coal spillage provisions took the form of a pit with steel shuttered sides from which gravity deflected coal spillage onto the chain conveyor which then loaded them back on the main belt conveyor.

The location of the fixed catches shown on the 'Rope Driven Initiator' sketch is contingent upon the total amount of stretch in the tub shackles which, when collapsing at the end of the run, will allow the second set of 8 tubs to gravitate back clear of the crossover such that the front set can pass on its reversal so to run under the belt to the loading position. This enlarged diagrammatic sketch illustrates the operation principle of the rope activated initiator which runs in channel iron guides below and between the track rails. The enlarged drawing of the travelling initiator shows how it engages with and pushes along the tub train, together with the steps to reduce noise levels as it returns to its starting out position, and also the provisions made for absorbing shock at either end of its run. Partly automated, the initiator automatically stops and then reverses at the end of its 'inbye run' and at the end of its 'outbye run' shuts down awaiting further push button starting for the next cycle. One complete cycle took about 1 minute giving the system a capacity of some 700 tons per hour, being operated by three men, effecting a total saving of ten men per 24 hours. It worked extremely well and opened up the colliery for the successful application of coalface mechanised mining.

Jim Hall, the Colliery enginewright and his son were of great assistance with regard to both the construction and application of the system; they were more than helpful in all the activities I undertook at both Radford and Wollaton collieries.

(OPPOSITE) PHOTOGRAPH 32. MECHANISED LOADING STATION. FULL SIDE
CONTROLLER: WOLLATON COLLIERY

The foot lever when depressed unlocks the tub stop and allows two tub axles to pass over
and then closes, having placed in its correct position the next tub to be filled. Even with
simultaneous handfilling from two or three faces the rate of coal flow was very high at
certain periods of the shift. Tubs might well be filled in 10 to 15 seconds, during which
spillage, being very high, formerly blocked the system. However, with the introduction
of the spillage arrangements shown such situations were now controlled.

Some four or five years later coalface mechanisation was established in a situation equal
to the demands involved. At about this period (1957) the American mining engineers set
up automatic loading stations which were completely manless in many of their mines. The
electric trolley locomotive placed the mine cars in precise situations in which the conveyor
was re-started if it had stopped awaiting a fresh supply of mine cars with capacities of up
to 10 tons. The system shut down when all the available cars had been filled. However,
where a failure obtained they really had problems. Although we had all the basic elements
with which to have automatic loading we preferred to have a couple of men as an insurance,
since the rate we loaded out 15 cwt. tubs (often at some 600 tubs pre hour) was at a
considerably faster rate than took place with the American 5 ton or more mine cars.

The system worked very well and proved a great asset towards the success of the
extensive coal face mechanisation which followed immediately after its installation.

5

Coal Face Mechanisation

Discovering the ultimate in coalface mechanisation, and getting rid of the collier's shovel virtually took the form of a crusade to quite a number of mining engineers throughout the country. Norman Siddall was equally as enthusiastic as myself; indeed there were many more of us throughout all the coalfields similarly motivated in trying to find a solution.

However we both came upon our different personal solutions from two contrasting approaches. He was a few weeks ahead of me relative to our respective discoveries, but he was much further ahead with a scheme for a comprehensive coalface mining system. The Midland Institute of Mining Engineers conducted its meetings (amongst other places) at the Ashby-de-la-Zouch Mines Rescue Station, Leicestershire and normal business meetings were usually followed by the presentation of technical papers given by the various members usually at monthly intervals during the winter. At the end of a particular session the members retired to a certain public bar of their choice in Ashby for an informal drink and general mining or other discussions.

The Trepanner and Hydraulic Powered Supports

It was following one of these meetings that Norman and his associates got down to looking for basic principles which would enable a continuous coal producing machine to be developed. The principle of the trepanning wheel-cutter came up for discussion. It was thought to have possibilities and a sketch was duly prepared on the back of an envelope showing a diagrammatic format.

Countless ages ago the trepanning tool had previously been used for cutting holes in human skulls to let out evil spirits or hot air, although I know of no recorded instance where it has been used on any politicians, particularly those who have in recent times created such a catastrophic effect upon the UK Coal Mining Industry.

Anderson Boyes Ltd (Motherwell) through its Senior Field Engineer Robert Thorpe, a fellow graduate of Norman's at Sheffield University, took the idea to Motherwell, where the company developed a workable design and subsequently constructed an experimental machine built round the basic idea. About this time Tom Seaman of Gullick Ltd (Wigan) developed an advanced type of hydraulic powered roof support which opened up mining possibilities way beyond the limits of normal imagination of that day. Norman brought the two together on an experimental basis at Ormonde Colliery in the No.5 Eastwood Area of the East Midlands Division. The installation, as experimental systems go, was tremendously successful.

The machine, called the trepanner, had a great advantage in that it did not unduly affect the house and large coal market to any appreciable extent (very important at the time). The collier's shovel, with all the manual effort it called for, was now redundant. The support system opened a vista of roof control and security to the workman formerly undreamed of. Comprehensive mining systems with their high rates of production and productivity had now potentially arrived. Mines were no longer something apart. They could be compared with the modern manufacturing factory, even to the extent of complete automation as a more advanced installation at Woolley Colliery (Darton) in the No.6 (North Barnsley) Area later technically proved under the direction of one Cecil Peake, Area General Manager at that time.

The Anderton Shearer Loader

During the period whilst Norman was getting his installation together, I recall following a full day at Clifton Colliery, getting back to my office at Bestwood about 5.30 at night. Everyone but Ron Swindall had left or was leaving. Sam, my clerk, had got everything prepared for signing and appropriate action, nothing contentious, no immediate commitment. Browsing through one of the monthly mining magazines I almost slipped off my chair. There it was, the basic development I was looking for, outlined in a short technical article. What was being described was a development by Anderson Boyes Ltd whereby one of their standard panzer mounted longwall machines could be modified by the application of a special gear box with its right-angle shaft equipped with five or six rotating disks about 3ft. in diameter. The periphery of each disk was fitted with equidistantly spaced cutter-picks.

An hour or so later I called in at Ron Swindall's office, showed him the article and told him what I intended doing about it. We discussed its effects on proceeds since the machine produced a undesirable percentage of fines or dust so degrading the overall revenue return,

'Where are you going to put it?' Ron enquired.

'Gedling Top Hard; we have a spare face there abandoned because of former roof control problems. It might just be the job we been looking for and not much will have been lost if it doesn't succeed. I might want some help in getting the Dowty hydraulic roof supports through on controlled revenue.'

'Tell me what you want and I'll phase it for you,' he said, now being really interested. It was not until 9.30 p.m. that we decided to pack up and go home.

Next day I rang Robert Thorpe of Anderson Boyes Ltd and asked him to call in and see me urgently. Actually he came the same morning about 11.00 a.m. Handing him the technical article,

'What's this, Bob?' I asked. He looked at it and laughed.

'I knew you'd be first,' he said, 'Actually it's an article that's slipped through prematurely.'

'Be that as it may, can I get the replacement gear box and disks?' I asked.

'It's not that easy, we're not ready yet. It will take a few weeks but as far as I am concerned you'll get the first conversion kit,' he replied.

'OK, Bob, that suits me fine,' I answered.

Then he started to tell me the background to the article and the basic development. The machine had been designed for the French potash mines. It had revolving drums embracing peripherally mounted cutter-picks and was mounted on a standard panzer face conveyor. James Anderton, the Area General Manager of the St Helens Area (North Western Division), was another of us on the trail of total coalface mechanisation, and had been shown the potash machine arrangements. Taking these as a base, together with his mechanisation department led by Tommy Lester, he evolved the Anderton shearer loader about which the article had been written, based upon his initial experimental developments which had proved highly promising. Incidentally, I understood James Anderton ultimately received an award of £10,000 from the National Awards Panel for his efforts, the repercussions from which have developed and extended universally way beyond the limits of conception at the time.

Naturally I had to clear the situation and my proposals with N.R. Smith, the Area Production Manager. I'd committed orders for equipment. To my amazement he turned them down because they would degrade the products. Protesting, I pointed out that the proposals were only for an experimental installation in the Gedling Colliery Top Hard Seam, where reduction of revenue would be but marginal. He subsequently made a personal visit to the St Helens Area, and on his return he emphasised the machine's unsuitability because of the increased production of fines. Naturally I was very disappointed, both by not having been given the opportunity of making such a visit, but more so in view of what I had in mind; however I didn't cancel the arrangements I had made with Robert Thorpe.

Within a few days, on a suitable occasion when N.R. Smith and I were in a planning meeting together discussing Gedling Colliery, I referred to my proposals together with the insignificant risks involved in pursuing them. Allan Hill was most supportive and ruled in my favour; I was given the go-ahead. A short time afterwards Robert Thorpe lived up to his promise and the equipment I was waiting for was delivered.

Explaining the situation to Robbie Scott and Gus Wilkinson at Gedling was no problem. It was the first machine of its type to be introduced into the East Midlands Coalfield, and maybe the second in the whole country. Colliery Managers are not unlike opera singers engaged to take a leading role in a new opera. They derive great assurance and pleasure in such situations.

In those early days the machines had a capacity of about 70 h.p. which limited many of the basic factors such as depth of web, reliability and maintenance which controlled potential production performance. The panzer conveyors had advanced but little on the original German design and construction. Powered supports were most primitive by current standards. We were probably at the same stage as the Wright Brothers when they flew their first aeroplane, thus pointing the way to the evolution of the Boeing 747 and Concorde. Reliability and maintenance with such an underpowered machines were especial problems. These, however, were subsequently overcome by technique variations, design concept changes and progressive development, not to forget the hard work and dedication by many nameless people over many years.

The location of the face installation was as intended in the Top Hard Seam at Gedling Colliery. My colleagues lost no time in getting the equipment installed and from the moment of starting it up I knew instinctively we were 'on a winner' and had found what we had long been looking for. In hindsight, the overall situation at the time represented in its embryo state a watershed in coalface mechanisation, ultimately leading to the annual megaton coal-producing installations of the current time. Then we were setting the seal on the format and direction in which coal production was to take, although this wasn't readily obvious.

The basic technique and organisation have not changed, other than to introduce bidirectional operations, i.e. the machine produces coal continuously as it travels through the face in both directions, The following simple explanations are intended to give some idea as to what basically modern mining practice is all about.

Now that you are familiar with panzer face conveyors, hydraulic props, longwall coalcutters, gate belt conveyors and other relevant mining matters and will have developed a good background understanding of mining technique, I am sure you will find the following not too difficult to follow and maybe of interest.

PHOTOGRAPH 33 Anderton Shearer Installation - I. Gedling Colliery, Top Hard Seam. Let's start simply with this early face, now fitted up and ready for working. Everything is quite familiar: the Dowty props (with the cabbage heads), the steel roof joists, the panzer face conveyor. What are the differences in this picture? Just a couple of little ones:

Firstly, the 6 to 9 inch band of stone on top of the coal seam, the cause of earlier hand-loaded difficulties (note the machine haulage rope line on the left hand side about the middle of the seam). Secondly, look carefully at the bottom right-hand corner of the picture. Do you see that rack toothed sword, with a long lever and an adjustable anchor staker? Remember the 'sylvester' in connection with the Samson stripper installation at Elsecar Main Colliery. That's it, the primitive means we had of pushing the panzer and equipment forward into the new track following the Shearer's extraction of a 2' 3" slice from the coalface.

PHOTOGRAPH 34 Anderton Shearer Installation - II. This is a beautiful clear view of the former standard machine which was converted to a power-loading machine in the manner referred to. The conversion gearbox with its six-disk pick cutting media is prepared and ready to start to cut the face. Note the part of the machine casing which protrudes down towards the panzer flights. It houses a large 'crown wheel' which drives the disks. The large steel deflector plate plough provides for the cut coal to slide down onto the conveyor during the production run and the plough lifts the coal onto the panzer on the machine's backward return clean-up run. The hydraulic props in the foreground are of Dobson Ltd manufacture. The machine haulage rope is now on the right-hand side directly over the conveyor spill plates.

PHOTOGRAPH 35 Anderton Shearer Installation - III. Here the machine is shearing off a 2″ 4″ slice of coal along the front of the face and loading it onto the panzer during the production run and loading it onto the conveyor. The troublesome band of soft stone above the seam can be seen to have fallen and been carried away. The three phase power supply cable is laid between the top of the spill plates and the front row of supports.

PHOTOGRAPH 36 Anderton Shearer Installation - IV. This is a view from the opposite (front) side of the machine looking down the face. The machine is in transit towards the main loader gate carrying out the clean-up run, thereafter being pushed over with the primitive tools mentioned earlier.

PHOTOGRAPH 37 Anderton Shearer Installation - V. The machine is ploughing across the face to its starting point which reduces the capacity of the panzer conveyor, since the machine is moving at 30 ft. per minute in the same direction as the conveyor chain; hence the congestion shown in the picture. Improvements did not come easily and involved many people who had to spend many long and arduous hours underground. In this connection I pay tribute to Anderson Boyes Ltd for the tremendous support in all situations they gave to us through Bob Thorpe and Forrest Anderson. It is not unusual in undertaking developments of this kind to experience embarrassing breakdown and delay, but in such circumstances a phone call to Sheffield or Motherwell invariably and rapidly bought spares, supplies or speedy investigation.

The machine was a phenomenal success, and both the Area General Manager and the Area Production Manager (now converted to the machine's potential value) undertook inspections of the working installation and considered it to be such a great proposition that immediate steps towards its extension throughout the No.6 (Bestwood) Area, East Midlands Division were initiated. James Anderton, informed of the progress we had made, invited me up to his St Helens Area to discuss with him our experiences. Together with Tommy Lester I visited his installation and during the afternoon spent a couple of hours in detailed technical discussion. I really think James was concerned that we were hard on his heels and likely to take the lead in the breakthrough he had made and wanted to know the true situation at first hand.

Colin Rudge, in my old Yorkshire Area, was but a little way behind us and pursuing the same objective. The Manvers Area workshops made the conversion from a standard machine to the Anderton shearer by constructing a gearbox with a modification which involved sections of the Meco-Moore cyclic power loader.

The application of the machine outlined in Photographs Nos 33 to 37 inclusive was limited to a seam height of three feet in thickness, so the prospects for its introduction at Radford Colliery were quite unlikely. In view of this I started working on modifications which would extend its range down to about 2′ 6″. The final design was as shown in Photograph No.38.

PHOTOGRAPH 38 Low Seam Anderton Shearer Loader. Drawing on my former experience with the Meco-Moore cutter loader, I came up with the idea of floor mounting, back to back, two Anderson 17" standard coal-cutting machines with converted cutting ends, interposing a short cross conveyor between, all on a common base plate. The machine was designed to operate bidirectionally, i.e, cutting and loading out in both directions. This was, I believe, the first ever approach towards this objective and is particularly important in mining thin seams to maximise production per yard of machine travel. The final design we worked to is shown in the photograph in its constructed format and can readily be understood.

Jim Hall and Chris Corden, Radford mechanical engineer and Group electrical engineer respectively, helped with its design, construction and early development which embraced the cross conveyor and the machine's modified cutting ends: i.e., new cutting end, gearboxes and drums.

Trials were carried out at Radford, where we encountered great difficulty because of the design of the conveyor. It constantly jammed with fine coal and could not transfer the sheared coal onto the face conveyor, whilst a second major difficulty persisted with the low peripheral speed of the cutting picks. We had grossly underestimated this during the design of the gearbox. Dust was also a problem.

However, we suspended further development in favour of extending what had already been proved successful and was being introduced at all collieries throughout the group, the No.6 Area and the East Midlands Division. This particular period was, I believe, to have been a watershed in the affairs of the Coal Industry's mechanised development, in that what Norman Siddall, James Anderson, myself and others had initiated created great interest within every coalfield. It certainly gave positive indications as to which way coalface development should follow. Robert Thorpe and Normal Siddall produced a technical paper on their experience with the trepanner and hydraulic powered supports at Ormonde Colliery. Taking Bobbie Scott along with me, a similar paper was produced on the experience we had obtained at Gedling Colliery relative to the Anderton shearer loader, and both were presented at different times during the same annual session of the Midland Branch of the National Association of Colliery Managers.

We each received a silver medal for our contributions, duly celebrated at the 1957 Annual Convention when Forrest and James Anderson of Anderson Boyes Ltd celebrated the occasion by a dinner at one of the top London hotels. Some two years later I received a second silver medal for further contributions made in connection with the development of hydraulic-powered supports within the No.4 Huthwaite Area.

TOMMY WRIGHT

At this point I wish to introduce a mining engineer who became a very close friend and colleague. Everyone knew him as Tommy Wright. His background was similar to my own, and his outlook both wide and imaginative. At the time I first met him he was Area Production Manager for the No.4 Huthwaite Area,

East Midlands Division. He rang me up just about the time we had successfully established the Anderton shearer installation at Gedling and were obtaining good results.

'Can I come and see that machine of yours at Gedling tomorrow? I hear you are doing great things with it and I'm interested. We have some excellent prospects here,' he said.

'Of course, Mr Wright, I'll be waiting for you,' I answered.

This was the original installation of its type in the East Midlands Division and features the initial conversion gearbox and rotating cutting disks mentioned. Over the years however it was subjected to intense modifications such as drums of varying sizes and shaped scrolls, being fitted with cutting picks of various types. Tommy duly came along and, together with Robbie Scott, I took him to visit the installation. As I said earlier, the mining conditions were not too good, but the machine was working soundly and dealing with difficulties quite well, Tommy was greatly impressed and, being an excellent practical mining engineer, he fully appreciated what we were trying to do. On the coalface he turned to me saying,

'Youth, this is not a soft job. What you've done is bloody marvellous; I've found what I want here.'

Prefixing a sentence with 'Youth' was Tommy's mark of pleasure and respect, to whoever he was addressing.

On return to his Area he ordered six machines in bulk. As I found later, that was his basic approach (quite in contrast to my Area Production Manager's early reactions). I have heard, from time to time, criticism of his approach to spending by much lesser men. In that respect he can be judged alongside that famous railway engineer Isambard Kingdom Brunel, who gave the south-west of England a network of railways opening up a vast economic potential. He too was subjected to the same criticisms. However, like Isambard Kingdom Brunel, Tommy Wright got things done with satisfactory results. That visit subsequently proved to be a landmark in the affairs of the EMD No.4 (Huthwaite) Area and had later a beneficial effect on my personal career.

Current coal face mining equipment has advanced in terms of design, construction, reliability and technical application to a degree unimaginable in those early years, this being dealt with in succeeding chapters. One interesting aspect is that the Anderton shearer loader set the pace of development. With greater capacity developed within the Anderton shearer, the face conveyors (and indeed in many cases the outbye transport conveyors) had to be developed to match to the equipment's shortfalls thrown up by experience and expanding production. As the system was extended to a wide expanse of coal seams of varying thickness and with a large variety of mining conditions, support systems had to be developed to meet the new situations as they were opened up, such that every aspect of the mechanised longwall system developed, at a corresponding rate, to meet the overall requirements of an integrated system built round its producing element.

PHOTOGRAPH 33. ANDERTON SHEARER INSTALLATION – I: TOP HARD SEAM,
GELDING COLLIERY

This was the actual coal face we chose for undertaking experiments with the Anderton Shearer Loader Installation. Formerly it had been worked by conventional machine cut handloaded methods but had to be abandoned due to the working difficulties already referred to. Usually when undertaking such developments one looks for the most promising conditions so to reduce the risks of potential failure. Our earlier experience of the excellent system of support to the working face we had developed with the panzer-mounted coalcutter was transferable to advantageous effect. In addition, for a substantial period of the men's working shift they were working under a highly protected area with considerable risk reduction and increased safety.

The problem of fines gave little trouble and the effect upon revenue was negligable, although we did have to handle an increase in large stone on the surface sorting tables much as expected. However the development did open up a large parcel of coal that had been set aside owing to the problems of manual working.

PHOTOGRAPH 34. ANDERTON SHEARER INSTALLATION – II

This photograph shows the Anderton Shearer Loader in the LH or main gate stablehole positioned and ready to start its shearing operation through the face which is about 180 yards in length, with the seam of the order of 34″ in thickness. In the foreground are Dobson hydraulic props being used with German type extensible bars. The combined steel loading plate and plough deflected the coal onto the panzer conveyor during the production run from 'left to right' through the face. On the return run 'right to left' the plough cleaned up the coal spillage left in the sheared track.

Dust was controlled by heavy jets of water projected on the rotating disks. Very shortly after this photograph was taken the disks were replaced by a drum with pick boxes fitted with cutting picks mounted on its periphery. Substantial development was undertaken over many years in connection with drum and pick design. The modern scrolled drum is quite sophisticated and associated with a cowl has eliminated the primitive arrangements shown.

With bidirectional operation the machine produces coal in both directions, the machine in this case having cutting or shearing elements mounted at either end.

PHOTOGRAPH 35. ANDERTON SHEARER
INSTALLATION – III

This picture shows the machine under actual shearing operations with the coal being transferred to the face conveyor. The efficiency of transfer in those early days wasn't good with the result that spillage was left behind and often had to be cleared with the shovel. Some idea of the amount can be gleaned from the photograph.

An examination of the support system indicates that the bar in the foreground has been advanced roughly at a distance of some 4 metres behind the machine. As the machine crosses the face these bars were advanced progressively.

Normally the men would attempt to complete two strips per shift so advancing the face upwards of 1 metre per shift with a production of about 300/350 tons. Working two shifts per day gave us 700/750 tons per 24 hours which was quite an acceptable return compared with conventional hand filling, in those early days, particularly so as the face productivity (tons per face manshift) was higher and the coal produced much cheaper.

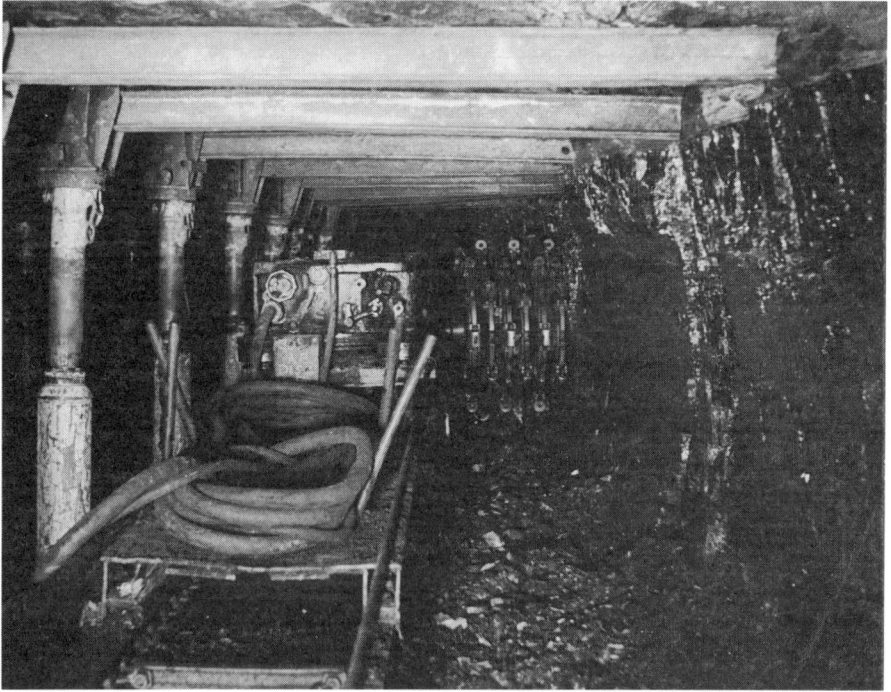

PHOTOGRAPH 36. ANDERTON SHEARER
INSTALLATION – IV

This photograph shows the shearer on its flitting and cleaning run from right to left across the face. It provides an excellent view of the front of the machine and shearer disks. The spillage deposited between the panzer conveyor and the foot of the seam has yet to be cleaned up by the facemen, following which the panzer will be pushed over into the newly cleaned track. With the panzer in its new position the extreme back row of supports are withdrawn and re-erected adjacent to and along the waste side of the conveyor.

The stone and shiny coal above, in the upper part of the seam, represents the source of the extra large material which falls in the new track behind the shearer as it cuts through the face, the effect of which is clearly shown in Photograph 37.

The electrical power cable handling arrangements are quite simple and consist of a sledge pushed or towed by the shearer, with the electric cable being fed off the sledge whilst cutting and loading the coal, and coiled back thereon during the fitting and cleaning run.

PHOTOGRAPH 37. ANDERTON SHEARER INSTALLATION — V

This is a leading end view of the shearer ploughing or cleaning the new track as it travels from right to left across the coalface. Here again the source of the massive coal/stone lumps can be seen along the roof of the seam.

The picture shows a state of conveyor congestion in that the panzer chain is unable to take the material away at a sufficiently fast rate; moreover, the effective speed of the chain is reduced by an amount equal to the flitting speed of the shearer, since chain and shearer are both moving in the same direction.

Such situations created problems at the transfer points (i.e. where one conveyor delivers onto another), with the material having to be broken by hand-picks. Such problems invariably generated solutions. In this particular situation mechanical coal breakers were fitted to the stage washers situated between the face panzers and the gate belt conveyors.

PHOTOGRAPH 38. LOW SEAM SHEARER LOADER, RADFORD COLLIERY
(DECEMBER 1955)

Mechanisation of Thin Seams:

This was a special development undertaken in an attempt to increase the range of the machine's applications to seams below 3'0" down to 2'3" in thickness. There were vast possibilities and potential throughout the UK were it possible to achieve this.

The basic machines were standard AB17 longwall coal cutters with the cutting end modified in a similar manner to the Anderton Shearer Loader. Both the gearboxes and the cutter drums were constructed in the machine shops at Radford Colliery. Lack of height below 3'0" required the machine to be run on the floor of the seam in its former manner. The basic problem was that of conveying the cut coal across the machine track onto the face conveyor, which in the trial situation undertaken at Radford Colliery was a bottom belt face conveyor.

To make the machine bi-directional two machines were placed back to back as clearly shown in the photograph. This was, I believe, the first time bi-directional shearing was considered or attempted. The trials threw up a number of problems: the conveyor constantly jammed owing to the internal compacting of 'fines' within the unit; belt speed required to be increased; the peripheral speed of the picks was far too low, being less than a half that of the proven machine; and finally, the production coal dust gravely impaired the men's working conditions, such that the experiment had to be shelved.

Had time been available I feel we could have developed something useful from our efforts. In the circumstances we were working hard towards the extension of the proven system throughout my Group.

6

Staff College Exposures
and Organisational Change

The Board's Staff College - Chalfont St Giles

At intervals throughout my career I was exposed from time to time to the National Coal Board's Staff College at Chalfont St Giles, Bucks, in a variety of situations such as syndicate nominee, occasional lecturer, and symposium delegate at National Production Conferences, Meetings and Assemblies. On one occasion I attended for the purpose of being subjected to syndicate questioning about the 'Role of the Group Manager in Area Management'.

Other than for Board policy meetings, gaining new experience in informative technical papers, new mining techniques and reviews of new equipment from my colleagues and field initiators, I never found any real stimulating or lasting inspiration from such exposure. Standard procedures on organisation and management are all very well in standard non-variable situations (if such exist) but throughout my experience in a number of the Board's Areas, the range of differences which prevailed was extremely wide and uncertain, a situation which was more than exemplified by the comprehensive range of performances which existed between the Board's 43 Areas at the time (see Appendix). There was, however, a standard objective common to all Areas, that of producing the tonnage of coal the nation needed, safely and at profit. In this respect, the better Area performances did not necessarily reflect the highest levels of management ability.

The Board's Staff College at Chalfont St Giles was ill-conceived in many people's opinions, including my own, for a number of reasons, namely, its location; its initiation under Dr Fisher (then of BBC Radio Brains Trust fame); and a succession of staff whose successes in the mining field were either yet to be discovered or insignificant. There were, however, exceptions, Things tended to change over the years under the influence of the teaching dogma practised. Though colliery managers had formerly made underground inspections almost daily, subsequent to the influence of the Staff College, their underground inspections became less frequent, on the basic premise that, with true delegation, it wasn't necessary. This was the work of the undermanagers. I would add, in fairness to the colliery managers, that the growing tentacles of bureaucracy also contributed, although the colliery manager's staff at individual pit level was substantially increased over the years.

I recall that the work I was undertaking for R.G. Baker then as manager of Earl Fitzwilliam's Elsecar Main Colliery, at a later date covered the employment of three persons. To keep abreast of the activities, changes and variations within a large colliery could take upwards of a month (four to five days a week) to examine the whole mine and, not unlike the painting of the Forth Bridge, the process needed to be restarted at the end of each cycle. The whole underground situation was in a constant state of change as new working areas were exposed.

There were many advantages involved with the high frequency of visits underground. There was close contact with both men and junior officials. Adverse situations between men or management could be anticipated and avoided. With deteriorating mining situations, remedial changes could be more quickly considered. The monitoring of essential mining development ensured that corrective action could be determined more swiftly and more readily applied. Indeed, the maintenance of high safety standards was more effectively provided.

Another feature about the Board's Staff College was the post-lecture adjournments to the bar at night. Some people were good socialisers, extremely witty, fluent at small talk, telling good jokes and showing great bonhomie, whilst the more intense amongst them preferred to talk and think seriously. Here was a potential yardstick for assessing social acceptability, for invariably people with authority were present. Under such circumstances the labelling of a tried or untried person as, 'He's a good mixer', or 'He has the potential of an Area General Manager', or 'He cannot delegate', could be rewarding or unhelpful to the individual.

The well-labelled person is placed and off he goes to be trained for the requisite labelled post. Floundering at the first hurdle he is moved forward to obtain further experience in another field. Blundering at his next obstacle the process is repeated. Subsequently, he emerges ready to take the job, for which his so called training has prepared him. The poor chap adversely labelled gets no such treatment. Whenever he faces an interview the label is there before him negating his chances, yet the source of his labelling may not have had the slightest idea of what the implications of his label really did mean.

I would like to refer to the following situation. Having sincerely attempted to recount my experiences of many years as accurately and carefully as I can without bias or animosity, I would ask the reader to reflect on the justification of the label: 'His level is no higher than that of a junior official', as applied by one person at a very high influential management level to myself at an emergent stage of my career. How far and wide such an opinion was conveyed I don't know and it is, in the circumstances, somewhat inconsequential as events unfolded, but it represented the seeds of potential personal damage, being both unwarranted and uncorrected by the subsequent opinions of many highly qualified and capable people. To me it would seem that such label-makers are incapable of moulding or shaping good raw material into useful creative effect and in some circumstances lack the characteristic of self analysis, particularly with regard to their personal levels of achievement. If they were, such labels would not be used in the first place.

During my period as Area General Manager of the No.9 (Neath) Area of the South-Western Area. I was asked to attend the Board's Staff College so to undertake discussions with a particular section of the delegates about 'The Role of the Group Managers within an Area'. At the same Session Bill Rowell, Area General Manager of the Scottish Alloa Area, was asked to do the same thing with another section of the delegates. Bill was a fine and highly competent mining engineer with whom I had worked from time to time during his period on Headquarters staff. We knew and respected each other very well. There were however tremendous differences between the two Areas. His was much bigger than mine. He serviced the electricity industry and other large consumers, whereas my comparatively low output of anthracite was most valuable as household, central heating and specialised fuel. The respective mining and labour relations situations too had important differences, not the least of which embraced deeply entrenched customs and practices maintained over countless generations within the anthracite coalfield.

The concepts of organisational delegation being projected at the College were based upon those of Henri Fayol and others of more recent date. Fayol was a French mining engineer who in 1916 had for many years managed a large French mining company and put forward his ideas about the organisation and the supervision of work. In 1925 he advanced a number of principles and functions of management which he had published. His central idea was that of 'unity of command', which stated that an employee should receive orders from only one supervisor. This, at the time, helped to clarify the organisational structure of many manufacturing organisations. Clearly this is a straightforward concept simple to interpret and apparently easy to apply. But over a period of years following discussions with a great many of those delegates returning from sessional courses at the Staff College, I found their interpretation of delegation to be more in keeping with, 'You told someone what you wanted doing and left him to get on with it, in his own way, in his own time, and awaited the outcome.'

The nature of the delegation format practised by some very successful people I grew up with and which I tried to follow took the following form:

1. First determine to whom matters needed to be delegated and whether or not he or she is capable of dealing with them. If not, make further enquiries to ensure their capability, or make another choice.

2. Ensure that the matters to be delegated are adequately understood by the person chosen.

3. Fix a time interval for completing the execution of such delegated matters.

4. Follow up the progress being made in relation to the specified time allocation.

It was usually the last two points that I found presented difficulty in that a degree of discipline was naturally applied, and resentment could arise where there was a failure to deliver within the specified time interval.

The enthusiastic recipients usually had no difficulty in accepting and working to time constraints and would instinctively try to prove an underestimation of their abilities by completing the assignment given to them earlier or better than specified. Moreover they would keep one informed of their progress, particularly when testing out ideas etc. developed along the way, in describing the difficulties being encountered. Most of the staff I had the good fortune to work with were of that ilk. Those not so motivated invariably had to be sorted out at the end of the allotted time, and usually went to great lengths concerning the alleged difficulties they had met and the impossible nature of what was being asked of them. Such difficulties were never referred back at the time they were met, or help sought relative to them. Further pressure tended to cause some resentment as manifested in such expressions as:

'He won't let me get on with the job.'

'He's not satisfied with nuts and bolts, he has to design them.'

Those making such observations usually do not have the courage to express them direct. If that had been so meaningful discussions would have arisen, in which case any real problems arising would have been cleared and the *bone fide* difficulties resolved. After all, such are but negative reactions. Referring back to the situation of Bill Rowell and myself at the Staff College, nothing was said at its conclusion or comparisons made whereby apparent conflict could be examined, but some three or four years afterwards, it was impersonally referred to openly at a full meeting of the College. Our approaches had been different and one of us didn't fit into the pattern of things being projected. These projects were often put forward by people who would run for cover if attempting the accomplishments we took in our stride.

I had quite a lot of correspondence with Jack Long, Managing Director of Long Airdox Ltd, mine machine manufacturers in the United States whom I knew very well and had met many times. He was deeply interested in British mining systems and over the years I had fed him with quite a volume of material. Jack was also friendly with W.V. Shepherd, then the Board's Director General of production with all-embracing responsibilities for the collieries. Jack told me of a meeting he had had with him during which I became the topic of conversation, when the Director General said,

'His problem is that he can't delegate,'

Jack's reply was, 'We've quite a few excellent people like Charlie in the States doing similar jobs - why don't you save the Board money by getting rid of all those making no contribution? If it is true what you say, we would.'

The apparent reply was a shrug of the shoulders.

Lest the reader feels I'm being particularly sensitive or paranoid on the subject, I'm not really. Annoyed maybe; for such remarks are an insult to all those excellent colleagues I and others were privileged to have led and to have been associated with over a long time. The successes and achievements described here are a great tribute to their personal contributions and supportive efforts.

I do not believe it is unreasonable to feel that staff controlled and staff oriented teaching establishments lacking enthusiasm, zeal and motivation for maximised company performance can get their priorities wrong and focus endeavours more upon the means, rather than the end for which the organisation was set up in the first place, and misjudge as 'out of step - lacking management ability' those operating purposefully, enthusiastically and effectively in pursuit of that latter objective in their own way.

Jack Long was a particularly good friend to me in those early days. He encouraged me in many ways, opening up many valuable visits to the American mines and advised me greatly on American methods and techniques. I found him tremendously progressive and enterprising.

On one particular weekend I joined him and his family at Greenbriers, West Virginia, where I had a most relaxing time with him and his children, ten pin bowling, swimming and following other pursuits. I recall playing with his youngest son Army my first ever game of ten-pin bowling, and with the luck of the novice made a 'strike' in the first bowling attempt (not repeated at any time since). I got on well with the children. They kept me alive in many of their activities, one of which was to judge their attempts at raising castle structures with playing cards.

Greenbriers is a fabulous country resort set in a most beautiful environment (the US President had a cottage there). Recently it was disclosed that a subterranean maze of many miles of tunnels, together with numerous offices and facilities, had been secretly constructed within Greenbriers' environment. The purpose was to ensure continuity of USA Government administration in the event of the Cold War evolving into a nuclear calamity.

In the mid 1950s, the Board recruited, for administrative training, a number of excellent youngsters, undoubtedly amongst the cream of the colleges and universities at the time. They were sent into the Areas as 'AAs' (Administrative Assistants). The one we received was a youngster, Peter Mullen, who was given a thorough basic field grubbing following a long period at Area, actually working in various departments. We received him at Clifton Colliery where, working to manager and undermanager, he fitted in like the proverbial glove and was well received and respected by both officials and men.

In those early days the analytical degree to which I subjected the managers in various manpower recovery situations was really intense, as for example:

If working on any improvement scheme in which labour savings were to be effected and a number were scheduled for upgrading to the coal face, such an analysis would require the name and check number of each man recovered, and his grading suitability determined. (Some men by reason of their health or former accident partial disablements had to be carefully examined for suitability for a potential replacement job, one which they could undertake without detriment to their individual welfare.) To get the required number of men upgraded could involve a detailed analysis of maybe two or three times the actual number of recovered personnel, in that it involved a fair degree of 'ringing the changes'.

Peter fell into one such situation. He applied himself with the aptitude and skill of one who had been versed in the approach for many years, Not being afraid to dirty his hands and face, he made many visits underground to ensure a smooth transition of each man to his upgraded or alternative post. Heavy overtime was undertaken without a murmur. Many were the nights all four of us would work. Sometimes it was irksome but as things fell into place we were thrilled.

He had the great respect and appreciation of us all. I believe he ultimately spent some time at the Board's Staff College. At least he was one example with great ability and who could generate confidence. In getting capital project approval for manpower saving schemes I never had a moment's difficulty with the Area Accountant. He would usually say,

'Do you have a manpower schedule, Charlie?'

'Yes - here's a copy,' which he would accept in good faith.

I was the only one, to the best of my knowledge, to carry out such a detailed survey, but it worked and over a long period throughout my career it enabled us to upgrade a large number of men into coal face production with an expansion of output tonnages, a situation of great benefit to those upgraded in terms of wages and to the National Coal Board in terms of performance.

Administrative Staff College, Henley-on-Thames

Whilst I was employed as Area Production Manager No.4 (Huthwaite Area) I was nominated by the Board to attend Session 36 (1959) at the Administrative Staff College. Don Severn, the Area General Manager, temporarily withdrew my responsibilities under Mining Legislation for three months of the course tenure, his advice being,

'Go, lad, get the best out of the opportunity without worrying about the Area.'

The Henley Staff College (not unlike the Harvard Business School, USA) is situated alongside the River Thames in a most beautiful environment and incorporates residential and recreational facilities. The seventy or so attending delegates were nominated from a wide range of companies, institutions, active services, industry and other activities throughout the whole of the United Kingdom and overseas and were those who undertook the financial responsibilities for their respective organisations. Some of the delegates had passed through Oxford, Cambridge and other universities. They, I believed, represented the country's potential supply of future leaders over a wide range of activities. Reassuringly, I found amongst them a small number like myself who, by the nature of their situation, had not had the good fortune many of the delegates had enjoyed by their birth or other circumstances. I was not alone.

Naturally I felt very apprehensive, for I believed that most of the nominees were really brilliant in view of their particular circumstances and background. I was selected by the Board along with one D.A.N.R. Simpson on Headquarters staff, whom I knew quite well. We'd met on several occasions, particularly when he was involved with the early Planning and Development of Abernant Colliery in the South Western Division's No.9 Area.

On the first Sunday night of the session we were called into the lecture hall to hear a talk followed by a general discussion which was given by one of Hoover's top executives. Douglas Simpson and I were sitting together. The lecturer outlining matters relative to trade and commerce was waxing forth with great fluency when suddenly he started to criticise the National Coal Board, the dirty and deplorable condition of its colliery surfaces, and the grievous state of its assets.

He really was 'going to town'. Turning to Douglas Simpson, I said, 'Are we letting this character indict our profession in this manner without some challenge?'

It needed courage in the situation to make such a challenge and took quite some time to summon it but it came. I stood up and addressed him as follows:

Had the speaker ever travelled by train through the industrial towns and cities of the Midlands or the North? If so, did he not observe that the criticism he was making against the National Board applied in no small measure also to many large factories, industrial works, British Rail and other enterprises alongside the main tracks of the rail network? Why was he singling out the National Coal Board for such critical attention? The National Coal Board were spending enormous sums of money on the reconstruction of colliery surfaces and the provision of miners' welfare facilities. As an example, Gedling Colliery in the East Midlands Division was in the process of spending over £250,000 on surface reconstruction offices, baths and amenities for some 3,000 men employed. Did his Company ever have to meet such situations at any of its manufacturing units?

I sat down drained, with all the symptoms of one in a highly uncertain situation. I could not nor would have wanted to disguise the remnants of my Yorkshire dialect. What would be the reaction of those delegates present?

The speaker's reply was that he hadn't intended his observations to be exclusive to the National Coal Board, and it was good to note that the NCB were developing a new breed of enlightened executives. It augured well for the future.

At the bar following the meeting quite a number of the those present at the meeting came up to me and said how pleased they were that someone had challenged these people. At the time Hoovers were not too popular in the retail trade because of the 'switch sales' technique they followed. I'd no idea what this was and in order to find out it was necessary to expose my ignorance on the matter. The delegates responded quite helpfully. I gained a little confidence but I needed a great deal more with which to face the tasks ahead.

Very much later I was told that it was usual to project the first speaker of the first Session in the role of an Aunt Sally, so that the ice could be broken early at the start of the course and so provide for real stimulus.

The delegates were arranged in a balanced syndicate format, in line with their individual vocations, skills and experiences, each syndicate approximating to the constitution of a regular company working board, with provision for the widest corporate coverage. The range of subjects fed to the delegates was extremely wide, fully supported by an extensive library, current magazines, articles and press publications. The subjects were fed at a most exacting rate throughout the full Session, almost to the final day. The course generated heavy pressures throughout and certainly called for effort. However, one was free to take from the course what one wanted without reference or question.

Visiting speakers were drawn from the widest possible range including public affairs, politics, industry, economics, manufacturing, finance and commerce and trade unions. A well organised system of visits to the City, institutions, oil terminals, public works, manufacturing industry, finance houses, etc. dovetailed and were integrated within the subject content of the course.

It was customary throughout the course for the Principal, Sir Noel Hall, and

Lady Hall to invite a number of the assembly for drinks prior to the 'report back sessions'. These were very hospitable and lavish occasions. I wasn't a heavy drinker and never have been, so that it didn't require more than a couple or three gin and tonics to get me beaming. It so happened that the only 'report back' task I did coincided with one of these fabulous invitations. I'd schooled myself not to go beyond the single gin and tonic. Moreover I'd carefully written down what I had intended to say at the report back session. After the first drink, however, something went wrong. My former resolve had slipped.

To this day I don't know why, but I do have a strong recollection that Sir Noel and Lady Hall had two children present who were most adept in ensuring everyone's glass, no matter what the beverage content, never fell below an imaginary plimsoll line, and undertook their topping-up responsibilities with a degree of enthusiasm and application worthy of the widest possible exposure. I began to glow, but what was worse, my limb control stabilising mechanisms had been badly impaired. Somehow I managed to be in position to present my report and when my turn came to speak I stood up glowing with the incandescence of a 1000 watt electric light bulb and stammered,

'Sir Noel, ladies and gentlemen. I've just come back from a lavish drinks party, most of you know how lavish. Before going I'd taken the precautions of writing out notes of what I was going to say here, now I find I can't read them.'

The place rocked with laughter, I had to improvise - I don't know how well but the audience members were really encouraging.

At the end of the course, on being interviewed by Sir Noel Hall, his first questions were invariably:

1. What did you think of the course and what ideas do you have towards making it better?

2. What did you personally get from the course?

The course was exceptionally well organised and called for deep application. It ran with great efficiency. The staff were excellent and possessed a wide visionary outlook. There was little I could suggest. The answer to the second question was quite easy. My confidence had improved dramatically, and exposure to many facets of business, organisational trends, financial undertakings, the City, economics and international GATT and other agreements had stimulated thought, interest and discussion. Here I felt individuality was not impaired; one took the things considered helpful and useful in enhancing one's make-up whatever one's area of application.

As a representative cross-section of the country's up and coming high executives there was quite a range of variation. Some people absorbed things quickly and easily, no doubt permanently. Others, like myself, had to slog. Certain members however created a life-long impression on my mind. Not only did I feel some were of great intellect but as natural human beings they were amongst some of the best I have encountered. J.P. Gregory (I believe him to be on the Cadbury's Board), C.E.H. Morris (if I recall correctly from Llanwern steel plant, Port Talbot), and G.P. Hogg, (accountant with Boots, Nottingham) are three that readily spring to my mind. I never met any of them afterwards.

Nationalisation of the Coal Mining Industry

In the very early days of my career, nationalisation of the mines was discussed in the pits, trade union and WEA (Workers Educational Authority) class lectures and political circles regularly, with the leaders painting pictures of a New Age, an Elderado and an end to all arduous effort, something which would happen almost overnight with political control in the right hands. However, it was not until the passing of the Coal Industry Act 1946 that the pits came under State control and the National Coal Board was set up.

In the event there was a great deal of disillusionment. It could not be otherwise. Many parts of the industry were in a particularly bad state having been starved of capital over many years and run down inordinately almost to the point of closure. However, there were some large and efficient companies within the coalfields which were well run and profitable.

There was no short cut to meet the time element needed to convert available capital into structural works and technical achievement, or to advance the potential results of such development beginning to emerge at the coalface arising from the foresight of the leading companies engaged in the application of the Meco Moore and other cyclic power-loading machines. Quite a number of these early machines were tried during the early 1950s.

PLAN I - THE COALFIELDS OF GREAT BRITAIN This plan published by the National Coal Board shows the coalfields of Great Britain and their constituent coal qualities in terms of classification or specification. The mining situations were variable among the respective coalfields, Areas, and collieries. Seams varied from a few inches in thickness to over 20 feet and from level gradients to almost vertical projections. Some locations were very heavily disturbed with geological faults, washouts, igneous intrusions and the like. The roof and floor strata, above and below the seams, embodied a wide range of difficulties. Weak roof shales were often difficult to control whereas working immediately under strong, hard sandstone was usually accompanied by excellent mining conditions. Floors of weak fireclay often allowed the supports to penetrate with immediate adverse effect to the working places. Firedamp and water were no less of a problem. Some situations were so heavy with gas that emission techniques were developed to collect and commercialise the gas for colliery steam generation or other purposes. Throughout my career in the Coal Industry great strides were made to control all such adverse situations and with the advent of powered supports, safe mining practice was achieved under some of the former most difficult of mining circumstances. The range of systems and techniques developed to meet the whole gamut of problems involved with coal extraction over the past forty years has been truly amazing.

The economic development of cities such as Sheffield was no accident. Located over an area of excellent coking coal and within easy access to the Cleveland ironstones, as demand for steel increased, growth was quite natural and commercialisation followed. With the development of high quality steels came the manufacture of other steel products; the associated plant and factories further contributed to the city's rapid growth and expansion. Sheffield steels were world renowned whilst manufacture of steel cutlery held a similar place on the international scene.

Similarly Cardiff, within easy access to large quantities of high quality steam, industrial, electricity and gas, and general purpose coals, was ideally placed both for coastal transport and export shipping. Newcastle, with its port facilities on the Tyne and its close proximity to the Durham and Northumberland coalfields, similarly played a predominant part on the industrial scene.

Under the direction of Parliament through its then Minister of Fuel and Power, The Right Honourable Hugh Gaitskell, CBE, MP, the National Coal Board was set up to run the Coal Industry. Its constituent members were at the time as follows:

FIRST NATIONAL COAL BOARD - BOARD MEMBERS (1 JANUARY 1947)

> Viscount Hyndley, GBE, Chairman
>
> Sir Arthur Street, OBE, KGB, CMG, CIE, MC,
> Deputy Chairman
>
> Mr J.C. Gridley, CBE
>
> Mr Ebby Edwards
>
> Sir Charles Ellis, FRS
>
> Sir Joseph Hallsworth
>
> Mr L.H.H. Lowe
>
> Sir Charles Reid
>
> Sir Eric Young

The organisational chart on page 166 shows how the Board effectively administered and controlled its coalfield assets which, within my experience, worked very smoothly in those early years, during which time a great deal was accomplished in revitalising many of the old collieries. The operational unit is the colliery (of which there were roughly a thousand), each with a colliery manager whose responsibilities at the time were defined in the 1911 Coal Mines Act and Regulations made thereunder, subsequent to which the 1954 Act was introduced after the pits were nationalised. The collieries were grouped in Areas under the control of Area General Managers, just as in the past groups were so contained within the jurisdiction of private companies. There were 48 Areas embraced within the eight Divisions conforming to the locations of the respective coalfields shown in Plan No.1.

In addition to the transfer of some 800,000 men and staff from private enterprise into nationalisation, some seven hundred employees were included who, in the past, had been engaged on central activities of one kind or another. Certain activities formerly the responsibility of Government Departments and other bodies were transferred within the new arrangements. The main change

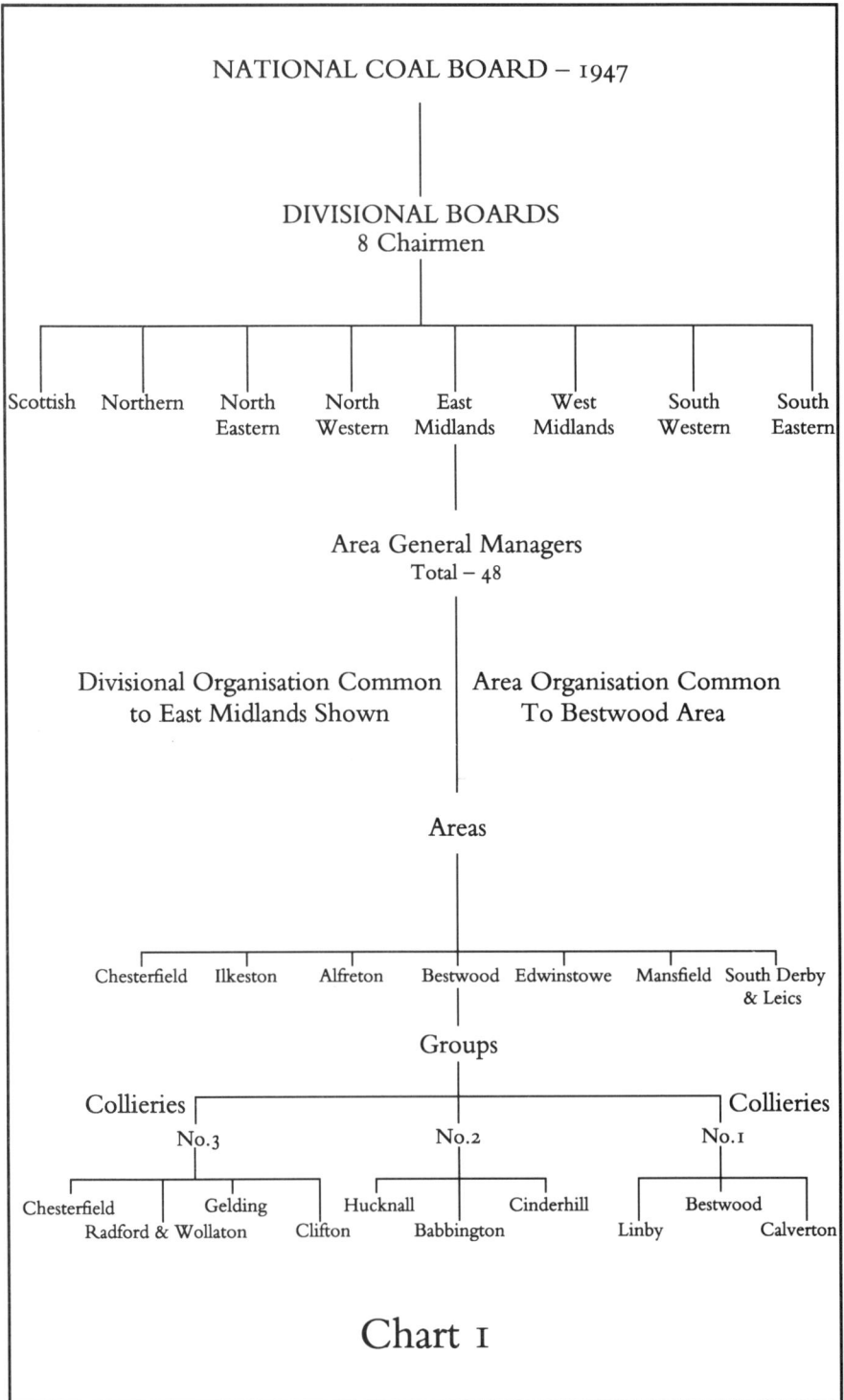

NATIONAL COAL BOARD – 1947

DIVISIONAL BOARDS
8 Chairmen

Scottish Northern North   North   East   West   South   South
                    Eastern   Western   Midlands   Midlands   Western   Eastern

Area General Managers
Total – 48

Divisional Organisation Common   Area Organisation Common
to East Midlands Shown   To Bestwood Area

Areas

Chesterfield Ilkeston Alfreton Bestwood Edwinstowe Mansfield South Derby
                                                                   & Leics

Groups

Collieries   Collieries
              No.3   No.2   No.1

Chesterfield   Gelding   Hucknall   Cinderhill   Bestwood
     Radford & Wollaton   Clifton   Babbington   Linby   Calverton

Chart 1

involved the replacement of some eight hundred boards of directors, mostly part-time and pursuing separate policies by eight Divisional Boards working within the limits of a national policy laid down by the National Coal Board.

PRINCIPLES OF PROPOSED ORGANISATION

The principles on which the organisation of the coal industry was based reflected the economical and commercial circumstances of the time. The marketing function had no boundaries. Coal produced in one coalfield or district could be sold in any other. Many inland markets such as London drew their supplies from more than one coalfield. The export markets similarly were fed with coal coming from various parts of the United Kingdom. Flexibility in marketing was to be essential if supply in varying locality situations was to be quickly adjusted to meet the pressure of directional and intensity of demand. There was need for the national co-ordination of selling and a national policy to regulate process and qualities. The function of marketing in this situation, I believe, underwent a great change, with a greater accent being placed on distribution than on an aggressive drive towards developing new and overseas markets.

For long-term planning, the industry as a whole must be dealt with as a single unit. With production costs varying from coalfield to coalfield and Area to Area, capital development had to be planned to favour those Areas where costs of production and delivery to each market was less than those from alternative sources. What is more, a national policy for prices and capital investment should have as its corollary a national policy for wages. Policies for coal prices, capital investment and wages are all facets of the industry's economic potential; thus the setting up of completely autonomous Divisions or Areas would be inconsistent therewith.

Concerning its administration and control provisions, there had to be a clear line of command from the National Coal Board to the coalface and from the coalface to the National Coal Board. The authorities in the Line were the National Coal Board (responsible to the Minister and Parliament), the Divisional Boards (each collectively responsible to the National Coal Board), the Area General Managers (personally responsible to the Divisional Board) and the Colliery Managers (personally and directly responsible to the Area General Managers). A basic premise which the Board had towards its organisation provisions was stated as:

> Though good organisation is essential if effort is not to be misapplied and energies dissipated in friction and frustration, men matter more than charts and good principles of management are no substitute for good managers. Organisation is only a means to an end, to enable the coal industry to contribute as much a possible to the standard of living of all people.

This I always felt to be basic to both progress and achievement and indeed subconsciously worked to it in that some decisions made and actions undertaken during my career may well have been fraught with personal risk and possibly over or on the edges of strict formal policy contravention. Departmental staff were covered at National and Divisional Levels. They followed the 'Line and Staff' format of the Organisation Area Level Chart as shown. Prior to nationalisation, the General Manager or Managing Director of a colliery company looked to his Board of Directors (usually representing the shareholders) for direction and guidance.

With the creation of the Divisional Boards, full time Directors were appointed with special responsibilities for specialised staff, and with a joint interest in the framing of Divisional policies and plans and supervising their execution by the Area General Managers whose performance they watched.

The Staff Departments usually found at Divisional and Area were: Production, Labour, Marketing, Welfare, Finance and Scientific, each advising the Area General Manager on matters within his specialist sphere. They also acted as consultants to the colliery managers and carried out their specialist duties on his behalf, e.g., keeping financial accounts, running the Area laboratory; undertaking functions of welfare on behalf of the mining communities.

Subsequently the Board sought advice on its organisational structure from the private sectors. Dr Fleck of ICI chaired a committee to that effect. Many of his findings were adopted, the chief of which was the appointment of two Deputy Area Production Directors (Planning & Operations) and giving Purchasing and Stores a departmental status. During the long experience I had with the Board following its inception, I found most of those I felt to be good managers were highly individualistic, enterprising, enthusiastic and capable of decisive action under the widest range of circumstances. Maybe these were the persons the Board had in mind although, in practice, they were often subjected to backlash in one form or another when procedure was contravened. However, this is no more than life in general calling for one to rise to and deal with the occasion, whatever its nature.

In hindsight, I have often pondered upon the Industry's organisational aspects. I believe the vertical structure has proved very costly and departmentalism at too many levels developed to an extent that it hindered, rather than enhanced, progress. E. Northcote Parkinson in his book *Pursuit of Progress (Parkinson's Law)* in those early years examined this facet in detail with a high degree of enlightenment and humour, the combined effect of which was to detract from the more serious issues involved.

NATIONAL COAL BOARD - FIRST YEAR'S (1947) PERFORMANCE

The Coal Industry's performance under the nation's ownership was embodied in the National Coal Board's Report and Accounts for 1947 and for every year thereafter.

TABLE OF COALFIELD PERFORMANCES YEAR 1947

DIVISION COALFIELD	TONNAGE LONG TONS	PRODUCTION COSTS		REVENUE	PROFIT/LOSS
		TOTAL	PER/TON	TOTAL PER/TON	
Scottish	22,185,069	43,014,148	1.939	44,614,174 2.01	+1,574,098
Northern	35,899,472	81,282,387	2.264	72,483,418 2.02	-8,795,131
North Eastern	38,310,290	74,000,749	1.963	74,623,771 1.95	+623,022
North Western	13,233,010	30,998,325	2.342	29,297,884 2.21	-1,570,675
East Midlands	34,683,982	57,291,028	1.653	64,801,230 1.87	+7,510,621
West Midlands	16,642,170	31,438,275	1.889	34,116,449 2.05	+2,686,092
South Western	22,418,864	59,385,596	2.649	48,648,935 2.17	-10,741,405
South Eastern	1,375,359	3,644,061	2.698	3,153,698 2.29	-490,526
United Kingdom	184,748,211	381,040,46	2.068	371,836,736 2.063	-9,203,905

The Appendix sets out the individual Area performances achieved during the first year of operation.

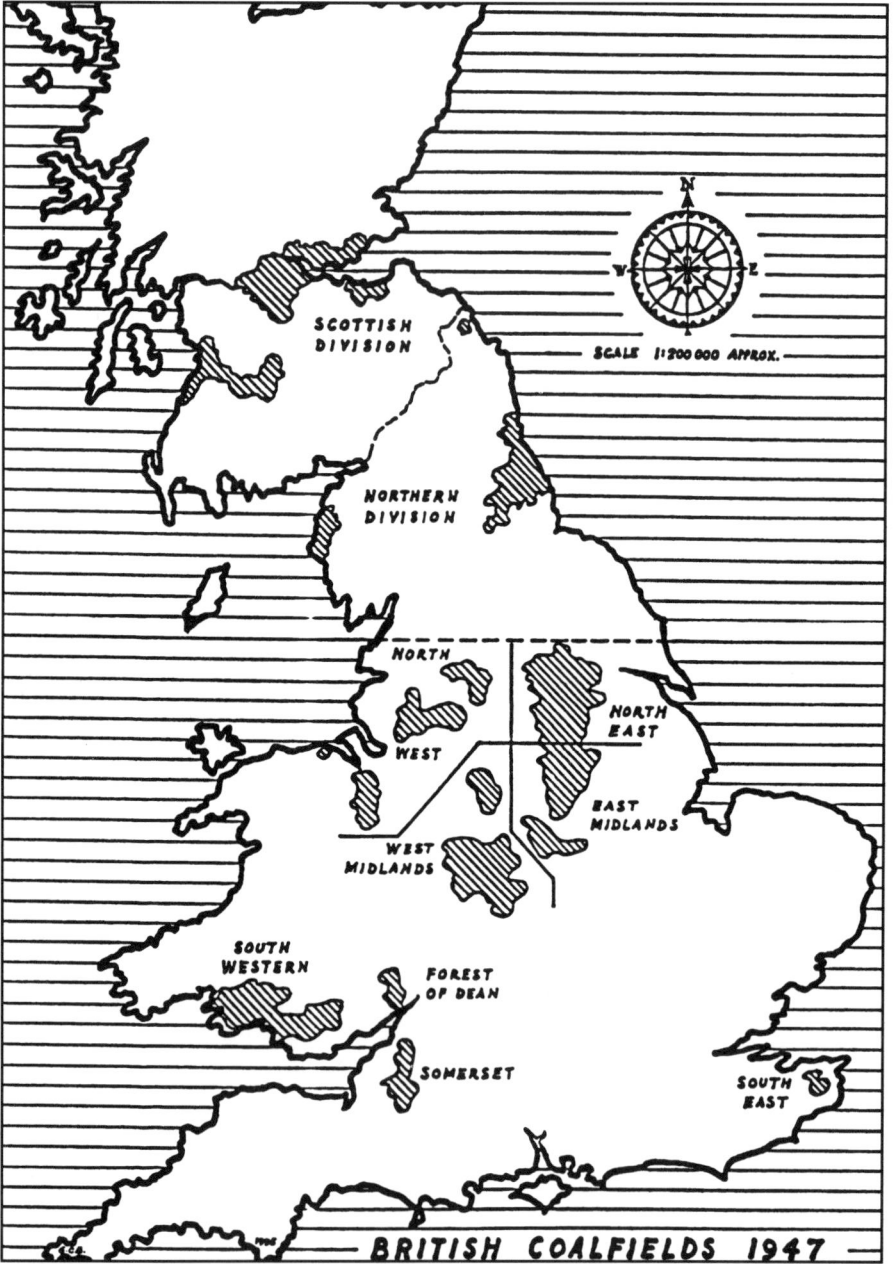

BRITISH COALFIELDS 1947

7

Area Production Manager

The Fleck Report did have an immediate and direct effect upon my career. One of its Board's recommendations was that the Area Production Manager of each Area should be provided with two additional assistants: a Deputy Area Production Manager (Planning) and a Deputy Area Production Manager (Operations). This was in a situation where other responsibilities they had formerly carried out for a considerable period during private ownership and early nationalisation had been, or were to be, withdrawn from them, for example control of industrial relations, purchasing and stores and transport.

Deputy Area Production Manager (Planning)

East Midlands Division

The outcome was that Allan Hill, the Area General Manager, called me into his office and explained the whole Fleck situation, following which he asked me if I would undertake the Area Production Manager (Planning) job alongside Walter Sharpe (No.1 Group Manager) who had been approached and had accepted the Area Production Manager (Operations) position.

The chance of a new experience, including the construction of Cotgrave Colliery within a green field site, I found attractive. I agreed and we were subsequently appointed. Harry Saviour, a North-Country man, was the Area Planning Engineer. He had received extensive training under the Bestwood Collieries Group (Col. Lancaster). He was an excellent planning engineer, one of the best I had met at the time or would meet in the future. The new post was restricted to mining engineers with first class Certificates of Competency, and so he was barred from making an application.

The Gedling project was now completed and operating quite successfully. During its execution Harry and I had spent a lot of time together and had developed an excellent relationship. To match Harry's skills in his post would have taken me several years. On discussing my new post with him I told him so. His reply was,

'Look, Charlie, you can bring something to the job I can't provide. The way you have handled the Gedling Project proves that [the scheme was operating better than had been projected] you can get things done.'

That is how we started together and during the short time we worked together I found him extremely supportive and very loyal.

PRINCESS MARGARET CUTS THE FIRST SOD - COTGRAVE NEW MINE

On 7 April 1954 Princess Margaret performed the ceremony of 'cutting the first sod' for the shaft-sinking at the new Cotgrave Colliery, following which she visited Calverton Colliery and spent some time underground. Harry and I were in attendance amongst a host of Divisional, Area and other dignitaries who had been invited.

Cotgrave, to me, was a real bonus in terms of experience. The chance to obtain the sort of knowledge it provided occurs but infrequently. I lapped up all I could. The Divisional Sinking Engineer, Arthur Wadsworth, although young, had acquired a substantial amount of basic experience in this field. He had a tendency to try and impress people as to how good he was, which was totally unnecessary for he really was a very capable and highly confident person. He had charge of both the Bevercoates and the Cotgrave shaft-sinking projects, discharging his responsibilities in a very able manner.

BOARD CHAIRMEN VISIT COTGRAVE DEVELOPMENT

At a later date, when the Cotgrave constructions and shaft sinking processes were in full embodiment, we had a visit from Sir James Bowman who a few days earlier had been appointed as the Board's Chairman. In addition, we had Sir Hubert Houldsworth still in post. We were given to understand that Sir Hubert had learned of the intended change from a public announcement made by the Minister of Fuel and Power. Sir James Bowman attracted the widest entourage during the visit from those present. I walked round alone with Sir Hubert who hardly spoke and appeared heavily stressed throughout the visit. To me it was a very sad occasion for, during the few times I had met him, I had been greatly encouraged by his interest and obvious devotion to the Industry. Maybe the higher one climbs in an organisation the more intense are the disappointments when matters of personal import adversely change.

Harry and I discussed at length the surface design and underground layout and shared concern about the best possible means of marketing the coal. For this, we obtained Board permission to engage Simon-Carves of Cheadle, Cheshire as design consultants both for the colliery and for the coal preparation facilities. As things turned out, for the former we supplied the ideas and they did the research, legwork and drawings. Concerning the latter, they were experts in this field, so we went along with them.

If I recall correctly, we sank two 20 ft. diameter shafts through about 200 yards of water-bearing strata, adopting a strategy of freezing a cylindrical block of strata some 35 ft. diameter solid which was down to a depth of 250 yards, and thereafter sinking the shafts through it, with a large freezing plant operating continuously during the whole process. The shafts were lined with a special chemical resistant concrete. Mechanical grabs were introduced for loading out the loose material produced by patterned explosive charges, this being loaded into skips and discharged at the surface. Shaft water we dealt with by the use of sinking pumps secured to and mounted onto the shaft lining.

Subsequently Arthur Wadsworth had a heart attack which he overcome but, being an intense tennis player and golfer, he pushed himself very hard and his second attack proved fatal.

Over the first twelve months Harry and I really got down to developing both major and minor reorganisations and reconstructions for Linby, Babbington and Hucknall Collieries.

Watching operational things developing in the Area daily began to have its effect. The extension of coalface mechanisation was moving quite well along the lines developed at Gedling Colliery, and other interesting things were happening, I was beginning to feel I was out of the stream of things and somewhat isolated. Allan Hill had by now retired, N.R. Smith had been appointed as Area General Manager. Walter Sharpe had been upgraded to Area Production Manager with Len Hogg, former No.1 Group Manager, assigned Deputy Production Manager (Operations). I was isolated. However, after about fifteen months or so in the job, out of the blue I received a phone call.

'Youth, what are you wasting your time for in that dead end job you are doing there?' said a voice immediately recognisable as Tommy Wright's.

'Why do you say this, Tom?' was my answer.

'Because I want you to come and join me as my Deputy (Operations). Put an application in for the job. It's coming up within the next two weeks.'

'OK, Tommy, I'll do as you say, thanks,' I said with a degree of pleasure I hadn't experienced for quite some time.

'Don't let me down, Youth, the job's yours and I want you,' he said before ringing off.

I applied for the job and met Tommy's AGM Don Severn on the interview panel (which included W.H. Sansom, the Divisional Production Director) for the first time. He was extremely pleasant and encouraging. No doubt he knew more about me than I was aware of. I was given the job. When the news broke, N.R. Smith was the first to know officially, even before myself. He sent for me.

'Why did you apply for the No.4 Area job?' he asked.

I certainly wasn't going to unburden myself before him in the circumstances, so I quite cheerfully replied,

'Because it's a step up.'

'What makes you think that? It is not so!'

'Look, Mr Smith, I believe it is so and you'll find I'm right. There will never be an Area Production Manager promoted from my current job. Such promotions will always be in a direct line from the Deputy Area Production Manager (Production) post,' I countered with a degree of conviction which even surprised me. Events proved me right. There never was.

'Will you give it further thought before coming to a decision?' he requested.

Having been overlooked by him for the same post within his Area in favour of Walter Sharpe's choice, with whom I had no quarrel, my future was in my own hands to shape the best way I could. Where loyalty to a person is not reciprocal one is free to act in accordance with one's best interest. One need have no pangs of conscience.

CALLED FOR INTERVIEW WITH W.H. SANSOM

It costs very little to be obliging in such situations, particularly when one's mind is firmly fixed. Two days later N.R. Smith came into my office and said that Mr Sansom would like to see me at Sherwood Lodge that afternoon at 3.00 p.m. When I entered Mr Sansom's office, his immediate greeting was,

'What's your difficulty, Mr Round?'

He already had approved my appointment to join Tommy Wright in No.4 Huthwaite Area. Cheerfully I replied, 'I haven't any difficulties; I'm extremely happy, Sir,' (which was quite true at that point).

'But your Area General Manager tells me you have some idea you are blocked in a dead-end job. What makes you think that?'

'Because if one carefully analyses the Operations job, it is more in keeping with the Area Production Manager's job than the specialised job of planning. In fact, Mr Sansom, one finds this obvious within a very few months of doing the job. One can easily get out of touch,' was my explanation.

'I've never considered the nature of the post in that manner, Charles, I am sure there is something in what you feel,' was his kindly response.

'You know, Mr Sansom, I was disappointed at not being considered for the Operations position when Walter Sharpe was appointed Area Production Manager in my Area. I was passed over by my Area General Manager, for I feel at least he could have tested my feelings at the time. I am sure I have served him well over the years.'

His final remark was, 'Charles, you have served the Divisional Board very well over the years since you joined us. Go to Tommy Wright, you have my blessing. Best of luck.'

With such acts of kindness and encouragement one becomes ideologically and loyally enslaved.

Thinking it over in retrospect I'd like to believe that our discussions had set something in train, for shortly after, the restriction of the First Class Certificate of Competency for the post of Area Production Manager (Planning) was removed and in due course Harry Saviour applied for the same post in the West Midlands. He got the job and I know that for a number of years he did give them excellent service. Unfortunately Harry had a problem: he liked a drink but when he went too far he lost control of himself. On every such occasion the next morning he would immediately go round to the persons he was with and ask about his behaviour and offer his apologies for any offence he might have caused. He had no recollection of how he had reacted under the influence. On one occasion a brush with the police involved the suspension of his driving licence for over twelve months. The morning before he committed suicide Harry went into his office, brought all current matters up to date, cleared his desk and left, never to return. Thus a very competent, able and nice person, in normal circumstances, was lost to the Coal Industry.

Deputy Area Production Manager (Operations)

NO.4 (HUTHWAITE) AREA - EAST MIDLANDS DIVISION

Thus I came to join Tommy Wright Area Production Manager No.4 (Huthwaite) Area as his Deputy (Operations) although only for a comparatively short time. I was there for only a few months but they were extremely happy ones, despite an experience the Coal Industry and those associated, both men and management, would have preferred not to have had.

My friend and colleague Ron Swindall, who followed events as they unfolded, was quite incensed.

'So that is how these people are going to reward ability, loyalty and unstinted support?' he said.

Despite my move Ron and I maintained contact wherever I was. Ron was the only person I ever told of the nature of my interview with W.H.Sansom.

I started my new job during the Suez Crisis, 1956. Petrol was rationed and I was still residing at Stoke Lodge, Gedling, Nottingham, which meant a round trip, Gedling to Huthwaite, of some sixty miles daily. Inadequate supplies of petrol dictated the need to take accommodation in the Saracen's Head, Sutton-in-Ashfield, for five days each week, returning home to Mary and the children at weekends. The support she gave me gave me during this situation was tremendous: never a complaint or remonstrance. It was good to be able to come home at weekends. John was progressing well at the Beckett School, West Bridgeford, Nottingham, and Dorothy had settled in at a local Council school in Netherfield. The presence of Harry, my gardener, made it possible for Mary to keep up with her whist drive and other social activities.

On starting work in the No.4 (Huthwaite) Area, my immediate concern was to get a good background knowledge of the operating collieries which meant spending a great deal of time making daily visits underground, sometimes with Tommy but usually with the Group Manager or the colliery manager. Eighteen collieries had been allocated to the Area, of which twelve were in Derbyshire and the remaining six in Nottinghamshire. The collieries had been allocated within four groups, each under the control of a Group Manager.

During the first month or so Tommy, with a couple of friends, would call in at the Saracen's Head from time to time and ask the hotel landlady or husband, 'Where's Charles?'

Invariably the reply was that I was my room working.

He would then get on the phone and say, 'Come down here, Youth, have a break; you've done enough for today,' following which we would call at another pub, sometimes until ten o'clock or after. When we went out together we would usually finish up at his home and Lily his wife would prepare us supper. Lily was a lady not unlike Mary: very strongly family-oriented, hospitable, no pretences, and one could always feel comfortable in her presence. Knowing I was living away from home she was usually concerned as to whether I was able to get something to eat, for many's the time I wasn't able to conform to the dining hours. However, the hotel manager and his wife were quite good in that respect and always made provision. They were a highly volatile couple and often flared up at each other. When they really got to know me I became a recipient for difficulties from both sides and had to try, judiciously, to smooth out their problems. However I got on well with them and they both were very good and kind about my welfare.

After about fourteen days in the Area, we were struck by an occurrence dreaded by mining engineers and men alike. Tommy and I were together underground making an inspection of Wingfield Manor colliery, a small Derbyshire Colliery producing about 1,500 tons per day. Having descended the mine at about 9.30 a.m. we were over 2½ miles inbye in the midst of the workings when a message came through at about 11.30 a.m. 'Would Mr Wright come to the nearest phone? he's wanted urgently.' The nearest phone situated on the main haulage road was upwards of half a mile away. We dashed off apprehensively, instinctively aware that something serious had occurred within the Area. Rarely is one called in such a manner. It had happened to me only once previously when, underground at Elsecar Main Colliery in 1946, I had received information that my brother John had been found dead underground at Wharncliffe Silkstone Colliery.

SUTTON COLLIERY EXPLOSION

Within minutes Tommy got to the phone. He picked up the receiver in the semi-darkness and amidst all the facial dirt he literally turned white and uttered two words: 'My God!' He put the phone down.

'Come on, Charlie, we've got a big disaster at Sutton Colliery.'

He couldn't tell me any more. We covered the distance from those underground workings at Wingfield Manor to the Manager's office at Sutton Colliery in our pit clothes at tremendous speed. Neither of us spoke, fearful at what we would find and would have to face.

Arriving at the Manager's office, Mr Donald Severn, Area General Manager, was there co-ordinating action. The Acting Manager Andrew Stone (Undermanager) and Jack Frith, (Group Manager) were instigating the emergency procedures. Amidst all the phoning and activities we learned that there had been a firedamp explosion and over twenty-five men were involved, some of whom had been killed. The circumstances were not clear. Instructions to withdraw all men had been given and most of them had already reached the surface.

The mine rescue team were moving underground on their way to the scene of the accident. Mansfield Rescue Station had been informed and stand-by rescue teams had been organised and were moving in. Tommy took over and coordinated all action and activities. He informed Mr W.H. Sansom, the Divisional Production Director, who was on the scene in a comparatively short time, bringing with him a stabilising ambience, which settled us all.

With the head lampman I initiated a comprehensive analysis of all men who were known to have been underground from the lamproom records, against which we checked those who had returned to the surface. We traced everyone, including those caught up in the explosion, and could account for every man. The names of all men whom we deduced had been working within the district in which the explosion had originated were highlighted for interview. The evidence of these uninjured men would be crucial and needed to be obtained at the earliest moment. Members of the Mines Inspectorate in the meantime had made their appearance. Until we had confirmation from the rescue team inbye as to the security of the situation underground no one, other than those already underground and the official rescue team, could enter the mine to recover those killed or to assist the numbers of men who were badly burned.

By about 4.00 or 5.00 p.m., we had clearance that the situation underground was secure and all men including the victims of the explosion were now on the surface. The inspection teams could now travel to the scene of the explosion's origin. I asked Tommy if he wanted me to go with the first team. We couldn't go together - no two top men in line could do so for obvious reasons.,

'No, Charlie, I'll go. You take the next inspection team in.'

At about 6.00 p.m. Tommy led the first inspection team underground, including the Mines Inspectorate and others who had direct concern. They were underground for about four or five hours. During the time they were underground, Mines Inspector M.V. Thomas and I were working on the list of names of the men whom we considered could well be in a position to provide evidence relating to the explosion. During this exercise I found the Inspector very helpful and cooperative. I have great respect for him. We met on many occasions afterwards.

During a lull I was able to phone Mary. Highly stressed and emotional, her first words were,

'Are you all right, love?'

She'd heard the news and had deduced I was likely to be embroiled. Harry told me that when she heard the news of the explosion and recognised Tommy Wright in the news bulletins her face had paled.

Continuing, she asked, 'Are you at the pit and do you have to go underground there?'

'Yes, dear, I am - nobody is allowed underground until it is declared positively safe.' I answered.

I may have to some extent reassured her, but she'd been in and around pits for a very long time and was fully conversant with the associated dangers. I certainly wasn't going to cause further worry by informing her, immediately I'd left the phone, that I would be leading the next inspection team underground. She wouldn't have slept for the next three nights (if she in fact did), for it was some four days into the future before I was able to able to see her again.

Together with Mines Inspector M.V. Thomas and others we went underground at about 11.00 p.m. Leading the inspection team from the front, I had expected to obtain evidence of the explosion's occurrence long before we did. At about 1,000 yards we first encountered an oily smell but no physical damage to the supports or to the equipment installed along the roadway we were travelling in. As we moved inbye we began to see discolouring and a blacking of the supports and the oily smell got stronger. It was probably some two hundred yards from the point of the explosion's origin that real evidence was seen in the form of loose material. Boards, pieces of timber, and small road roller pulleys were disposed as though they had been picked by the explosion's blast and carried forward. As the blast became spent, it deposited these along the roadway, the lighter material being carried the greatest distance. I was again surprised to note that the conveyor system had not been disturbed to any great extent, though some conveyor structure covers had been displaced where material carried by the blast had struck them.

When we got to the coalface we discovered that the greatest damage had occurred to a new Anderton shearer mechanised unit in an advanced stage of being tooled up for operation. It was at the roadhead or entrance to the coalface. Over the panzer conveyor drive unit was a huge cavity within the roof strata. We judged it to be more than 20ft. in height, some 15ft. wide and about the same in depth, an approximate volume of well over 100 cubic metres. A heavy concentration of firedamp was present in the void, which called for a small ventilator fan later to be used. On the coalface there was no damage or disturbance. However, we did find methane of less than 1% in the general air current. The return gate indicated no signs of disturbance or after-effects from the explosion.

The electrical terminal box on top of the drive motor was broken but there was no other apparent damage to the equipment. The sides of the roadway adjacent to the void had been greatly blackened and the area disturbed by random deposits of material including timber.

The conclusion which we arrived at was that firedamp (probably within this cavity) had been ignited electrically by electrical sparks from the damaged terminal box. However, these assumptions needed to be clarified by experiment. There was no evidence to indicate that coaldust had played any part in the explosion. Our opinions closely confirmed those of the first inspection team. The situation was referred to the Safety In Mines Research Board who later determined that firedamp within the void could have been ignited by a stone falling from the cavity thus damaging the electrical terminal box connections. They proved that a large stone, on falling, brought with it a fuse-like cord-film of firedamp and when the stone damaged the electrical connections the sparks produced ignited this, forming a fuse which fired the whole body of gas within the gas-filled cavity.

We arrived out of the pit about four o'clock the following morning with a tremendous number of relevant matters awaiting urgent attention before us. M.V. Thomas and I had the long task of collecting evidence from the men and other sources wherever we could find it, which meant a close examination of all past safety reports relative to ventilation and electrical equipment.

Both Don Severn and Tommy Wright instructed me to give the matter my exclusive attention and not to consider or get involved in any other matter. In a situation of this kind this has great merit, for no person can readily forecast the nature of the 'fallout' which may arise. As events unfolded, within a few months I had been able to carry out the assignment. I was glad, as it proved most valuable later. M.V. Thomas and I drew up a list of some seventy or eighty men we wanted to interview and accordingly set about the task. It took a very long time. During the intervals, whilst waiting for the men to be brought to the surface or fetched from their homes, we examined numerous Deputy's statutory reports, ventilation and dust records and electrical test details relevant to the whole situation. The quantity of ventilation in circulation through the coalface was somewhat low as compared with an established mechanised power loaded unit, but work was proceeding with regard to this, with the final stage of development being in the process of completion. From our interviews with the men at the scene of the incident, we established that those who had been on the intake side (furthermost from the explosion's origin), had felt a blast of hot air flowing back against the normal direction of air flow. Others nearer to the epicentre of the explosion observed a flash and heard a low rumbling noise accompanied by a rush of hot air. For those adjacent to the scene of the detonation, the flash had a longer duration, the noise was more intense and heat was very fierce. The degree of burns the men suffered was greatly influenced by the working garments they were wearing. The victims dressed only in shorts were burned in excess of 70% of their body area, but those with a light covering, vest or shirt, received a remarkable degree of protection. The five victims, of whom two were young boys of about fifteen or sixteen years of age, died from burns in excess of 65%.

During the period in which we were taking statements from the men I managed to slip away for a couple of hours and visited the living victims in the Burns Unit of the Mansfield hospital to which they had been taken. That visit was indelibly impressed on my mind and from time and time since this terrible incident I have relived the occasion in hallucinations during sleepless periods. Both the men and the boys were in intense pain and suffering great affliction. The nurses and doctors were constantly administering succour as much as they could and they too showed obvious signs of distress. Three of the badly burned men and the two boys died.

The visit that day brought to me most vividly the nature of the responsibilities we were carrying, not only by statute but more importantly personally, to the men we employed and supervised.

During the afternoon of the third day, on the completion of our task, I returned home. Mary was doing her domestic chores when I walked in. I will never forget the look of relief when she turned and saw me enter the lounge. Normally she was never demonstratively emotional, but on this occasion her reactions were unmistakable.

Criminal Proceedings Initiated by HM Inspectorate

The Mines Inspectorate initiated criminal proceedings against Tommy Wright, Area Production Manager, Jack Frith, Group Manager and Andrew Stone, Colliery Undermanager, who was Acting Manager. Teddy Maiden, the former Manager (having been transferred to a similar post at Pleasley Colliery a short time previously), and myself, who had but some two weeks recently moved into the Area, were exempted from these criminal proceedings. This was a great help to my colleagues for, being free from the stress and worry of such direct involvement, I was able to help prepare their defence without emotion or diversion.

The case was heard in the Criminal Court at Mansfield, and ran for several days. We had very helpful support from both Divisional and Headquarters Levels. The jury's findings were in favour of the prosecution; Tommy Wright was fined £50, Jack Frith was also fined £50, and Andrew Stone £25. They were thus branded as criminals. The defence lawyers lodged an appeal which was granted and a further hearing was included within the Calendar for the next Quarter Sessions in Nottingham.

In the meantime Donald Severn instructed me to continue with what I was doing, no matter what the circumstances. From the surveyor's scale drawings of the roadhead area embracing the explosion's point of origin, I was able to get the joiners at Hucknall Colliery to construct a scale model of the relevant roadhead/cavity portion of the roadway. These craftsmen did a great job with this reproduction, subsequently accepted as a court exhibit under the identification 'Mr Round's Model'.

At the Quarter Session hearing in Nottingham, we had excellent support from Dr Willet, the Board's Authority for Safety, and W.H. Sansom, the Divisional Production Director, both of whom were called as defence witnesses in addition to Tommy Wright, Jack Frith, Andrew Stone and myself. Tommy Wright occupied the witness box for some four hours. He, in turn, was followed by Jack Frith who was questioned for some 2½ hours, Andrew Stone for about two hours. My role in the witness box was to confirm the model, explain its close association with the real thing, answer questions thereon and deal with issues arising from the work M.V. Thomas and I had undertaken immediately after the explosion.

Dr Willet really rose to the occasion, but it was W.H. Sansom who was the star witness. He represented the classic way in which one should behave in such situations. To all questions he offered short and to the point lay answers, his technical observations were couched in simple terms, he ventured no information and never wandered, keeping his cool with remarkable ease. The Prosecuting Council was most uneasy. Had he been able to shake this fine person the ultimate outcome could well have been different.

The jury retired and were out for quite some time. Tommy, Jack, Andrew and I were all standing together in a group, wound up tight, highly stressed and apprehensive. W.H. Sansom came over to us and said,

'You four remind me of the miner who fell down a thousand yard mine shaft. After falling fifty yards he said, "By, that was clumsy." Falling a further two hundred yards he uttered, "I'm sure to break a leg." At a depth of five hundred yards he muttered, "It's going to be far more serious. I shall be lucky not to break my back." When he got to within ten yards of the bottom, he said, "I shall soon know" . . .'

We laughed in great relief with stress and tension assuaged. That was the position we were in. Within five minutes we were admitted back into the court-room to hear the jury's verdict to the effect that all defendants were cleared with the legal costs carried by the Mines Inspectorate who had initiated the prosecution.

The kind personal letters received from the defendants, Don Severn and W.H. Sansom, thanking me for my contribution are something apart - never discussed, certainly not expected - but when they came, I felt really privileged, proud and good to have served my colleagues in their difficult times.

Having been closely involved with this situation, I gave a great deal of time and thought to the problem of quick detection of firedamp in large cavities or high places. Taking the lamp into the cavity needed scaffolding of some sort. Moreover there were risks involved in doing this. Thinking in reverse I encountered the concept that if one cannot take the safety lamp detector up into the firedamp accumulation, would it be possible to induce the air/gas contents of the cavity by means of a tubular telescopic probe, into the lamp at a much lower level for test purposes? After making a series of sketches I passed the problem over to the Area Scientist, Leslie Wood, and one of his assistants, Peter Fowler, to work on. It proved to be quite an involved enterprise. It was shown that, by sealing all the safety lamp's air inlets, this could by achieved by simple induction. However, the results were not reliable and it was easy to lose the lamp's test flame. Peter Fowler came up with the idea of using an aspirator and this worked admirably.

PHOTOGRAPH 39 shows the final situation we arrived at which proved to be workable. Normally firedamp is detected within the general body of the air by reducing the flame to a small yellow bead; the gas burning produces a small blue cap on top of this. The size of the cap varies according to the amount of gas burning within the lamp. An equilateral small blue firedamp cap would signify about 2.5% of firedamp present in the atmosphere. At about 4.5% the firedamp explodes within the lamp. The lamp's internal gauzes prevent interaction with the external atmosphere. Here we induced an air sample with an aspirator from the upper open end of the telescopic tube and fed it into the lamp for measurement. We had to undertake a substantial modification to the lamp's inlet air system so that the lamp could operate normally and instantly be converted for use with the probe. The magnifying glass was fitted to make the detection of low percentages of firedamp, in normal situations, easier. Patents were applied for on 25 July 1958. We received an award of £100 each from the National Awards Panel. The subsequent development of a small, more convenient, electronic gas detector reduced dependence upon the safety lamp so that it never took off. John my son was aged sixteen years at the time of the explosion. During the work I was doing and the Court Hearings, John said to his Mum,

'Tell Dad I don't want to go into coalmining. It will last Dad's time out but not mine.'

I'd mapped out a career for him and really wanted him to follow me. He had really raised the matter at the right time in view of what we were involved in. I rang Norman. His response was strong and to the point.

'Let the lad find out what he wants to do himself. It's his own life he wants to live, not a continuation of yours.'

At the weekend I talked to him. 'John, Mum says you don't want to go into mining. What do you want to do?.'

'Did you know what you wanted to at the age of sixteen Dad?' he responded.

'I'd no choice, son, there was nothing but the pit and at sixteen I had already been working two years. Money was short. Moreover I had no qualifications to work with.'

'I like chemistry; that might be worthwhile,' he ventured.

'Right, shall we agree on that? You can have all the help I can give you,' I said.

Thus he had my blessing and chose his own future which opened up for him a degree of success I could not in my wildest dreams have imagined at the time. His prophesy about the life of the Industry proved correct, although on what grounds he expressed it he didn't say or was unable to reveal.

On being appointed Area General Manager of the No.4 Huthwaite Area Donald Severn had initiated a schedule of annual colliery reconstructions designed to improve the winding capacity of the colliery shafts, increase all forms of transport capacity underground and reduce the attendant non-productive manpower requirements. A typical scheme would be around £100,000, a sizeable sum in those days. It would embrace the mechanisation of the coal-tub handling arrangements, both in the pit bottom and on the surface. A number had already been done and New Hucknall and Teversal Collieries were in the process of reconstruction at a cost of £80,000 and £100,000 respectively. It involved the introduction of high capacity mine cars as replacements to the old small tub rolling stock.

Silverhill Colliery was in the process of preparing for a new electric winder embodying the use of a twelve-phase alternating current system designed to minimise operating electrical coal-winding costs.

I found it interesting to observe Tommy in action. He knew the Area in absolute detail and had developed a communications network which fed him continuously with information of all kinds: things that were going wrong; wagon build-up and demurrage costs; men leaving early; neglect of maintenance, etc., intelligence he often supplemented during his visits to the local.

His planning and progress meetings could be a mixture of intense stress punctuated by flashes of humour; or quiet and purposeful during which time he manipulated his staff members in the direction he wanted them to go. Such was not a planned Staff College induced procedural approach; it was his natural way of doing things. There was always a degree of uncertainty which kept people on their toes. At one point in the meeting he could be pressurising a Group Manager with regard to lack of progress in a stone heading and instructing him to install a loading machine, when the phone would ring. He'd listen and then turn to the Area Mechanical Engineer and for everyone to hear would say,

'There's a young lady downstairs with a new baby in her pram asking for Charlie Buchan [Area Mechanical Engineer]' - or any of us for that matter.

By such means he'd naturally control the stress level of the meeting in progress.

In view of the Staff College's doctrines about delegation I tried to find out how my contemporaries and superiors handled their situations. The variations were quite considerable, but most of them acted in a natural way according to their personalities and former experience, certainly not dogmatised by the infiltration of Staff College academic analyses. Tommy's approach was much similar to my own except that I did absorb detail to a greater extent. He too set time limits on all his delegated matters and followed them up.

I recall that on one occasion we had a major problem of roof control in a newly installed mechanised coal face at Bentinck Colliery. The local Management needed support and assurance, so that I spent four consecutive days with them trying to find a solution and to get things operational. I was usually down the pit and inbye by the time Tommy got to work at about 8.45 a.m. On the fifth morning however, as I was preparing to leave the colliery manager's bathroom (changed to go underground with Tom Jameson the Colliery General Manager), a message came through,

'Tell Mr Round not to go underground; Mr Wright wants to see him at Area.'

Changing back into my civilian clothes, I went back to headquarters to see him. As I walked into his office he said,

'Sit down, Youth, I'll be with you in a minute.'

He continued with his morning mail and from time to time various members of the staff would call in to see him. After I'd been there about an hour and a half I was impatient, feeling my time was being wasted.

'Look, Tom, what do you want me for? I've been here for nearly two hours and there are things to do.'

He turned towards me and with an infectious disarming smile said,

'Charlie, I know what you have to do and what you've been doing this last four days. You've made damn good progress with that Bentinck job, but this morning you are having a break and joining me for lunch at the Swan, in Mansfield.'

At lunch he told me, in the confidential manner he told many other people, that Norman Siddall, then Area General Manager of the No.5 Eastwood Area, was moving to become Area General Manager of the No.1 Bolsover Area which had declined in performance under the stewardship of its former Area General Manager who had been transferred to the National Headquarters Staff as Director General of Production. I felt delighted that Norman Siddall had been appointed to the post, knowing it could be but a matter of time for him to sort out the situation.

Within a couple of days Tommy started making contact with the Mines Inspectorate, NUM and NACOD (National Association of Colliery Overmen and Deputies) Officials.

I asked him, 'Why all this activity, Tom?'

Quite openly he informed me, 'These people can be very influential and helpful in many situations,' neglecting to mention the specific situation he had in mind.

Within a month of the lunch we had together, Tommy was, I felt, deservedly made Area General Manager of the No.5 Eastwood Area, and I, next in line to him, was called upon to undertake, temporarily, the duties of Area Production

Manager for some four months, during which I really began to get to know and appreciate the fine sterling character of Don Severn.

As a soldier in the First World War he was gassed by the Germans at Peinamunde. This seriously damaged his lungs and left him with a permanent asthmatic condition which deteriorated with age. Some days he was in excellent form whilst on others he stoically and quietly suffered. It soon became apparent that here was another fine person to work to, one in whom I could invest my total loyalty and give unstinted maximised effort. As a mining engineer he was highly competent and capable and his business acumen was superb, being immediately able to pinpoint in both balance sheets and statistics the vital elements that really mattered, often to our personal embarrassment during Area Management meetings.

During these three or four months I was temporarily acting as his Area Production Manager without realising that he was shaping and carefully guiding me where necessary and allowing me to act for him from time to time at Area consultative and safety meetings as well as in other situations, on all occasions quietly advising and encouraging me. He was in no hurry to get a permanent incumbent in post. I was quite happy that all that he and Tommy had set in motion was progressing most satisfactorily.

When the post was advertised all he said was, 'Charles, you'll be submitting an application for the position,' to which I affirmed that I would.

The interview panel consisted of W.H. Sansom, Donald Severn, the Divisional Staff Director and a Headquarters Representative, E.J.Kimmins. Prior to nationalisation both he and G.C.Payne had been Colliery Managers at Manvers Main Colliery and Barnborough Colliery and employed by the Manvers Collieries Company Ltd. Amongst the candidates to be interviewed, other than myself, was one Frank Darley who had served under E.J. Kimmins as Colliery Manager during his period as Sub Area Manager in the Yorkshire Division's No.3 (Rotherham) Area. Frank Darley was then working in one of the Lancashire Areas. E.J. Kimmins was pressing very strongly for his former colleague's appointment to the post.

'How serious are you, Mr Kimmins? Round's record since he joined us in this Divison has really been something. Over the past three months or so he has handled the Area Production Manager's job admirably and with great merit. Morever, how do you expect us to hand over the job to one who is accepting an OMS vastly inferior to the 45 cwts per manshift this lad is obtaining?' was Mr Sansom's observation.

At that time real power for local matters was vested in the Divisional Boards. Mr Severn informed me of this only after he had received confirmation of my successful appointment to the post.

'What you have told me, Mr Severn, is no surprise. There are many people about who consider the Wrights, Rounds and others of the industry should never emerge from the coalface beyond the lower management levels'.

With regard to the Headquarters representative's contributions to the Industry, I have little knowledge other than that, as a young student, I recall he was involved in a large capital project involving the merger of Manvers, Barnborough,

Kilnhurst and Wath Collieries. Any other major contributions or personal achievements he made or undertook within the Coal Industry, before or subsequent to the state ownership of the collieries, have escaped my attention, despite the fact that I have always tried to keep abreast of all mining developments.

Area Production Manager, No.4 (Huthwaite Area)

When I took up my new appointment Mr Severn, in a relaxed mood, was discussing his policies with respect to various aspects of Area operations.

'Charlie, if Little Jimmy had been ready, I would not have brought you into this Area. I've always worked to a policy of training my own people for all staff posts.'

My response was very simple.

'Tell me, Mr Severn, are you sorry little Jimmy was not ready?'

'Quite the contrary; I am more than delighted the way things have turned out. You brought us something we needed at the time we needed it and currently I'm happy and relieved to have someone who I know and can trust, one who I am sure will do a good job for me.'

At the time I first met him, little Jimmy certainly wasn't ready, despite having been nurtured and encouraged by Mr Severn throughout the various stages of his career. He subsequently followed me into the post I'd just been given. Maybe Tommy Wright and myself in our different ways contributed a little to his further development.

In her usual manner Mary took my new appointment in her stride. Regardless of whatever progress I made, she always kept her composure and approachability. Together with the children we moved to 'Greengables, Wingfield Road, Alfreton, Derbyshire'. Don and Mrs Severn with Allan, their only son, had lived there previously. It was a beautiful house set in a relaxing location, almost adjacent to the Alfreton nine-hole golf course. Don told me the house was built to the prize design of the Ideal Homes Exhibition of 1926. There was a substantial area of land, both back and front, which Harry our gardener made excellent use of. John travelled by bus to the Beckett School at West Bridgeford, taking just over two hours for the round trip. He developed the ability to do his preparation and homework on the bus, which allowed him time to develop in the sports of tennis, golf and swimming. Dorothy transferred to the Girls High School, Alfreton, where she settled in very well.

PHOTOGRAPH 40 shows John and Dorothy at about this time. They have always been grand youngsters. There was, however, quite a humorous incident concerning them when they were five and three years of age respectively. We were then living in Cherrytree Street, Elsecar. One night they had been particularly fractious with each other and Mary had put them to bed at about six o'clock instead of their usual time of 7.30 or 8 00 p.m. When I walked in from work later, there was such a disturbance in their bedroom that I went up to investigate. Both were crying and between his tears and sobs John complained, 'She waited whilst I got to sleep and then she hit me.'

However, they have always got on well together. Mary did an excellent job on their development and upbringing.

Mary was exceptionally happy; she liked the new home and in a very short time found a venue for her whist drive activities in Sutton-in-Ashfield which she attended along with Mrs Peter Harley, the mechanisation engineer's wife.

The uncertainties associated with an acting post having been cleared by my appointment allowed me to approach the tasks ahead with my old levels of enthusiasm intact and supplemented by a desire to ease, as far as I could, the physical difficulties of my Area General Manager. Some days his health was so bad that he would struggle along the corridor as far as my office, come in and take a chair, pull out his aspirator and for five or even ten minutes try to overcome his breathing difficulties. I knew not to burden him with anything at such times. Lest the reader think such problems impaired his ability to carry out the high responsibilities of his post let me state unequivocally that he carried out his duties meritoriously and taught me a great deal in the process of doing so. When I needed his advice he was always generous and forthcoming, his humour ready at all times to diffuse awkward situations. He was a highly competent person and more than matched any of the Area General Managers I had worked under and was later to work with.

Coalface mechanisation was being extended at all pits and an experimental trial with Gullick power supports was showing good prospects with fair progress being made. However, we were experiencing appreciable damage with disruption of production under the conditions in which they had been applied. In the little village of Pinxton was a small engineering company, Herbert Cotterill Engineering Ltd, which undertook plating and constructional steelwork. The company did a fair amount of work for the collieries within the Area from time to time. Often a number of us would meet together at the engineering shop for a coffee on a Saturday morning and discuss with Herbert Cotterill the problems we were having with the powered supports. In use they were distorting badly, giving us great problems of maintenance with frequent output losses.

We redesigned the support bases and canopies, stiffening up where any weaknesses were found. Herbert's little company did the work for us speedily. Moreover, there was negligible transport involved between the engineering shop and the pits as opposed to sending the supports back to Gullick's (Wigan), some three hundred or more return miles away for repair. Over a period of a few weeks we had built into the redesigned and strengthened units an improved working life. As an aid to efficient production and greater safety, they were proving their worth. The Area General Manager knew what we were up to and with his blessing we undertook a programme of powered supports installation on each of the mechanised faces that were equipped with hydraulic or friction-type individual supports. Getting them on capital charge raised difficulties, but I asked Mr Severn to allow me to purchase individual hydraulic legs, valves etc. in bulk, as a controlled revenue charge, to which he agreed. Over several months we built four or more complete installations, the Herbert Cotterill Engineering shop constructing the bases and canopies in one section and doing the fitting and testing in another.

PHOTOGRAPH 41 POWERED SUPPORT MODIFICATION is a surface view of the modified and strengthened Gullick powered-support loaded for transfer into the mine. Provision has been made to fit extensible roof bars of German origin. Had Gullick's, the original manufacturers, followed up with a reasonable degree of industry the progress of their installations within the Area, the experience we gained would have been available to them also and could have been incorporated in their newer designs.

PHOTOGRAPH 42 is of an installation of 180 of the modified units in the Blackshale Seam at Langton Colliery. Details of the 1 metre extensible roof bars shown at the top of the picture are to be quite clearly seen.

It took quite some time for Gullick's Ltd (Wigan) to discover our dramatic upsurge in the purchase of hydraulic legs and associated hydraulic control equipment, and still a longer time to realise that we were actually assembling complete installations. Gullick's protested strongly to the Board's purchasing and stores department, who sent down a high-powered party including a senior purchasing and stores representative, together with John Adcock, Chief Mechanisation Engineer (whom I knew well) and others I didn't know. Advising Mr Severn of the projected situation he said,

'Get your facts together, sit down and prepare your case. If they start giving trouble let me know.'

The party of about five or six came to see us, each member provided with files from which they raised their charges reinforcing one another's statements blackening all that we had done, including that we had been contravening Board's instructions which was quite a serious matter. Who was responsible in the first place and on whose authority did we carry out the manufacture of powered supports? These were matters of serious import to the Board.

My party consisted of Peter Harley (very active in the British Association of Colliery Management at the time), the Area Purchasing and Store Manager, and a representative from the Finance Department. We had agreed to sit quiet and refuse to be drawn until they had exhausted all their allegations. This we did although it gave the meeting the impression we had no defence. Regretfully, the whole situation would have to be reported to the Board, who would take a serious view as to what we had done.

Opening up, I accepted full responsibility for the situation which had arisen because the equipment we had been supplied with had proved inadequate to meet our mining conditions. Gullick's had failed us also, in that we hadn't received the proper support and service an experimental situation required and should have been given. As such, we had been thrown back on our own resources. Moreover, had the manufacturer's representatives given us a reasonable degree of interest and service they would have known what the situation was and they would not have met our bulk orders for the hydraulic legs, valves, etc., without which we couldn't have done a thing. 'Further,' I added, 'concerning the Board instruction we have contravened, could I see the copy you refer to, I can't find ours.' There wasn't one remotely close to what we were discussing. The Head-quarters party went very quiet particularly when I added,

'Now, when you report this incident to the Board, kindly include in your report that we have reduced the price of a support by some £180 per unit, or upwards of £32,500 per installation.'

On learning this, they were really shaken and extremely quiet. As Blaster Bates, the West Midlands explosive demolition expert, often said in parody of Queen Victoria's immortal expression,

'They were not remotely amused.'

In fact, over the last half hour they were with us we were to be congratulated on our initiative and had done a splendid job for the Board. Informing my Area General Manager about the situation he laughed, but I had a gut feeling we hadn't heard the end of the story and the last laugh would be with them. It was. A new National Directive was issued which made it impossible for anyone to repeat what we had done and moreover Gullick's bought out Herbert Cotterill Engineering Ltd, retaining Herbert as manager for a few years.

W.H. SANSOM RETIRES

Towards the end of 1959 W.H. Sansom, the Production Director, East Midland Division, retired from the National Coal Board's service. Disappointingly, my circumstances were such that I regretfully was unable to be present at his retirement dinner and presentation which we, as Area Production Managers, arranged as a mark of great respect. I had amongst the No.4 Area Surveying Staff a draughtsman who was an excellent artist. Having prepared an address to Mr W.H. Sansom I asked him to convert it into an illuminated address on behalf of us all. This he did, following which I obtained the signatures of those within the Areas and Division who were close to Mr W.H. Sansom. The illuminated address was presented to him by one of my colleagues at the dinner. It really was most beautifully prepared in colour and was indeed a work of art.

PHOTOGRAPH 43 reproduces that address. After the retirement dinner I received the following letter from our greatly respected East Midlands Production Director-Departmental Head.

Fairfields Grantham.
Dec. 13th 1959

My dear Charles,

I was very sorry indeed that you were not able to be present at the presentation last Thursday - I know you would have been there had it been at all possible.

Mr Noble passed on to me your exceedingly kind letter which has given me a great deal of pleasure. I would like to have expressed my thanks to you personally, as I have done to others for the most useful and handsome gift you were all so kind as to present to me. It will always remind me of your kindness and help. I was particularly touched to receive the illuminated address which I am not surprised to hear was your idea. I think it is most beautifully done - a work of art - and the wording on it although it touched me so much gave me a great deal of pleasure. I shall treasure it all my life and I hope that my son will do so too.

I would like to take this opportunity to thank you for your friendship and for all that you have done to make my job such a happy one.

I have of course taken the greatest interest in your career since you have been with the EM Division. You have made wonderful progress, progress which has been due to your own hard work, ability and great enthusiasm for your chosen profession. I sincerely hope that you may progress even further.

Thank you again for all you have done and for the exceedingly kind things you say about me in your letter in which I know you are sincere.

My best wishes for your advancement in your profession and for your future happiness.

My wife joins me in sending our kindest regards to you and your wife.

Yours sincerely

W.H. Sansom.

Our Production Director was a humane person of great ability and as with others I have mentioned one was greatly enriched by exposure to him. He could and did inspire one's endeavours and one could rely upon his support and guidance, no matter how intense was the problem that had to be faced. He was a good man.

The manifestations of fate are unpredictable and in this connection I again met E.J. Kimmins. He came into the Area to make a visit to Newstead Colliery, an extremely well organised and excellently managed pit run by John Collinson and Ray Gregory, Manager and Group Manager respectively. The colliery had excellent standards of safety, workmanship and performance, and its productivity at 3.06 tons per manshift was well above the national average at the time, output closely approaching the 1,000,000 annual saleable tons mark with a profit level of almost £1,000,000 or £1 per ton.

Before going underground with my visitor I was told that Mr Rex Ringham, the Divisional Chairman, wanted to see me at 2.00 p.m. at Sherwood Lodge which, travelling through Newstead Abbey Park, is not too far distant. In my experience of conducting members of the Board's staff and others in visits and inspections underground, the visitors invariably can be classified as:

1. Those with a genuine desire to determine the logistics of whatever is being undertaken, and to partake of the mining experience being obtained, readily sharing their own on an informal and enthusiastic basis.

2. Others whose motivation would appear to be that of trying to convey to the local Area, Group or Colliery the importance and superiority of themselves and their station together with the outstanding abilities they possessed.

With the big people: Lord Robens, Lord Ezra, Sir E.H. Browne, Lord Halifax, Wilf Miron, W.H. Sansom, A. Bowker, Sir N. Siddall, D. Severn, Alfred Kellett, R.G. Baker and many others already mentioned, and, in general, the American mining engineers and operators, there is no attempt at self importance. They act with pure interest and meaningful contribution and invariably fit into the first category.

The others often adopt a superior, arrogant, snobbish demeanour with the immediate effect of establishing distinct divisions. Protocol is a useful tool as it helps to induce levels of importance within the parties. A third but none the less potent approach is that of niggling over minor matters of inconsequence. I would not deny that during my career members of my staff could well have been justified in placing me many times in the third category (and no doubt did), on the basis of niggling born out of impatience.

Our visitor on that day could well have been classified in the second category on more than one count. Compared with the job I was doing, he was on that day but a bureaucratic encumbrance trying to justify his presence. There was no specific purpose to his visit that we knew of, yet, during the whole time spent on that visit, I felt his whole approach was to try to show there had been no relative change between us since those early days of nationalisation in the No 3 (Rotherham) Area Yorkshire Division on 1 January 1947, when I was starting out as a young unproved Colliery Manager at Rotherham Main Colliery, fifteen years before.

Mr Rex Ringham and Interview

Incidentally calling in to see the Divisional Chairman, Rex Ringham wanted to inform me that Sir James Bowman, Chairman of the National Coal Board, wished to see me at Hobart House about a transfer to the post of Area General Manager of the South Western No.9 (Neath) Area. I do not think my visitor would have been in the least envious had he known, particularly as Tommy Wright had previously turned it down.

I recall as a young student looking up to the Colliery Manager at the great apparent distance between us and thinking, 'This man must have the qualities of a genius, be highly educated and be a person of real vision and enterprise.' After one qualifies and becomes successfully versed in the practice of fair or good managership, looking sideways one's thoughts are changed:

'There is nothing special about me, and there is nothing special about him. How did he get there? I wonder what his route was? I know what mine has been.'

Things and people are not always what one thinks or imagines they are, no matter from what position they are viewed.

My career in the East Midlands Division No.4 Huthwaite Area had been a very happy and rewarding one. I had two regrets upon departing: I was leaving behind Don Severn for whom I had developed great admiration, fondness and respect, and I had to leave John my son behind to complete his studies so that for the first time our family was to be fragmented. Both Mary and I were really sad. However, we hadn't realised John was grown up and, at the age of seventeen, had a much higher degree of confidence than I had had at twenty-one. He was more than capable of standing on his own feet and opening up his future.

There is little doubt about the fact that public school training does develop a degree of confidence in their pupils which Local Authority council schools fail to achieve. This is strongly evidenced in all my four grandsons, Michael, Peter, David, and Dorothy's son Paul, currently attending Wellington Public School, Somerset.

Sir James Bowman - National Coal Board Chairman

My interview with Sir James Bowman lasted about thirty minutes and was quite straightforward. He informed me that the Board wanted to fill the post of Area General Manager - South Western Division's No.9 (Neath Area). They had submitted a list of potential candidates to Mr D.M. Rees, the Divisional Chairman, and he had made strong recommendations that I be given the post. I told Sir James that I had never met or worked for D.M. Rees.

'That is immaterial. He knows all about you and has studied your record. It's you he wants.'

I accepted the post, knowing that others had turned it down.

Before leaving the Area, Don Severn arranged a party for Mary and me, the staff having bought us a large cabinet of fine quality cutlery. In his farewell address he said to the staff,

'I am sorry Charles is leaving us to work in the Welsh Valleys. I have no doubt he will do very well there but before he goes, I would like to publicly acknowledge my debt of gratitude to him. More than any of you realise he has been of great strength to me and on more than one occasion gave me unstinted help when I needed it most. I shall miss him greatly.'

Before opening up the next stage in my career I would like to complete, so far as I am able, Don Severn's situation following his retirement.

Following my departure Little Jimmy Wright, as Don used to refer to him, was appointed to my vacated post of Area Production Manager which he carried out until after Don's retirement. Mr and Mrs Severn were living at the time in the Area General Manager's house at A-Winning, a small Derbyshire village near Alfreton. He was replaced by Harry Bennett, formerly Area Production Manager of the East Midlands South Derbyshire and Leicester Area.

On 31 March 1963 Mary, Dorothy, Harry and I motored up from Ammanford in West Wales to Alfreton, Derbyshire, to attend John's wedding to a delightful young and very pretty little Alfreton girl, Sue Radford (who has over the years, like Mary was to me, been a pillar of support to him). Sue's uncle Sid I knew well during my time in the No.4 (Huthwaite) Area. He worked in the Mechanisation Department. My first question on seeing him was to enquire after Mr Severn.

'Didn't you know?' he said, 'Mr Severn had a bungalow built adjacent to Greengables. He now lives there.'

I broke away from the celebrations at the earliest moment and drove down to my old address. Mrs Severn came to the door and I could see that she was in an extremely anxious mood and quite stressed, most unusual for the lady as

I knew her. She was a Magistrate of many year's standing and was normally quite strong.

'Is Mr Severn at home?' I asked.

'No, Charles. Allan my son has gone to fetch him from a small country club in Derbyshire. He'll be moving in in about a hour.'

'Moving in, Mrs Severn. What do you mean?' I asked.

'Since you left many things have happened,' and with tears in her eyes, she opened up to me with deep emotion.

Don retired and they made tremendous efforts to find suitable accommodation. They had been unsuccessful and since Don had always liked Greengables he enquired of Little Jimmy Wright (then in residence) if he would agree to the ceding off part of his frontage so that they could build a bungalow. It would have been no hardship and it wasn't his property anyway. Mary and I had lived there some three years or so and, in a similar situation, we would have agreed and indeed felt it to be a privilege in view of the help and support he had given me during the short time I was with him.

Jimmy refused and both Mr and Mrs Severn were distressed. However, the house was vested in the National Coal Board's ownership. After appealing to Wilfred Miron, the Divisional Chairman, the transfer of the portion of land required was authorised; thus the bungalow's construction was set in motion. Jimmy's reaction was to build an ugly 6 ft. solid wooden fence along the separation boundary. In the meantime Mr and Mrs Severn were being put under pressure to leave the Board house they occupied in the village of Blackwell. Harry Bennett, the former Area Production Manager of the South Derby and Leicestershire Area, his successor, had a large family and was travelling long distances. He had a big problem.

To ease Harry's situation Mr and Mrs Severn moved into a small country club a few miles from the village of Blackwell, Derbyshire, to await the completed construction of their new bungalow, which took several months. At the time I met Mrs Severn she had just entered the bungalow early that day to get the placed warmed up and ready for Don taking up his new residence.

I had been with her for upwards of two hours when Don came in with his son Allan. He had deteriorated badly since I last had seen him and was in real difficulty. It took him quite a while to recognise me, following which he said, 'Sorry, Charles lad - I'm knackered, I've got to go to bed.' Those were his last words to me. Within a matter of days I was informed later he had died and had been interred. Unfortunately I wasn't informed at the time and was unable to pay my last respects to another fine and great person who contributed much to my development and character.

The above is the gist of Mrs Severn's account. I have no reason to disbelieve what she told me and I am sure there are many friends and acquaintances of Mr and Mrs Severn who would readily testify to her honesty of expression. Like Donald her husband, she was well respected by all who knew her.

Several years later Greengables was taken over as a Cheshire Home and for quite some time now has cherished and supported those human beings less fortunate than ourselves. Having contributed in the form of entertainment to one such Cheshire Home near Swansea during that part of my career in Ammanford, West Wales, I found that the physical and mental plight suffered by the inmates, although in the hands of fine human and dedicated people, certainly tugged hard on my heartstrings. I am of the opinion the property could not have been put to better use.

Returning to the wedding ceremony I was both sad and depressed. Mary and my sister Doris were together when I entered the room.

'What's the matter with you?' they both asked.

It must have shown. It would seem there are a number of different ways in which one can express gratitude and appreciation to a mentor who has contributed greatly to one's career.

Shortly after this incident I had a visit in Wales from Sir Andrew Bryan (formerly Chief Inspector of Mines) who then investigated claims on behalf of the National Awards Panel. Apparently Jimmy had laid some strong claims as to his contribution to the development of powered supports, I couldn't recognise them particularly, as there were quite a number of us involved and Jimmy's contributions were little different to those of others. We had no case with which to make any such claim in that we were merely strengthening and beefing up the original Tom Seaman designs - and there was nothing radical involved therewith. I don't know how he fared with his claims.

SIR GEOFFREY AND LADY BARNETT

Prior to leaving the No.4 (Huthwaite) Area, East Midlands Division, I had established close contact with the Managing Director of Hardwick Industries, Mr Wilfred Allsop. At Gedling we had helped him to develop successfully the Dobson hydraulic prop with which we ran into early difficulties. The props were prematurely unloading with loss of support to the roof, quite a serious situation. However we kept the problem under wraps, and the Company and ourselves quietly worked on the difficulties to the point of resolution.

The same problem occurred with an identical installation in South Yorkshire but this was given wide publicity, much to the embarrassment of the manufacturer. Because of this, Wilfred Allsop, Managing Director of Hardwick Industries, came over to see me at the No.4 (Huthwaite) Area Headquarters with design drawings for a new type of powered support. This was basically a 'goal-post frame' support as opposed to the Gullick 'box form' construction which was in its early state of development.

'Charles, you helped us tremendously with the problems of our hydraulic prop. I have with me designs for a new powered support - we'd be most grateful for your interest and help in proving its feasibility.'

Peter Harley (Area Mechanisation Engineer) and I spent more than two hours subjecting the designs to comparative analysis relative to the appreciable experience we had built up with the then current Gullick units, widely applied and fully operative.

The design looked very promising and had a number of encouraging new features. We agreed to undertake trials at Newstead Colliery. The installation was duly delivered and installed and trial operations were started. Subject to the usual applied modifications, fairly normal in such circumstances, the system worked exceptionally well and his particular design was established, following which a number of installations were manufactured.

Roughly three weeks before taking up my new appointment in Wales I received a very nice handwritten letter from Sir Geoffrey Barnett, Chairman of Hardwick Industries Ltd, inviting Mary and myself to join a small dinner party in the Victoria Hotel, Nottingham. Mary and I accepted though she was apprehensive; she tended to screw up tight at the prospect of meeting high personages she didn't know or hadn't previously met. However, she did have an uncanny knack of instantly identifying nice people. Those having an affected or snobbish approach she would spot quickly, become uncomfortable, turn-off and remain quiet. During that time, both on the American scene and within the UK, companies were tending to look at the wives of potential high executives and during weekend parties in some exclusive surroundings would carry out an evaluation of how supportive such wives would be to the post of the high executive, by making an assessment of their level of education, behaviour in high circles, social qualities and ambition in the furtherance of their husband's careers. There may have been other facets. However we never knowingly had to face them, and probably would never have reached the first hurdle had that been so. I recall a film of many years ago which featured this concept starring, if I remember rightly, Cary Grant.

Mary, to me had qualities which were priceless. She had a strong moral code, deep love for her family and a simple uncomplicated approach to living. She was able to create a home atmosphere of happiness and security that visitors often sensed and remarked upon. I really did have a rock-like haven to turn to when in difficulty. What greater support to a husband is there?

Having accepted Sir Geoffrey Barnett's invitation we went along to the Victoria Hotel where a small room had been set aside, beautifully appointed and adorned with flowers and shrubs together with lavish provisions of food and drink. On entering the room Sir Geoffrey came across to us and introduced us to the small party present: Wilfred and Mrs Allsop and Lady Barnett. Inwardly I wondered how Mary would react. She knew of Lady Barnett having seen her countless times on TV in 'What's My Line?' and other shows and now, here she was in person. However I was amazed at her dignified ease and obvious pleasure in discussing with Lady Barnett her background, and talking about John, Dorothy and her family environment. Lady Barnett appeared quite en-thralled and talked to her at length, pursuing her with many questions. Mary was really at ease. The situation unfolded as though they had known one another for years. We had a beautiful and indeed memorable night, the company being both charming and hospitable. On our way back to Greengables, I asked how she had managed to be so easy and comfortable in the circumstances.

'You can talk to nice people,' she replied and that was it - no elaboration. Norman Siddall over the years had memorised a number of her uncomplicated quotations - that was one he missed.

Wilfrid Miron - Deputy Chairman East Midlands Division

Before taking up my appointment in Wales, Wilfrid Miron (then Deputy Chairman of the East Midlands Division) invited me over to his office. Having been born in or near Llanelli he knew West Wales intimately and for over two hours he explained, advised and encouraged me tremendously, all of which proved more than helpful when I got there. Wilfrid was a highly competent person, a brilliant and humorous after-dinner speaker. On one occasion he made media and parliamentary history by referring to the then MP for Kidderminster as a 'carpetbagger'; the detailed circumstances which gave rise to this statement escape me, other than that it concerned ill-informed criticism of the Coal Industry. Despite pressures from MPs he stood his ground, much to the admiration of us all.

One piece of advice which was of tremendous help and importance and created great happiness for Mary, Dorothy, Harry and myself was,

'Get to know the people – join in whatever is taking place and become involved.'

As will be seen, this we did in no small measure so that we were all able to look back on the time we spent with our Welsh friends and colleagues as one of the most rewarding parts of our family history.

On the retirement of Rex Ringham, Wilfrid was appointed the Divisional Chairman and remained so until the Divisions were abolished in 1967. During this time, the East Midland Division made substantial progress under his leadership. Many were the times, when dealing with Area problems and relationships in very difficult situation, that I would recall a great deal of his advice to helpful effect; moreover I remember most vividly his observation, 'The situation in West Wales differs greatly from that in the East Midlands. Don't lose heart if things don't come quickly and progress is slow. The situation can be changed down there and I am sure that if anyone can effect beneficial change you can.'

That, together with similar views expressed by Norman Siddall, helped me to negotiate my way through the darkest of situations. I really enjoyed working for the East Midlands Board and to paraphrase what Mary said,

'One can derive great happiness by unstintedly giving one's all to capable, competent and understanding good people.'

That part of my career spent in the East Midlands Division tremendously extended my experience, developed my character, built my confidence and exposed me to the influences of some very fine people for which I am truly thankful. The whole family, including Harry Wharton our gardener, was very happy there.

Not to mention the body of the East Midlands miners would be most remiss of me. As a body of excellent, hard-working and progressive men they have no equal within my compass of experience; whatever success obtained within the Division in the final analysis was theirs.

The manner in which they were later betrayed by the politicians of the 1980s and 90s following the support they gave the Government in one of its most testing periods is more than difficult to understand.

PHOTOGRAPH 39A AND B. JOHN
AND DOROTHY ROUND'S SCHOOL
DAYS

John in his Beckett School (West Bridge-ford, Nottingham) uniform. At the time he was just over 11 years and had just started his new school, his fifth school in almost as many years. However the school did an excellent job with his academic training and character development, following which he went on to develop and sustain highly cred-itable achievements in his later life.

A snapshot of Dorothy at the age of about 9 years during a vacation. She at the time was attending a private school in Burton Joyce, near Nottingham; later she attended the Girls High School in Alfreton, Derby-shire. In the first five years of their school life they were substantially disturbed with many school changes; however neither of them suffered too badly in the circumstances.

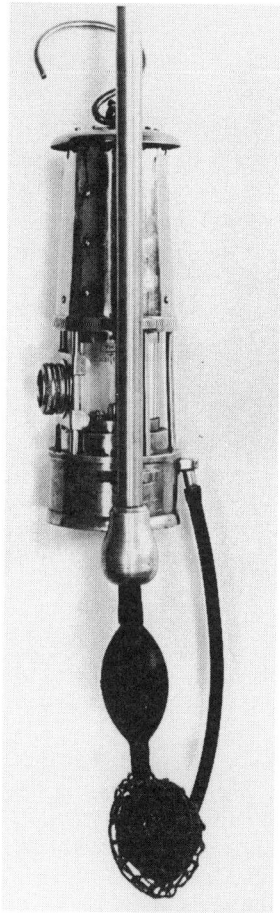

PHOTOGRAPH 40. MODIFICATION OF
OFFICIAL'S SAFETY LAMP (RWF)

Normally firedamp is detected within the general body
of air by reducing the flame to a small yellow bead;
the gas burning produces a small blue cap on top of
this. The size of the cap varies according to the amount
of gas burning within the lamp. An equilateral small
blue firedamp cap would signify about 2.5% of firedamp present in the atmosphere. At about 4.5% the
firedamp explodes within the lamp; the lamp's internal
gauzes prevent interaction with the external atmosphere. Here we induced an air sample with an aspirator
from the upper open end of the telescopic tube and
fed it into the lamp for measurement.

PHOTOGRAPH 41. GULLICK POWERED SUPPORT: NO. 4 (HUTHWAITE AREA)
EAST MIDLANDS DIVISION

The original Gullick hydraulic 'box type' support after a short period of commissioning occasioned heavy maintenance. The canopies fitted on the top of the hydraulic jacks were subject to heavy distortion, whilst the bases twisted and bent. Lateral thrust pressures gave rise to abnormal damage configuration of the vertical hydraulic legs which provided the basic support.

Stiffening both base and canopy improved the situation greatly, in that the roof control system at the face involved 'Groeschel' extensible steel bars. These were incorporated and fitted over the side legs.

To control the problem of dirt getting into the chock and also to provide for extra stability to the vertical jacks, protective steel plate was added. The performance of the supports was dramatically improved following work undertaken on them.

Access to and replacement of the horizontal advancing jack was made easier. Rubber block inserts were fitted to provide a degree of flexibility to the individual vertical supports. After a great deal of effort on a modification trial, further modification, retrial basis, we did get to the stage in which the installation functioned quite well, following which we then built our own units, a situation which gave rise to Headquarters investigations.

PHOTOGRAPH 42. MODIFIED GULLICK POWERED SUPPORT: LANGTON
COLLIERY, NO.4 (HUTHWAITE) AREA

This is a photograph of the original modified installation in the Blackshale Seam at Langton Colliery. The seam was about 3' 8" in thickness, the face being about 200 metres in length. The manner in which the extensible steel bars were fitted to hydraulic supports is shown very clearly along the top of the photograph. Mining conditions were not the best, particularly the soft floor aspect. This gave trouble in the supports cutting deep therein as they were advanced pushing mounds of dirt before them, as may be seen. However the installation worked very well and represented a substantial leap forward in those early days.

Usually the men gave us 2 strips per shift with about a 24 inch slice, i.e. approx. 4 ft. advance per shift up to 8 ft. per working day yielding 750 tons. Face productivity was of the order of 8 tons per faceshift.

Following the experience gained with this installation, four or five complete installations were built on the basis of spare part assembly which when completed were transferred to other collieries quite successfully with a progressive improvement in the overall Area situation.

PHOTOGRAPH 43. RETIREMENT TRIBUTE TO W.H.SANSOM

Production Director, East Midlands Division, National Coal Board
1 January 1947–31 December 1959

8

Area General Manager
Appointments

No.9 (Neath) Area, South Western Division

Background Information

The No.9 (Neath) Area usually referred to as 'The Anthracite' covered parts of the Amman, Swansea, Dulais and Neath Valleys of West Wales. At Vesting Date, 1 January 1947, when the collieries were taken into national ownership there were 66 collieries operating and producing a total output of 3,828,965 tons with an overall productivity of 14 cwts per manshift. Financial losses at the time were of the order of £4,307,486 per annum. This situation of heavy financial loss had been prevalent for many years.

In 1959 the annual output for the Area was approximately 1,358,000 tons. Face and overall production were 40.5 and 14.8 cwts per manshift respectively, proceeds being £5.455 per saleable ton. Total losses for that year were £3,367,000

Early in January 1960 I was appointed to the post of Area General Manager. In comparison with the former Areas I had previously worked in, the tonnage produced was small, but its size in relation to the difficulties encountered was quite a different story. The total output was being won from eighteen small collieries, the majority of which were little more than holes in the ground, badly developed with a minimum of capital expenditure. Varteg, Pwllbach, Abercrave and Onllwyn's were fair examples.

During the early years of nationalisation the Swansea Area was split in two with the formation of the No.9 (Neath) Area in which the rate of annual losses exceeded £3,250,000 annually, a situation which had prevailed for many years. Cefn Coed and Abernant Collieries were subject to some of the most difficult mining conditions I have ever experienced. Heavy dust yields underground had taken their toll in miners' lives over the years, whilst the firedamp yield at Cefn Coed was extremely high, ultimately being captured and used for steam raising and heating surface buildings.

Industrial relations were very difficult with the men being adept in exploiting management weaknesses. Over many years they had established what was known as the Seniority Rule. It had been honed and shaped by the men over a very long time to the point at which it was so flexibly adverse to management control that it made the task of the managers onerous and often despairing. In simple terms: if one was driving a large engine and another man was driving a smaller engine, any vacancy arising with the former had to be filled by a transfer from the latter, regardless of his suitability. This basic principle was applied to almost all facets of labour activity; moreover it was extended to unemployed personnel and indeed between pits. Its ramifications were virtually impenetrable. To the men it was sacrosanct, to the management it was anathema.

I applied tremendous effort throughout my career in Wales in trying to come to terms with this situation, by endeavouring to get the original formatted agreements, of which there were many versions quoted but few in accessible print, but to no avail. However, some of the work put in did pay off on one very important occasion as later recounted. Such was the basic background to starting my new post (of which I was thankfully ignorant at the time), although I don't think it would have influenced my decision to take the post. The job was packed with potential experience of all kinds.

Temporary accommodation was found for me in the Castle Hotel, Neath, which then contained such an array of rugby shirts in all colours, sizes and denominations as would have been a priority of purchase to any organiser or participant in a Notting Hill or Rio de Janeiro Carnival. I used to thread my way through them to get to my room each night. I really ought to have counted them. They could well have been a potential source of winning bets at any Welsh public bar later on.

Our permanent home was to be Wernoleu, Ammanford. It was a large house with two big lounges, a most sizeable dining hall, kitchen and snug, seven bedrooms and two bathrooms: an old house, but quite nice. It was set in upwards of two acres of park land. Mary transformed it from a dark dismal place into a very attractive light and airy domicile and we were all happy there. However, it was more than three months from taking the post before we moved in. On that day, some kind soul called with a large beautifully cooked *sewen* for our lunch and refused to tell us from whom it had been sent. I was sorry we were never able to thank them personally for such a thoughtful and kind act, I do so now.

On the Monday morning following my arrival the Deputy Chairman, Albert Walsh (who was later drowned whilst swimming in the sea during a holiday vacation), called in the Castle Hotel to guide me into Ammanford some seventeen miles distant. He apologised for the absence of the then Chairman, D.M. Rees, who was in hospital with a collapsed lung, indicating that his date of return to duty was uncertain. After spending the morning introducing me into the Area and to the Area Staff he left before lunch. At this point I was now facing the biggest test of my career up to then, although I wasn't aware of it.

Philip Weekes was the Area Production Manager. I don't think he had himself been there very long at the time I took the Area over. He really had, and has, a wonderful personality, and excelled in small talk, social graces, stories, humorous discussions and recounting past occasions at the public bar or in other similar environments. It would have been a great shame for him to have lost such prowess slogging away in the Anthracite, but his guardian angel moved him on to an identical but easier post in the No 4 Aberdare Area, where, for a further short while, he worked under Tommy Wright (who was of the same opinion as myself), before shortly afterwards being moved again to a post at the National Coal Board's Staff College at Chalfont St Giles.

Tommy Wright's Appointment as Area General Manager No.4 (Aberdare) Area

The circumstances of Tommy Wright's transfer into Wales were the subject of a great deal of media coverage at the time and are briefly recounted thus. Tommy had for a long time had great interest in the sport of clay pigeon and game shooting and, whilst Area General Manager of the No.4 Eastwood Area, he was accused of using the Board's manpower and resources to further that interest on behalf of himself and other associates. The detail of the situation I do not know, other than that Tommy maintained he was but a small part of the much wider organisation he had inherited. The Board's discipline involved his transfer to the post of Area General Manager of the No 4 Aberdare Area in the South Western Division. This was quite a stiff punishment for him as he never really settled down in Wales. At the time of the media's outburst about Tommy's case, the novel *Lady Chatterley's Lover* by D.H. Lawrence was being tested in the Civil Courts for pornographic content. Giles, cartoonist of the *Daily Express* produced a sketch embracing both events. His cartoon pictured an underground coal heading being driven by a couple of miners using an Anderson Boyes AB.17 coalcutting machine. The underground official, with blasting equipment, a shot-gun secured to his belt and a couple of large explosive containers filled with game hanging over his shoulders, is striding up the heading towards the two miners, one of whom is in the act of greeting him with,

'Good morning, Oliver! How's Lady C. this morning?'

The drawing really was fabulous and could well have been taken from an actual photograph. I tried to get a copy for insertion at this point, but the Editor of the paper ignored both my letters.

The first few weeks I spent my time visiting the collieries, assessing the mining situations and looking at possibilities for manpower saving, application of coalface mechanisation and surface reconstruction; and getting to know the NUM Agents, Branches and leaders. The mining conditions in some of the collieries were quite good and, with the East Midlands Coalfield's approach, would have yielded very good profits, especially for the Anthracite. Yniscedwyn (black vein seam), Pwllbach, Abernant, Dillwyn and Cwmgorse (red vein seam) Collieries were cases in point. My final assessment, particularly with regard to tackling the problems of the Anthracite, were:

1. The ideal solution would be to start the construction of four or five new small collieries and close down the unpromising ones as the new ones approached completion and came 'on stream'. There would be many advantages in undertaking this: shortened communications; provision of a modern transport system as replacement for rope haulages; controlled manpower projections; high ventilation standards; and as a basic background for coalface mechanisation.

2. To embark on a programme of suitable coal face mechanisation, with particular reference to fines in that the difference between duff (fines) and French Beans, the smallest heating product (about 1/8"), was upwards of £5 per ton lost revenue.

When it was possible to have discussions with Mr D.M. Rees I put my views to him. He told me that although this was the logical and a commendable thing to do, Headquarters Approval was distinctly remote; a great deal of money had been ploughed into the construction of Abernant and Cynheidre Collieries and the reconstruction of Cefn Coed Colliery which, with the fulfilment of the promised forecasts, would have readily met the Board's anthracite needs.

Later, however, before I left the Area, I was able to go an appreciable way in this direction. Abernant, Cynheidre and Cefn Coed had definitely failed to meet their commitments. Mr Alfred Kellett, who had replaced Mr D.M. Rees who had retired on the grounds of age and ill health, helped me initiate Treforgan and Bettws Collieries, two new collieries. Both of these were passed on to my successor, the first in advanced constructional completion stage, and the second as an early Stage II preparation, with a projected total capital and revenue cost of some £1,267,843. With his further help we were also able, by driving 700 yard drift from the surface to inbye workings, virtually to bring Blaenant into the new colliery category.

During my career in the anthracite we were able to close eight of the original eighteen collieries without any real incident or labour difficulties. I don't think anyone even noticed.

Subsequently, during a personally financed trip to the United States, I found that a similar situation had obtained in relation to their anthracite (hardcoal) business in Central Pennsylvania during the 1920s and 1930s, but the business went into severe decline. Today it is virtually gone except for two special operations. These remain because the Congressman of the district got a bill passed requiring that all American installations in Europe were to be heated by hardcoal. The situation, I was informed, had been strongly and adversely influenced by hardcoal Welsh miners who had emigrated out there. In fact, the famous or infamous American miner's leader John L. Lewis was born in or near Pontadulias. Swansea, West Wales, the approximate centre of Welsh hardcoal production. He built up the American Mineworker's Union to become a very powerful and violent organisation.

Early in November 1960 my old friend and colleague Norman Siddall wrote me one of the most uplifting letters of encouragement I ever received during my long career in the Mining Industry. It couldn't have been delivered at a more opportune time. My spirits were low and I was becoming depressed. Occasionally that used to happen, although I always tried to appear cheerful and confident, even when I wasn't. The gist of his letter was that it takes time, upwards of four years slogging in his case, to change failure into success, and one had to weather the testing period of pouring money into pits with no return, whilst the critics sit back and convince themselves you are making a balls of things. Say nowt, your time will come, mine has done, and this was not half the problem you have, was his final observation.

He had, in four years, raised the No 1 Bolsover Area performance from a productivity of less than 29 cwts to over 36 cwts per manshift with corresponding profit rise from £628,000 to £4,530,000. His predecessor, W.V. Shepherd, must have found it difficult to digest such a phenomenal change. Maybe this was the point from which he viewed all Norman Siddall's associates with some misgivings. In my case, events of the future retrospectively examined could well support such a view.

Having settled down in Wernoleu, we had a number of visits from the Ammanford police amongst whom was WPC Isobel Richards (WPC3) who had served a great deal of her time in Birmingham before coming back home to Ammanford. Isobel was an avid whist-drive fan; she and Mary became great friends and followed their simple pleasures at all the police whist activities in Ammanford, Llanelli and elsewhere. Mary was highly respected in police circles up to and including the Chief Constable. Whenever I was away she always felt secure with Harry Wharton as part of the family and police surveillance which was unsurreptitiously provided. She knew them all and was always ready with a beverage of some sort when they made their inspections. Dorothy took a commercial course at the Ammanford Technical College following which she became a bank clerk at Lloyds Bank. Pontardawe. She stayed with them for over four years, later being transferred to Sheffield and Rotherham before she married John Fordham.

PHOTOGRAPH 44 shows Miss Richards with Dorothy at the time she had just passed her car driving test in Llanelli. Isobel exerted a fine helpful influence on Dorothy during her teenage years.

John whom we had left behind in Alfreton, completed his studies at the Beckett School, leaving with nine subjects at 'O' level and three 'A' level passes. He found himself a job with a firm of dyers and finishers in Nottingham but had to leave because of a degree of colour blindness. I met the manager of the company before he left, and he explained to me,

'We like John; he's an excellent lad and can stay working with us as long as he likes, but it wouldn't be fair to him. His colour blindness affliction would be a serious handicap to his progress in our industry as, for a simple example, we have to be able to differentiate between 300 shades of white.'

Leaving this company, he joined a detergent manufacturing company, Diversey Ltd, which at the time had Canadian roots. He started out in the company's laboratory in the little town of Riddings, Derbyshire and undertook some research for them. The results pleased his employers. Being newly married with a young child on what were comparatively low wages, he pressurised the company for a salary increase. The outcome was that he was later offered the choice of three positions:

i)   join the company's research establishment in London.

ii)   take up a post as plant manager of their Beirut factory in the Lebanon.

iii) become plant manager of its Irish subsidiary in Dublin.

He phoned me and, after explaining the situation, asked me my opinion as to the post he should accept. My advice was as follows:

i) Unless you have the aptitude for a career in research which I don't think you have I'd forget London.

ii) Beirut and the Middle East are politically unstable and uncertain in view of the Arab/Israeli situation.

iii) With the Dublin job you'll be a big fish in a little pond with everything in your favour: expansion as the company develops; less internal competition etc. I would take the Dublin job.

That was my advice (maybe tinged with the fact that both Mary and I would have been uneasy and concerned had he opted for the Middle East appointment) upon which his comments were,

'I have decided to take the Dublin job, Dad - I wanted to test my decision against your views and comment before getting committed.'

Subsequent events proved he had made the right decision, although at one stage things were going badly in that the relationship with his immediate boss was strained and he decided to leave.

In one of the Dublin Sunday papers was an advertisement for a plant manager for a renowned company within the area. This attracted his attention and he applied for the post, and was called for an interview. Casually I asked him,

'John, did you tell these people of your colour blindness? It would seem to me in that particular situation to be a matter of some importance.'

'No, Dad - it didn't come up in the interviews,' he replied.

About a month following his interview he was called to a second interview during which he informed the panel of his mild degree of colour blindness. Colour and colour differentiation was however most important in the company's line of business. Regretfully he was told it would be a great problem both for him and them, and reluctantly they would not be able to consider further his application. They indicated that they were deeply impressed by his honesty. He took his disappointment manfully and I was proud of him. To the best of my knowledge that is the way he has conducted himself through his career. Both he and his wife Sue are highly respected and very popular within the Dun Laoghaire area of the Irish Republic.

APPOINTMENT OF AREA PRODUCTION MANAGER

With the transfer of Phil Weekes, I was in need of a replacement. Getting a good candidate from England wouldn't be easy, as the best candidates would be difficult to persuade. In any case, I always fought shy of importing staff from one of my former Areas, having worked on the principle of making the best of what resources were available within my new environment. To do otherwise created divisions within the staff, with all the problems that entailed. Roger Griffiths had been passed over many times because, I subsequently learned, of his grave disloyalty. It was indicated to me, very much later, that his lack of loyalty had an adverse effect on my predecessor's position but I gravely doubt this, both in fairness to him and from a deep analysis of the prevailing situation after taking up my appointment.

No real progress had been made within the Area over a number of years. There was a mistaken belief that the construction of Abernant would resolve the problems of the Area, although there were many years involved in its construction. It was way behind schedule. Anyhow, the job had to be filled. Advertisements were sent out but attracted a very poor response.

Griffiths had been selected as a former management trainee by the Powell Dyffryn Company and had been processed through their system. He was quite a good mining engineer but had no mechanised experience. Moreover, he did have a streak of ruthlessness which the men and officials within the Anthracite said was characteristic of Powell Dyffryn Company's training and methods. This he used from time to time.

However my Staff Manager, Evan Evans, had formerly held a Senior Post with Powell Dyffryn. He was a very competent and most likeable person but had been pushed into a backwater by my predecessor, probably on account of his former Powell Dyffryn connections. We came to the joint conclusion that Roger was the best of what was available, so we appointed him.

He was a very industrious person, and wasn't afraid of work or of making underground visits. Moreover, given instructions he would work to them. There was a very great contrast between him and his predecessor. During the first eighteen months or so we all worked exceptionally hard introducing coalface mechanisation, and in this respect real local talent emerged. Bernard Evans was in charge of the Mechanisation Department, which he handled with great skill and real organisational ability.

The basic format for mechanisation adopted was the German coal plough and the A.B. shearer. By the end of 1965 upwards of 50% of the coal produced was power loaded. Had the enthusiasm and efforts of the Board's local staff for the development of power loading been reasonably matched by the men, the adverse anthracite situation would have been successfully transformed virtually overnight. The criticism we received within the South-Western Division of creating a situation of 'mechanisation indigestion' because of the rate we set up the processes needed to be considered against the situation we were in.

The abominable low performances of the anthracite producers had existed for many years over many generations. They were and are still glibly explained away by the alleged abnormalities of working anthracite. It is quite true that some of the most difficult mining conditions I ever met were found in Abernant, Cefn Coed and to a much lesser extent Wernos Collieries. In the main, there was a substantially high percentage of average situations in which fair application would have given highly creditable results, particularly in view of the comparatively high margin of proceeds the products enjoyed. There appeared no other imme-diate alternative than to close everything down and restart, but even this also would have needed to embrace the development of mechanisation.

One adverse factor which cannot be overlooked was that of coal dust. Because of many years of inadequate attention, at one period in the Anthracite coalfield deaths due to pneumoconiosis exceeded one a day. Anyone who has seen a youngster a little over seventeen years in an advanced stage of pneumoconiosis retains an enduring imprint on his mind, not unlike that of the burned victims of a firedamp explosion. However, in this respect the Anthracite led the field in the developments of technique, which greatly contributed to the control and elimination of the disease.

The low productive performance which prevailed was engrained in the working and industrial practices, adversely influenced as they were by the Seniority Rule. The prospects of achieving positive improvement by the most brilliant of engineers or administrators in a free society were, I strongly believe, dependent upon effective and total change, which in the circumstances prevailing were most unlikely.

Problems of Underground Roadway Stability - Abernant Colliery

I referred to the abnormalities of Abernant which were exclusive to the Peacock Seam at a depth of some 900 yards or so. The Peacock coal had many of the colours of the spectrum with a sheen like silk and, with a fixed carbon content of over 99%, was closely allied to the diamond. It was fabulous coal, but in accordance with one of the unwritten laws of nature, 'If you want the best, boy, you gotta work and pay for it,' as might be expressed in the American vernacular. Acknowledgements to the American TV advertisement, about that time, in which a lady is seeking directional advice from a workman on the sidewalk in New York.

'Can you tell me the way to Carnegie Hall,' she asks of him.

'You gotta practice, lady,' he advises.

So it was with the extraction of the Peacock seam. Opening up the pit bottom was fraught with the release of vertical and lateral forces of such intensity that in a matter of hours closure of the excavation could be observed taking place. Re-excavation and the insertion of pressure resistant linings tended to slow things down a little. These linings took many forms: timber, heavy steel archers, reinforced concrete and mixtures of all such materials, and indeed others such as sand. Despite the wide range of mining practices we applied, we didn't reach Carnegie Hall.

PHOTOGRAPHS 45, 46 & 47 give some idea of the adverse mining circumstances which prevailed. For months we worked at the problem of getting stability, i.e the resistance of the inserted lining balancing the pressures to which it was subjected. I went to Germany to study similar problems in the Ruhr and bring back solutions, but even the most advanced they had, which took the form of a pipe-like structure with the pipe walls of over 6ft. thickness consisting of crushing timbers and massively reinforced concrete, failed to meet the situation

PHOTOGRAPH 47 was similar to the German versions and whilst we did get a measure of success the cost was highly prohibitive and, equally important, the construction time was far too great in relation to the rates of coalface advance projected for profitable working. In the end, we did obtain some degree of success at a prohibitive cost of over £200,000 for a 200-yard length of tunnel, with projected workings calling for upward of 100 miles of potential tunnel construction.

In a later section of my career, I recall one of the Board's civil engineers in the NCB's Bolsover Area, Derbyshire, in an attempt to destroy a bone-fide property damage subsidence claim, trying to frighten my client and maybe myself with inadequately explained and understood references to 'lateral thrust' and by probing in relation to specially prepared reinforced property foundations at a depth of a few feet, with neither explanation of purpose nor end result.

The extraction of the Peacock seam had to be abandoned. We came up the shafts and opened up the Red Vein seam at a depth of 400 yards, which offered a vast range of successful possibilities subsequently left to my successor.

SOCIAL AFFAIRS AND WELFARE

Within a few short weeks of starting in the Anthracite I was caught up in social affairs and welfare activities. On the second Sunday following my entry into the Area, I had agreed to accept and discharge the function of Chairman of a public concert (whatever that was; it was well outside my range of experience). I was terrified. Had Mary been with me at the time she would have been even worse. The occasion was a concert to be held in the Miners Welfare Hall, Gwaun-Cae-Gurwen. It had a capacity of upwards of 700 (on the night it appeared to be of the order of 5,000 in the least), having served as a local cinema. The place was packed from top to bottom and there wasn't a soul I knew in there. On being introduced to the various dignitaries present I was, with great courtesy, conducted to the seat of honour reserved for such occasions. The MC started off by welcoming me into Wales but more particularly into the West Wales community (I had the feeling of 10,000 eyes penetrating into the depths of my soul. Whether or not it showed in my demeanour I did not know - the intense feeling I had was quite sufficient at the time), and he continued by giving to the audience as much of my background as had been discovered up to that point. They were very quiet and most respectful.

The concert started with a march written by T.J. Powell, the Welsh Sousa, a lively pleasant march called 'Castel Coch', played by the Gwaun-Cae-Gurwen band augmented to about forty players. The concert format followed the general pattern: overture, selection, artist, novelty piece, second artist, light number, instrumentalist and showpiece before the interval; after which the pattern was repeated. The lady soloist was a contralto, from the Swansea Valley. She had a most beautiful voice, matured by very competent vocal training. She sang four songs. The two which I remember were and are very popular throughout the valleys: 'Softly Awakes My Heart' from Saint-Saen's *Samson and Delilah* and 'What Is Life?' from Gluck's *Orpheus*. The quality of her voice and her beautiful phrasing and expression, on top of the mixed feelings I had within me, almost got through to me; whether tears fell from my eyes or not I do not recall, but unashamedly I was almost reduced to that level.

The interval came all too soon. What was expected of me I had no idea. Nothing had been prepared - I didn't know what to prepare - and here I was being conducted to the stage to stand in front of a packed audience with a pair of legs much less secure than when walking underground. My nerves were taut like the strain of a heavily loaded winding rope. I had to speak from the heart, I could do no other.

I just stood squarely in front of the audience and talked to them. I didn't address. I just quietly spoke as to an individual, starting with my background, that of my parents and the problems they had faced, my family development, my elementary and technical schooling, my brass band connections and the great fondness I had for all kinds of music, and finished up by saying that I sincerely hoped I could fit into the community and contribute to its welfare to the best of my ability.

The talk was received very well and the audience expressed its feelings unmistakeably, being both kind and encouraging. Conducted back to my seat the dignitaries on either side leaned over and spoke their words of encouragement. I debated long and hard with myself for the rest of the concert as to the correctness of my approach. I came to the conclusion that it is better to have clear information given direct from the fountain head than to have it furtively salvaged discoloured from its drains.

Some two weeks after this concert Mary came down to stay a week with me at my lodgings in Ammanford. She wanted to get some idea of the house we were to move into and get her thoughts together of the requirements necessary for making it our new home. We took the opportunity to visit some of the most beautiful countryside and places the Welsh people are fortunate to have, including Rhondirmwyn (location of the legendary Welsh Robin Hood). I had promised Wilf Miron I would go there. On the way back we cut across from Llandeilo to Carreg Cennen Castle, landing there about seven o'clock in the evening. We called in the local for a quiet drink and, taking note of the atmosphere and people who were beginning to fill the lounge, we settled down. After about twenty minutes a party of four youngsters in their mid-twenties came in and settled at our table. Unlike in most English pubs, the party enveloped us and within minutes had determined our origin. Immediately we were christened Mr and Mrs Yorkie.

The Welsh people were very quick at applying nicknames to people (probably because firms would have so many employees of the same name) and it was necessary to have additional qualifying descriptions or a more readily recalled identity. 'Jones the Plank' identified a joiner; 'Morris the Milk' was used for a milkman. I was told of an imported colliery manager from Lancashire who, having spent some two years in the Anthracite, commented, 'Thank God I haven't been nicknamed.' From that point and throughout the rest of his career he was referred to as 'John Thank God'.

The young party were celebrating that one of the girl's brothers was up from London for a short holiday. However, Mary and I became the focus of their attention and hospitality. The boy from London went across to the piano and started to play the introduction to an old Welsh melody which I knew to be 'Watching the Wheat', from a euphonium solo formerly recorded by Rowland Jones with the Bickershaw Colliery Band during the 1940s. The pub lounge immediately burst into song and for the next three hours artistry and music pervaded our whole beings. Incidentally, Rowland Jones was born in the locality of Gwaun-Cae-Gurwen and subsequently became a very popular operatic singer; I met him several times during that part of our life in Wales. During this benefit concert for Mr and Mrs Yorkie (that's how Mary and I regarded it), our young friends introduced us to the Welsh opera *Blodwen* by Joseph Parry.

Both the boy and his sister were excellent vocalists. First the young lady sang most beautifully 'O Dywed I'm Awel Y Nnfoedd' - Blodwen's aria, anxious about the fate of Sir Howell, the condemned knight. Then both brother and sister sang the famous love duet of Howell and Blodwen, 'Hywel Be Ti'n Geisio Yma'.

The last item really bowled Mary over - never before or since did I ever see her react as she did that evening. The next day in Ammanford she found a record featuring Rowland Jones and Lorna Elias in the two roles. What a wonderful evening we had. On the way back to my lodgings she said, quite out of the blue, 'I can be happy here.'

This was a great relief to me for she really meant it, and on our return to Yorkshire she always said that the time she spent in Wales was amongst the happiest parts of her life. We never found out who our young friends were that evening; however, they never will be forgotten.

A few weeks after we had moved into Wernoleu, Mary and I were invited as guests, along with other civil dignitaries, to join the Ammanford Annual Carnival and to make the juvenile fancy dress award presentations. PHOTOGRAPH 48 features this occasion. Maybe the delightful and charming young lady who won the contest is married now with children of her own.

SIR GRISMOND PHILLIPS, LORD LIEUTENANT OF CARMARTHENSHIRE

Taking our place on the platform we watched the procession and talked with many of the other visitors present. After about a hour a very tall, straight and immaculate gentleman approached me.

'Mr Round, please excuse me, but I would like to have a short chat with you if you don't mind,' he said.

He was no ordinary person. There was a natural elegance, ease and poise about his whole bearing. He was the type of man you feel happy to meet but there are not many such persons around. Mary had already sensed this and was completely at ease.

'Yes, sir, by all means. How can I help?' I replied.

'My name is Grismond Phillips,' shaking hands with Mary and myself. 'I'm from Carmarthen, but I understand you are from Yorkshire. Do you know of a little village called Wentworth?'

'Yes, Mr Phillips, Mary and I know it well. I was born at Hoyland Common, about three miles from the village of Wentworth and Mary was born in Ecclesfield four miles away.'

He continued, 'Are you familiar with a very large private estate, the seat of Earl Fitzwilliam, Wentworth Woodhouse?'

'We walked the grounds a number of times when we were starting out together, Mr Phillips,' Mary answered.

Continuing, I said,

'I managed one of Earl Fitzwilliam's former mines, Elsecar Main Colliery.'

We were all sitting together, fully relaxed, with the rest of the visitors and guests watching us intently. Neither of us had any idea who this nice person was. He was really interested and, with a large smile on his face, he turned to us both and said,

'And I married one of his daughters - the eldest Countess Maud.'

Mary and I looked at each other with obvious astonishment yet fully relaxed and comfortable. Noticing the effects of the information he had imparted he said to us both,

'Look, we have responsibilities to our hosts. I've really enjoyed meeting you both; we shall get together again very shortly. Do enjoy this occasion.'

It was sometime later that afternoon before we became aware of the nature of things when another guest asked us,

'Have you known Sir Grismond, the Lord Lieutenant, very long?'

Truthfully I replied, 'We have connections which go back a very long way.'

Sir Grismond was quite right. We met several times afterwards. Some two weeks after the Ammanford Carnival, Mary and I received a handwritten note inviting us to have tea with him at his home near Carmarthen. Mary really looked forward to the occasion as she had been happy in his company. He was kindness personified and opened up everything in which we showed interest. There was a wide variety of absorbing artifacts, works of art, pottery, replicas from overseas. The whole place was brimming with things of great fascination. What riveted Mary's attention most was a series of young children's short notes and drawings; they were hung on the wall of his lounge.

'I'll get them down for you so that you can really see them.'

I may not accurately describe the detail. Had Mary still been with us she would have been able to do so, although she never discussed our experience that afternoon, or the subsequent ones which followed, with anyone.

The short notes were addressed to Uncle Gis; some were signed Lillabet, and others differently. The drawings contained flowers, trees and natural other things identified in the same way. These he had received many years earlier from the young Princess Elizabeth and Princess Margaret, both of whom we understood were extremely fond of him and always referred to him as Uncle.

Early in 1962 Mary decided to try her hand at rearing four turkey chicks, as we had plenty of space and suitable outbuildings. We had decided to invite management and their wives to our twenty-fifth wedding anniversary at the year end. She nurtured the turkeys as pets and gave each one a name to which it responded almost like a pet dog. All four flourished to grow to over 28 lbs each. Christmas Eve came and with it our guests. However, the turkeys weren't available; she hadn't the heart to have them killed at that point, so that we had to bring in professional caterers from Ammanford. On that occasion, we entertained over 120 of our friends and working colleagues. Immediately after dinner some twenty of the Gwaun-Cae-Gurwen Band unexpectedly turned up and voluntarily played for dancing throughout the evening and night in the largest room we had. It was quite a memorable occasion which gave us a great deal of pleasure for many weeks afterwards.

The turkeys!! - they were quietly disposed of throughout the following year but Mary would never touch a morsel of their flesh. This experience signified the end of her smallholding efforts and ambitions.

The next time we met Sir Grismond Phillips was at Christmas Eve the following year, in connection with our Silver Wedding anniversary. He graced us with his presence at our home on that occasion. After he had settled down I referred back to the occasion when we first met him at the Ammanford Carnival.

'When we met you at the Carnival neither of us really knew who you were; later that afternoon we felt remiss and a little unhappy at not having recognised your status.'

'Nonsense, Mr Round; you will recall it was I who met you both and it was so bracing for me to have people to talk to so uninhibited and direct in their observations. It was I who felt privileged,' he replied. 'The way you and Mary talk so directly to one another is really refreshing to me for I know there is strong reciprocal depth of real human feeling between you both,' he continued. 'One has only to cross your threshold to encounter an atmosphere of happiness, human warmth and contentment.'

'It's Mary's Castle,' I said.

'And she keeps it well,' was Sir Grismond's response.

'I suppose I must do, Sir Grismond. He doesn't help much,' was Mary's observation.

Sir Grismond Phillips was to me one of the very few of nature's real gentlemen I have had the privilege to encounter during my lifetime. Such persons are easily detected when you meet them. They don't tell you - they don't need to - they are just there and you know it.

Activities Associated with Elderly Citizens

The No.9 Area Photographic Department was run by an outstanding person, David Jennings, an excellent photographer, joiner, fitter, turner, plater, electrician, audio-man and plumber. Whatever was required of him he could accomplish to very high standards. Over the years, he had built up an extensive library of slides, taken throughout England and many countries in Europe. I had at the time assembled some excellent audio and radio recording equipment and was busy, when I could find time, making and circulating tapes to America and Tasmania. I called in to see him at his photo-laboratory in Neath one morning and outlined an idea I wanted to promote.

In simple terms, what I wanted to do was to create an interesting two-hour tape-recorded programme reinforced by coloured slides and to present it to the Old Age Pensioners' group in Ammanford. He fell in with the idea and gave it enthusiastic and unstinted support. We put the first concert on at the Ammanford Old Age Pensioners' meeting place. I spoke about the music and David described the slides as he projected them. The response was tremendous. The local press publicised our efforts and, within a month, appeals for a similar show flowed in from the Dulias, Neath, Swansea and Ammanford Valley's Old Age Pensioners' Groups. We were performing two tape concerts a week throughout the winter.

During the summer, we took the recording equipment into the various Miners Welfares throughout all the valleys, recording talented children and adults, singers and instrumentalists (there was no shortage of high standard candidates) for use during the following winter. Tremendous goodwill flowed from these kind old people. They were appreciative to the point almost of real embarrassment. It was difficult to refuse taking a dozen eggs from a pensioner. It called for great care and sensitivity not to hurt them. The way we avoided such problems was to have a list of worthy causes and tell potential donors that such were more in need than we were. It did not however stop them plying us with their special baking fare for refreshment during the programme intervals. This went on continually for some four years, and when my time came to leave the Anthracite the Pensioners' Groups sent delegates to see if I would agree to them getting up a petition to Lord Robens asking that I should be allowed to stay in Wales. In fact, twelve months after leaving the Area, the Ammanford OAPs had me chair a concert, following which a lovely meal was given, during which they presented Mary and me with a Royal Albert coffee set. We had no inkling this was to take place but we both enjoyed the weekend and indeed all the time we spent with them. They were very good people.

Whatever I wanted to do as a project for the old people I could undertake. Each year we organised two large concerts in the Gwaun-Cae-Gurwen Welfare. I would chair a meeting of NUM Branch leaders together with Vincent Richards and other members of my staff who organised the events. The OAPs were fetched from the different valleys. Buses were stewarded by the NUM personnel, and the men at the different collieries covered all costs. I, like Gus Wilkinson, personally got tremendous pleasure out of working to bring some degree of pleasure and happiness, if only for a few hours, to these old stalwarts.

Brangwyn Hall, Swansea - Brass and Voices

One evening I called the conductors and band secretaries of the Ammanford, Wernos, Gwaun-Cae-Gurwen and Ystalafera Bands together. I had Vincent Richards alongside me; he had quite an organising flare. I simply said that we ought to put a massed brass band and choral concert on in the Brangwyn Hall, Swansea, and feature a national artist. They looked at me wide eyed and speechless, awaiting further explanations.

'When do you want this to take place?' asked Vincent Evans, the secretary of Gwaun-Cae-Gurwen band.

'18 January 1964 is the only date available,' I said.

'But that's a Rugby International day at Cardiff Arms Park; nobody will go!'

'Those who buy the tickets will; we've got to get in early and get as many tickets sold for the concert as quickly as possible well beforehand,' I told him.

'Who's going to conduct the massed bands?'

Sensing rivalries, 'I'll get Harry Mortimer,' I said somewhat audaciously. 'I'd like the four band conductors to develop and agree a programme, arrange practices and joint rehearsals as soon as possible. Vincent and I will contact the

Ammanford Choral Society to get their agreement and fix a programme in conjunction with what you come up with.'

It took another three or four meetings to get them fully confident, but we managed it. Vincent Richards worked at the event with the same enthusiasm I did. I had been motivated by a desire to put on a similar event to that of my old friend Jack McKenning, the Yorkshire Divisional Welfare Officer in the City Hall, Sheffield. He had Grimethorpe, Carlton Main, Frickley, Wharncliffe, Silkstone and other colliery bands to draw from: the cream of the colliery brass band world. Every year the whole event, of which the concert was but a part, was organised and supervised by Jack with Ray Jenkins of the BBC a resident conductor, a task he had then performed some fifteen times or more. Ray had a large following and was greatly thought of in and around the City of Sheffield.

Harry Mortimer agreed to join us and stayed with Mary and me at Wernoleu over that particular weekend. We had great fun. After the concert the following Sunday morning he sat checking his Littlewood's Pools in the kitchen. Mary was cooking lunch. He turned to Mary and said,

'I shall never get rich doing the job I'm doing,' and, lifting his pool sheet, continued, 'But when I hit these I am going to buy the Brangwyn Hall. It's a fine place'.

However, so far as I know the Hall is still run by the Swansea Borough Authorities. Harry was very popular amongst those bandsmen. He had great ability and brought out the best both of the individual players and the bands either separately or augmented. The programme we adopted is set out below:

MUSIC OF BRASS AND VOICES
THE COAL INDUSTRY SOCIAL WELFARE ORGANISATION
(NCB No.9 Area, South Western Division)
Brangwyn Hall, Swansea, 18 January 1964

Choir, Audience & Massed Bands
Cwm Rhondda
Blaenwern

Massed Bands
Slavonic Rhapsody No.2

**Choir and
Gwaun-Cae-Gurwen Band**
Hallelujah Chorus (Handel)

**Gareth Morgan and
Gwaun-Cae-Gurwen Band**
Rondo: Horn
Concerto No.4 in Eb
Major (K.495) (Mozart)

Choir & Gwaun-Cae-Gurwen Band
Grand March ('Tannhauser')
Hail Bright Abode (Wagner)

Gareth Morgan (Euphonium) **and
Gwaun-Cae-Gurwen Band**
Lucy Long (Godffrey) (Friedmamm)

Massed Bands
SELECTION - My Fair Lady
(Lerner and Loewe)

Choir and Gwaun-Cae-Gurwen Band
Tidi-a-rhoddaist
(Arwel Hughes arr:; H.Morris

Massed Bands
Sousa on Parade arr: Palmer

Massed Bands
Excerpts from
Symphony No.5. (Beethoven)

Choir, Audience and Massed bands
Rhyd-y-Groes
Pembroke

Choir: The Ammanford & District Choral Society
(Conductor: Hywel G.Evans)
GUEST ARTIST: **Gareth Morgan** (Euphonium).
Massed Bands: The Gwaun-Cae-Gurwen Band
(Conductor: Haydn Morris)
The Ystalafera Band (Conductor: Lewis Williams)
The Ammanford Silver Band (Conductor: Emrys Henry)
The Wernos Colliery Band (Conductor: Eric James)

The Brangwyn Hall was named after a celebrated artist known for his First World War and post World War paintings. He was commissioned to produce a series of murals for a special situation, but for some reason, having completed the murals, the commission was cancelled. However Swansea Authorities rescued them and designed the Brangwyn Hall to accommodate them in its 'concert hall'. The authorities were far-seeing, in that they preserved for the nation a real work of art.

The event proved a tremendous success, such that for the next two annual concerts ticket demand exceeded supply. The last concert I ran was chaired by Mr Alfred Kellett and Lord Robens. My successors had no real interest and so the event died.

During my Gedling days, Norman Siddall had initiated my interest in art by taking me to see a collection of Sir Alfred Munning's paintings hung in the Royal Academy in London, later loaning me his collection of Lowry prints of the industrial north and urban areas near Manchester. Some thirty years ago my colleague Irvin Spotte of St Pauls, Virginia, USA sent Mary and me at Christmas a large print (limited edition) of Ray Harm's 'Wild Turkeys' a natural and true to life painting in which the detail is tremendous. This has recently been passed over to my son John, now having the status of a family heirloom.

Subsequently, he sent a further painting, 'Woodchuck' by a second American artist, Sally Ellington Middleton. This too is a delightful print. Its quality and detail really are something. Both paintings were the result of a great deal of painstaking observation in the wild over many days and countless hours.

Having failed in our attempt to work the Peacock seam, the development of the Red Vein was sufficiently well advanced as to warrant the partial closure of Varteg colliery. We needed men and the supply from this source was thought to be appropriate. The number required was about two hundred. Each man had to be processed with reference to the Amended Seniority Rules discussed earlier.

The Colliery General Manager was one Joe Harrison, recommended to me by my Divisional Chairman. Following a couple of days inspecting his colliery in the north of England, talking to his NUM Branch leaders and noting the reactions of the men, I was satisfied I would need to undertake a great deal of research to find anyone better. Joe had no idea why I was spending so much time at his colliery until I told him later after we had appointed him. Joe responded very well but, not unlike myself, had already found that there were many adverse facets of working within the Anthracite that were way out beyond the range of a mining engineer's normal experiences and which made management thereof frustrating, extremely hard and often very depressing. Together, with my staff, we grappled with this basic transfer of men, most of whom were misfits with regard to their placing. It could hardly have been otherwise in the circumstances. Something had to be done, but there was no apparent solution to the impasse which developed.

Divisional pressure was naturally on us. We had many of its personnel visiting the colliery, more concerned with reporting our apparent failures than with analysing the basic causes from which these breakdowns originated. We had little respect for them, knowing that in the same situations which Joe Harrison, my staff and I were facing they would present a very sorry picture. This was proved to be true on more than one occasion when the Divisional Chairman wanted a more positive approach from them on the basis of real commitment, living and helping with our problems for days, rather than for minutes or comparative short periods and reporting on our so-called management failings. They were always glad to be called away or to manoeuvre their departure to more pressing duties.

Trevor James was the NUM Agent for Abernant. He and I had developed an excellent personal relationship which matured later into a very close friendship which still exists. Unfortunately, since the death of his wife Glennys he has suffered poor health. Trevor is a very intelligent man who, as a young person, held strong left wing views. He had travelled a great deal and read prodigiously. He was the one who introduced me to Kafka, Koestler, Jack London and others. In his later years he had matured considerably, coming to terms with the world as it was, not as in those early years he and others thought it should be.

Getting him to come to my office one morning, I told him (which he already knew) that we were having one hell of a time with those Varteg men who had been transferred to Abernant. The Amended Seniority Rule wasn't working and was of no value. Some of these men were saying they should not have been moved from Varteg. Ripples of disturbance were going through the colliery and upsetting the men previously there. Continuing, I told him that we had con-formed to the Amended Seniority Rules, with the result that many men had been slotted into jobs which they contended were their entitlements, but in practice not only were they unfitted to perform them but their personal safety was at risk. I believe this to have been the first challenge of the Seniority Rule on the grounds of safety.

'I realise that, and am concerned that you can't compromise with safety issues; neither can I. Let us think about it,' he suggested.

We did just that for the next two hours. We covered everything we felt to be relevant to the situation but were getting nowhere. Then I said,

'I've half a mind to call the men's bluff and send them back to where they came from.'

'If that's what you feel there is little I can do to help the men, particularly in view of your concern for their safety,' he replied.

We called a meeting of the men's leaders where I stated that any man that wished to go back to Varteg Colliery was free to do so, and that we would make arrangements forthwith. Two-thirds of the men elected to go back, which was far more than I expected. However, it didn't worry me for the situation could not have been sustained as it was. Moreover, it opened up a new source of recruitment for Pwllbach and Cwmgorse Collieries from both of which we drained the men until the final closure of the two mines.

DIVISIONAL AND HEADQUARTERS ACCOUNTABILITY MEETINGS

We were subjected to very tight accountability sessions, at monthly intervals and randomly. In the early stages we were called to the original premises located in Tiger Bay, Cardiff and later to the new Llanishen Headquarters. Both D.M. Rees and subsequently Alfred Kellett ran them intensely. There was no nonsense and when one had suffered a particularly bad monthly performance he really had to be in possession of all facts and appropriate explanations relative to the steps taken (or to be taken) to change the situation.

On one occasion I followed Tommy Wright, who had been subjected to very heavy pressure and was leaving his meeting with real anger. Before I went in he told me the circumstances. Apparently the Divisional Board's Finance Director had been highly critical and somewhat offensive. In anger, Tommy had told him to put on some pit clothes and join him for a week in inspecting his collieries, rather than to sit in judgement within his sumptuous office not having the slightest idea of the basic problems involved in running an Area. I followed him in. The Board Members, led by Albert Walsh the Divisional Deputy Chairman, were all quiet and subdued and to me it was obvious that the former meeting really had been heated. My meeting on that occasion was somewhat innocuous.

I recall that, several months after Mr Alfred Kellett was appointed as Divisional Chairman, things had been continuously rough and Divisional pressure intense and after a particularly raw experience at a Divisional Area General Manager's meeting at Llanishen (Divisional Headquarters) I was smarting. Whilst being driven the fifty miles or so back to Ammanford I started to analyse the situation somewhat along the following lines:

'Look chum, you were not alone in having it rough!

You are no more the recipient of heavy pressure than your Group Managers and Managers are. They are getting much more from yourself. The pressure you get is intermittent, the pressure you apply is almost constant!

What's your complaint? The Chairman's not only got the problems of your Area; he has in addition those of six other Areas of which a number are performing but little better than your own.

Grow up!'

I began to get things into perspective by the time I got home and I was back to normal, quite cheerful and thinking about the next day. I found it useful from time to time to indulge in such self-analysis.

Lord Robens and W.V. Shepherd together with the Director General of Finance with other support conducted the Headquarters Accountability meetings which were much less frequent. Generally I found His Lordship helpful, encouraging and understanding of the problems we encountered. I never lost faith that we could solve the problem of the Anthracite and worked at things accordingly. Twelve months before I returned to my old Yorkshire No.3 (Rotherham) Area His Lordship drew me aside following an Accountability meeting and said,

'Charles - I'm sending you back to take charge of the South Yorkshire Area but it will be in about twelve months time. Things are going badly in the Area. A short time ago, when a Member of the Divisional Staff was taking me to see the Area General Manager, he tried to usher me into the "broom cupboard", the door he had just opened to the Area General Manager's Office. Carry on as you are but say nothing to anyone. You will be transferred when I am ready.'

Some twelve months later I was returned home unscathed but greatly enriched by my sojourn in the Anthracite. Lord Robens kept his word.

ALBERT NAYLOR - VISIT TO HIS YORKSHIRE HOME

We continued to struggle. Many things were turning over in my mind constantly but one thought persisted:

'Go up to Yorkshire and see Albert Naylor.'

'Go up and see him this weekend - I won't be able to go, I'm fixed up with Isobel Richards; we are invited to the Chief Constable's home near Carmarthen,' was Mary's advice.

The next Friday morning I took the chauffeur with the Humber Hawk I had at the time, and we travelled up to Elsecar, arriving there at about 5.00 p.m. After a meal, I left my chauffeur with relations and went to see my old friend. His younger son Arthur had, a year or so earlier, bought him a new bungalow and set him up in quiet comfort in the village of Milton, about a mile from where he formerly lived. There was no difficulty in finding the house. My knock on his door was answered by a lady I didn't know. I told her I had come up from Wales to visit my old friend and had just arrived.

'Don't tell him; I want to surprise him.'

She told me he'd been very ill over the past three weeks or so, but she led me into the lounge and never spoke.

'Is that you, Charlie lad; I knew that you would be coming up to see me today,' a faint voice observed from the far end of the room.

'It is, Mr Naylor, how are you feeling?' I answered.

'I had a stroke a short while ago and have lost my sight totally; come up nearer, lad, so that I can talk to you. There is a lot I want to know.'

He was really ill but there was no self-pity, and he was still the same strong character I had known. I had to give him a total account of what I had experienced after leaving him those many years ago. Every detail he pressed for. Just before I left him, he said,

'Charlie, I am so pleased to have helped in your training. You have more than lived up to my expectations.'

'Without that guidance and case-hardening you gave me, Mr Naylor, I couldn't have survived,' I truthfully observed, 'I'll be up to see you again soon.'

He shook my hand with such warmth I was overcome with emotion. Two weeks later my old friend died. I was up for his final departure, for I owed a great deal to him.

MERCHANT SEAMAN NO. B/P 626691

Some two years or so after I received Norman's uplifting letter, Alfred Kellett, the Divisional Chairman, came over from Llanishen to Ammanford to see me, why I never found out.

'Mr Round, I've come to have a general chat with you. I know of the difficulties you are having and the way you are facing up to them. I have no complaints, maybe a few suggestions,' was his opening remark. We sat and talked for about an hour about my family, the pits, my interest in music, brass bands and so on. When we were both fully relaxed and comfortable, quietly he said,

'I'd like you take a good long break away from phones, people and pits. Go somewhere where no one can worry you about anything. You need and deserve to do so, both in the interest of what you are doing here but more importantly your health. Have a word with Mary, then give me a ring later.'

Thunderstruck I managed, 'But Mr Kellett, I haven't the time. There is too much to do here.'

'Look, Charles, it can wait. You'll be able to handle things a lot better after a good break.'

It wasn't usual for Mr Kellett to use Christian names, but by doing so on this occasion he demonstrated to me that he was sincerely motivated by his honest concern for my personal welfare. Later, talking it over with Mary that night, she was all for the idea.

'You need to accept, you're getting worn out. Go, we will be all right whilst you are away.'

I rang my Chairman on the following Monday and thanked him for his kind interest, telling him I would be happy to follow his generous suggestion. 'Leave it with me,' was his reply. In less than a month I was embarking at Troon, Scotland on the 4,100 ton Merchantman SS *Uxbridge* with a scheduled voyage to Casablanca for loading out 5000 tons of phosphate for delivery to Cork, Ireland. The Chairman had arranged it through the Divisional Marketing Department.

The ship was at the time undergoing a major modification in the Troon shipyard. It had been split apart and another 1,000 tons of capacity spliced into its middle section. The work was some twenty days behind schedule, so I had to spend that extra time at Ayr until it was ready to undergo its acceptance tests. Ultimately, I was able to depart on my voyage as a merchant seaman. The

Captain, George Fiddler, was a seventy-five-year-old Scotsman who had sailed almost everything from puffers between the Scottish islands and sailing ships, to tankers and big merchant ships all over the world. He hailed from the Orkneys and, as a well-read man, had a tremendous library in his cabin (which I shared with him).

To meet the shipping regulations I had to sign with the rank of Supernumerary Pursar B/P626691 for the foreign voyage. I remember the Customs Officer coming aboard at Troon.

'What's your crew, Fiddler?' he asked.

'Twenty-four souls and a Coal Board official,' the captain responded.

'None of your bloody nonsense, Fiddler, I'm in no mood for jokes,' the officer yelled impatiently.

Captain Fiddler looked at me and said, 'Tell this silly sod who you are, son.'

Captain Fiddler fetched a bottle of Crabbs 12-year-old best malt whisky whilst I explained my situation. We had a little session before the Customs man left us and from their joint conversation it was easy to determine they had been friends for many years.

My Divisional Chairman knew what he was doing for me quite well. The first day at sea was terrible; I couldn't talk about coal or collieries as nobody would have been interested, I couldn't phone; my thoughts alternated between Mary and the Area. I really was trapped. Over a glass of whisky that night, I explained my ordeal to Captain Fiddler.

'We'll sort that out. Who's your favourite author?' he enquired,

'I like the French author Emil Zola.'

'You'll find about six of his novels over there,' pointing with the stem of his pipe to his library, 'Pick one out and tomorrow night we will discuss what you've read.'

Next morning after a plain but wholesome breakfast I contacted the bosun.

'Look, can you find me something to do until lunch?'

He gave me huge can of yellow paint and a brush and started me off painting the ship buttresses which proved to be the limit of my seamanship. In the afternoon I stayed in the Captain's cabin and read, whilst the Captain invariably slept.

True to his word, after dinner, over a glass of Crabbs 15-year-old (this being the best Scotch whisky one can get in his opinion) he said,

'What book are you reading, Charlie?'

'Zola's *Nana*.'

'You're a bit young to read that sort of stuff. What will your parents say? I haven't encouraged you, remember that!'

'All right I will.'

'What's it all about?'

'A French prostitute seducing a French count.'

'Goodness, it gets worse. The lad's depraved. Better ask the First Mate to keep an eye on the crew.'

'Why ?'

'I don't want their noble minds corrupting - that's why.'

He was real fun but he knew those books from cover to cover, all of them, and was quite an authority on the Peninsular Wars. This sort of thing went on throughout the voyage but having formerly read and found Franz Kafka very thought provoking, I looked amongst his books and found two: *The Castle* and *The Trial*. He'd read them and was more than surprised to find that I had too.

'What others did he write that you know of?' Fiddler asked.

'If I have remembered correctly or not I don't know - but strongly recall two others: *Metamorphosis* and *In a Penal Colony*,' I replied.

We spent two or three nights or more discussing *The Trial* and *The Castle*. I think we both deeply enjoyed the experience.

Occasionally I could get him to recount his war experiences and, like a young grandson at grandad's knee, I listened intently to all he said. He could paint colourful pictures in a very matter of fact way. He had been in heavily destroyed convoys and had crossed the Baltic to Russia in the most abominable conditions, only to be treated by the Russians with the same suspicions as given to an enemy. On one occasion, whilst hugging the Spanish coast, his ship had been bombed but with only minor damage. He crawled for refuge into harbour only to find his way blocked by other sunken and damaged vessels. There were no heroics. 'Just a job we had to do,' was his general observation.

CASABLANCA

It took about ten days for us to reach Casablanca, but we were held out at sea for a further twenty-four hours. I was sent ashore by launch and given the ship's papers, as required by customs, and instructed to contact the ship's chandlers in the customs buildings, everything having been organised by ship-to-shore communications. It took quite some time to sort the papers out and even longer to contact the chandler. In fact he found me. I had started walking into the port when a Land Rover pulled alongside and an Arabic voice said,

'Are you from Captain Fiddler's ship?'

'Yes,' I replied.

'Get in, please, I will take you to where you are staying overnight. Have you got my papers? Please let me have them,' he said.

On reaching the chandler's stores he settled me down with a bottle of refreshing beer for it was quite warm weather. Outside there was a great deal of activity. I went to the door to see a huge procession marching up the road with bugles blowing, flags waving, people yelling and singing. They were coming out of offices and shops and assembling on the sidewalks, cheering and yelling vociferously. My host came alongside me and said, 'We are celebrating our Independence from the French.'

He took me to my hotel and, speaking French, introduced me to the proprietress, informing her when to awaken me next morning. The next morning I was out in the French part of the city early, had breakfast and wandered down to a main square, on the far side of which was a large wall with a number of entrances through it. Going through one of these I walked on to find a sort of menacing atmosphere about the place. It was becoming more dirty and shabby as I proceeded, and I started to become anxious and apprehensive almost to the point of fear.

Retracing my steps, I hurried back to the main square. Later, on telling the ship's First Mate about this experience, he told me he knew where I had been and that normally they themselves would not have ventured through the wall in fewer than groups of at least six. Apparently it was the Kasbah where life was considered quite cheap and murders often occurred.

I moved on to the main post office. I wanted to buy Dorothy an assortment of postage stamps but I couldn't get the counter clerk to understand by pointing to the different price denominated stamp sheets. I was alarmed to find a gendarme on either side of me, and a lot of excited conversation which I did not understand going on all around me. However, a young girl of about twenty-three stepped forward and spoke to the people involved in Arabic, following which I was in the centre of a lot of smiling faces. She turned to me and said,

'I have just told them you are English and wanted a mixture of postage stamps. They had misunderstood and thought you were going to take them without making payment.'

She told me she was a Jewish girl engaged to be married to an American airman in a month's time. I then told her that I wanted to buy something for my wife and daughter but I hoped I would not have the same problem as had just been encountered.

'Oh no, you won't, for a minute's walk from here are some very nice shops. I'll help you get what you want. There is a special way of shopping out here and if you are aware of it, it can be most helpful.'

She took me to the shops where she selected a couple of beautiful dressing gowns, bargained with the people and advised me and through her I made my purchases. I don't know who she was; moreover, she wouldn't let me acknowledge her kindness. Mary and Dorothy were delighted with their gifts. I hadn't the heart to tell them that the choices were those of the young lady about to be married, although I did recount this experience to them.

I took a taxi back to the docks and boarded my ship, which was being loaded with phosphate and was about two-thirds filled.

Sailing back loaded was quite a different experience from sailing in on ballast. The ship was very low down in the water so that, to get to the galley, one had to walk on raised planks. Crossing the Bay of Biscay was quite rough. Captain Fiddler was in a hammock. His experience showed. I was in a bed, rolling from side to side with the discordant rhythm of the waves and a sore head as a result of it banging the sideboards.

Arriving in Cork, I had to report for discharge at a Merchant Seaman's office. Captain Fiddler had made out my papers which had to be endorsed officially. My ship's papers, as shown here, indicate Captain Fiddler's highly developed sense of humour which contributed greatly to the success of the Divisional Chairman's kind gesture which certainly did much to condition me to meet future problems. I travelled by normal ferry from Cork to Fishguard and arrived in the early hours of the morning on the quayside. Mary was waiting for me.

She was a lovely sight and wasn't I glad to see her. We had been apart for over six weeks in total.

In retrospect that trip proved of tremendous value. I was toned up, in better shape and freed from stress. Alfred Kellett had proved to me he was a kind and generous human being of great insight. I state my thanks to him now. During a visit to Washington, County Durham, on one occasion I met an old mining man who told me that Mr Kellett's father had the same human qualities and was noted for his many acts of kindness, and that all his sons had been subject to a strong fatherly discipline. All of them were taught to be honest, fair and industrious.

Dulias Valley Collieries

Following one Divisional Accountability Meeting, I was held back to discuss with the Chairman and G.S. Morgan, the Divisional Production Director, the temporary splitting off of the Dulias Valley (Blaenant, Cefn Coed, Onllwyn I & III) Collieries, which would involve the concentrated supervision of Roger Griffiths with help from the Divisional authorities. I wasn't happy about this. I wasn't satisfying myself either, regardless of the Divisional Board's requirements. Something was called for and it wasn't being supplied. I had to accept. Strangely, I was never relieved of my legal responsibilities for these collieries during the whole period of the exercise which lasted several months.

Next morning I called in my Staff Manager Evan Evans, the former Powell Dyffryn man. He knew the background of both Roger Griffiths and G.S. Morgan with the old company extremely well. I explained the situation to him.

'You're not worried, surely,' was his first observation. 'Roger Griffiths will cause you some embarrassment but none that you can't handle. He was constantly involved that way with everyone he has worked for. Now look at things this way. Both G.S. Morgan and Roger Griffiths have been associated with those pits in one form or another for years. Do you not think that if they had the "magic touch" it would have been manifested years ago? Things up there have not altered - they never will. Relax, let them burn their fingers,' he advised.

I could not fault his logic. He settled me down with that quiet, confident smile of his which was always available, no matter what the occasion.

My Area Production Manager had never been over popular in the Area. He had his sycophants but there were some Managers and indeed Group Managers who genuinely feared him. In this new situation of reporting direct to the Divisional Production Director, thus indirectly through to the Divisional Board. he developed and expressed ideas about his position way beyond the actualities of the situation. His imagination ran riot and his statements, from time to time, indicated he really hadn't thought out his vulnerability and security very well.

It does not require any great ability to report up the line the things and problems being encountered that had been going on for years. The real test comes when you have to state your intent on how you propose to deal successfully with those difficulties, and later to account for the progress (or lack of it) you are making. To compound this particular situation, the support being received from the Divisional Authorities was no better than, if as good as, that available within the Area, with the added disadvantage of something less than total commitment. He wasn't denied any support from the Area Staff. Such a course would have been stupid. I wanted those pits to blossom just as much as the Divisional Board did. Like most humans I do have a degree of pride.

In the early stages there was a slight improvement as some of the steps which had been initiated were having marginal effect: nothing dramatic, but it did embolden Roger to make even more outlandish statements. Throughout the whole exercise, he neither rang me or undertook any discussions about the situation. So far as I was concerned that was no problem, I had the sympathy of the staff who didn't understand, nor were informed of, the basic nature of the exercise. They fed back to me not only the gossip but the nature of events at all times.

After a few weeks, during which the performances of these particular collieries fluctuated with no apparent sign of breakthrough, it was obvious to me the experiment would not succeed. A few months later the collieries and the Area Production Manager were returned to the fold. On his return to Area Head-quarters with Evan Evans, the Staff Manager alongside me, Griffiths was called to my office. In strong terms, he was informed that his approach throughout had been despicable, his disloyalty defied description, and his next step out of line would be his last. I would get rid of him by any means warranted. Roger was no coward. He made no excuses, but he looked up and quietly said,

'You'll have no trouble from me.'

He was true to his word, for the next year or so before I was transferred back to Yorkshire he applied himself quite well. I feel that the experience he had received had a most salutary effect. Moreover he came out of the exercise discredited by lack of success. What his private thoughts were afterwards he never revealed to me but then, thoughts that remain unexpressed, or are not assigned action, are neither constructive or damaging.

Dorothy, in the meantime, had developed a nice disposition and, like Mary, was well accepted wherever we were called upon to appear. She fitted into all situations with ease and confidence. Isobel the policewoman was most helpful in many situations, and taught Dorothy how to drive safely so that her daily twelve mile journey between home and the Lloyds Bank at Pontardawe, her place of work at the time, gave us little worry, She became quite friendly with David Jennings's daughter Elizabeth of similar age and was quite happy throughout her life in West Wales.

My sojourn in the Anthracite lasted a little over five years. The first and last year's results were as tabulated below.

ANTHRACITE - COMPARATIVE CIRCUMSTANCES 1960 - 1965

Pits	Number	Face	Overall	Total (tons)	Per Ton	Total Loss
1960	18	40.5cwts	14.8cwts	1,396,113	£5.485	£3.367,000
1965	10	50.8cwts	17.7cwts	1,249,265	£6.705	£2,429,800

Percentage Mechanised Tonnage – 1960: nil, 1965: 51.60%

During the above period Abernant Colliery was in the process of tooling up. Men were being transferred and working faces established. The mining conditions in the Red Vein seam were quite good but we were having difficulties with roadway construction. The men were not adapting themselves and this gave rise to difficulties. If this situation was likely to persist, consideration had to be given to an alternative mining technique and, with this in mind, I set in motion steps to secure a American type continuous miner to drive the roads mechanically in the form of coal-headings so eliminating the problem of cutting and handling stone as part of the roadway drivages.

LORD HALIFAX VISITS ABERNANT COLLIERY

Being a new colliery under construction we had very many visitors. One who stood out way beyond all others was Lord Halifax, then Minister of Fuel and Power. As I said earlier, when you meet a fine personage you know it instantaneously and it registers. He was by demeanour a pious man who started his day by reading passages from the Scriptures. As a result of War service, he had lost both his legs and moved around on artificial steel ones. He had a fine physique and at 6ft. 4ins. he stood almost head and shoulders above most men.

We were conscious of his disability and tried to ease his passage wherever he went but no, he had to be treated like any normal visitor. On the surface he climbed vertical ladders to obtain better views of the constructional works taking place. Underground, he refused temporary travelling conveyances laid on for him and crawled through the coalface chatting to the men and officials with genuine interest. All of them were fully relaxed and easy in his presence. One was inspired and uplifted to be in this brave man's company. I did learn from Lord Robens on one occasion that Richard Wood (Lord Halifax), the Minister of Fuel and Power I refer to, was one of the most helpful Ministers he had worked with, in that he had a genuine interest in the Coal Mining Industry and that he was a fine person.

His kind nature shone through during all the dealings and discussions he had with each and every one of us, from the Divisional Chairman to the ordinary miners he met on the coalface. His actions and assentations signalled him to be a highly intelligent person of stature. PHOTOGRAPH 49 is a record of the occasion.

LADY MEGAN LLOYD GEORGE VISITS AMMANFORD

During my career in West Wales I became Chairman of the Ammanford Technical College Board of Governors for a period of some three years and as such I was able to get Lady Megan Lloyd George and Lord Robens respectively to address the College on two successive occasions.

My first meeting was with Lady Megan whom I found very lively, quick witted and widely experienced. It was through the influence of my Area Industrial Relations Officer Islwyn Thomas that Lady Megan was accepted as a prospective

Labour MP. He at the time was Chairman of the Carmarthen Labour Party, and made the casting vote in her favour. Since that incident he and his family had become close friends of the lady. She was subsequently elected to Parliament and held the Carmarthen seat for many years.

On the day following her address to the College, Islwyn and his good lady, and Mary and I took Lady Megan home to Criccieth, North Wales, where we spent several hours in discussion and taking in the widely varied contents of her home and garden. David Lloyd George, her father, had received gifts, artifacts and plants from many parts of the world. Amongst them was a large silver galleon from the naval authorities following the First World War, a bound book of caricatures from Punch and other publications featuring himself, and many paintings. Whilst we were looking at her possessions she carefully handed me a wrapped article and enquired,

'Charles, do you know what that is?'

Unwrapping the object it was quite obviously a fountain pen, which I stated in answer.

'Yes, Charles, it is, but not an ordinary fountain pen. That is the pen that signed the Versailles Treaty in 1919 which was imposed by the Allies on Germany following her defeat in the First World War.'

She had been there with Lloyd George, her father, on the occasion of the joint signing, at the time being a young girl of about eighteen years. Apparently both Soviet Russia and America were not party to it, but she didn't tell us the reason why they did not participate.

During the course of our discussions Lady Megan turned to me and said, 'Charles, for many years the Anthracite Coalfield has done badly and lost a great deal of money - why?'

I thought for several seconds and replied,

'Lady Megan, I have upwards of 8,000 lawyers, educationalists, mining engineers, geologists, preachers and politicians. If I had but 1,000 Nottinghamshire miners they would make the place sing!! and –'

Wittily she cut in, 'But Charles, they sing beautifully in the valleys already and they make the best politicians. We have to get our politicians from somewhere and should take the best, don't you think?'

However I was able to give her a short summary of the situation as I had interpreted it to be, which she appreciated, adding that mining wasn't the best kind of work for men to undertake. We visited the location of her father's grave and the cottages in which he was brought up. She was a very interesting person, quite charming and most hospitable.

Sir Humphrey Browne

I met Sir Humphrey Browne, the Board Deputy Chairman, for the first time, following a request from Mr Alfred Kellett to pick him up one night at about 6.30 p.m at the Swansea Airport and drive him to the Dragon Hotel, Swansea. I was there in good time. He arrived privately in the Board's plane precisely on time. We shook hands and walked to my car. Without guile or motive I said,

'Sir Humphrey, instead of staying at the Dragon Hotel could I offer you the hospitality of my home? Mary and I would be glad to have you.'

He reflected and for several seconds and then said,

'I appreciate your kind offer, Mr Round, but I can't. One has to be careful not to offend people in such matters. Were I to do so, it might create embarrassment to Mr Kellett your Chairman. I'm sorry, but why don't you come and have dinner with me at the hotel?'

I joined him for dinner in his reserved room. Again I was fully at ease. It was a most enjoyable occasion. He had determined my fondness for music and that was the only subject we discussed.

'Who is your favourite composer, Mr Round?' he enquired.

'Wolfgang Amadeus Mozart,' I answered with enthusiasm.

'I thought it might be, but you will find later on that your tastes will widen and embrace many others. Might I suggest you try the works of Sir Edward Elgar and Delius, you will find their works most enjoyable,' was his advice.

He then recounted how his interest in music had developed firstly in early baroque followed later by the romantics, Mozart, Haydn and others of that era. In later years his tastes graduated to Elgar and Delius. However I have found mine have not varied greatly over the years and I still get my greatest inspiration from Mozart, J.S. Bach, Haydn, Beethoven, Handel, and others of the period. I believe with Mozart his range of composition covers the whole gamut of human emotions. One particular composition of his has helped me through many an unpleasant and difficult situation. That is the beautiful soul-exposing clarinet quintet in A Major, K.581, which in all its majesty has everything with which to soothe the passing torments of a hectic and active life.

The four hours of dinner and discussion with Sir Humphrey was quite different from anything I had formerly had in any similar situation. I left him close on 11.30 p.m., greatly uplifted in spirit and well being, with the strong belief that here was another man of great natural quality. Next morning I had a call from Mr Alfred Kellett.

'Mr Round, Sir Humphrey Browne has asked me to call and thank you for your kind offer of hospitality and apologises for not having been able to accept it.'

'Many thanks Mr Kellett, Sir Humphrey invited me to have dinner with him which I did. I hope I haven't caused you any embarrassment. When I asked him to stay with us I did so out of pure welcome without realising the basic position involved; I had no other motive.'

'I realise that but we're all tied up with protocol, Charles, don't worry,' was his final assurance.

My old Planning Engineer Dan Jones retired some two years or so after I had taken charge of the Anthracite. Dan knew the area quite well and had developed quite a bright young planner, Roy Boneham, who hailed from Mansfield, Notts., who replaced him. Calling Roy into my office, I gave him an assignment to go through the Area and come up with projections for new potential small colliery constructions, with yields of about 1,500 to 2000 tons per day designed for low manpower levels.

Abernant's projection was between about 3,500 and 4,000 tons per day and to me was unlikely to get near that figure. Moreover, new, well laid out small units, carefully planned in relation to projected manpower, offered better scope for control. Where labour relations were difficult, strikes at a small colliery would not be so damaging as one at a colliery twice or three times its size. Another aspect of the situation was the strong family ties that existed in the separate valleys. Two small villages about a mile and a half apart were Cwm-twrch Uchaf and Cwm-twrch Ichaf (Lower and Upper Cwm-twrch) and, from talks with the old people, I found that both sections of the village were fiercely independent and, for instance, passionately resisted boys seeking girls in the adjacent village and vice versa. Small colliery units I felt would have a much better chance of developing strong internal close-knit ties than large ones with men recruited from a number of valleys in the sort of situation prevailing within the Anthracite Coalfield of West Wales.

Roy came up with four or five possibilities which narrowed down to three:

1. Shortening the underground communications at Blaenant Colliery by driving a sloping tunnel from the surface, which would have the effect of concentration and thus give us the advantages of a new colliery.

2. The Varteg Colliery men having been returned from Abernant would shortly be redundant. We needed a unit to exploit the adjacent reserves Varteg could not profitably extract. The location of the proposed new colliery was the old Treforgan village site, so the Treforgan Colliery was set up and the access drifts started with the minimum delay.

With the Divisional Chairman's interest and support, we got these two plans through and approved on an emergency basis, by-passing many of the planning procedures which would have been most delaying.

3. Bettws Colliery on the outskirts of Ammanford was about a mile and a half from the Area Headquarters, the proposed colliery taking its name from the adjacent village. The remaining time available to me within the Area precluded against me pushing this development beyond a Stage II Submission.

All three collieries were completed after I left the Area. During my last two or three years within the No.9 (Neath) Area I had great support from Alfred Kellett and this I gratefully acknowledge. I believe he had the most difficult job within the National Coal Board, one he discharged with great credit in the very difficult circumstances under which he was operating.

Before moving on to the next stage in my career I would like to pay sincere tribute to the people of West Wales. They were kind, hospitable, highly intelligent and stimulating to live with. They brought things out in me I never knew existed. Mary was similarly affected and found her life very happy and rewarding amongst them.

On returning home, many were the times I was asked,

'Did you regret having to spend part of your career in Wales?'

My only regret arose from my inability to infect the mining fraternity with some small measure of the enthusiasm I have always had for my profession and also not to have been able to turn heads to looking forward into the future, rather than perpetually into the past.

Area General Manager
No.3 Area, Yorkshire Division

The thought of returning back home to take over the No.3 (Rotherham) Area naturally gave me the greatest of pleasure and, having acquired a wide measure of confidence and experience during the fifteen years I had been away, I was tremendously assured that I could make further contributions to the Area's success as my predecessor had. The Area had a total annual production of some 5,500,000 saleable tons produced from ten collieries having 14,500 wage earners on the Area's books and a staff of between six to seven hundred. The revenue was of the order of £5.615 per ton or some £27,725,000 per annum. Coalface mechanisation, almost exclusively the Anderton shearer loader and powered supports, were in the process of being introduced and needed to be successfully extended. Throughout the Area, mining conditions were very good when compared with those I had encountered in parts of Nottinghamshire and West Wales.

It was in the early months of 1965. I was back home, the Area in which I was born, and was again working alongside my old mentor R.G. Baker, still Deputy Chairman of the Yorkshire Division. (The Chairman was William Sales, who had formerly spent some time in the East Midlands Division.) He found us a very nice house which the Board had bought from a surgeon (Mr Livingstone, who subsequently removed a varicose vein from my left leg): No.9 Broom Lane, Rotherham. It overlooked the Herringthorpe Valley playing fields and was close to the new Leisure Centre. It had plenty of room internally, and sufficiency of garden with a large greenhouse. My old gardener Harry was still with us and happy with the prospects the new facilities offered to him,

Dorothy had been transferred to the University branch of Lloyds Bank in Sheffield. John was still working for the Diversey Corporation in Ireland and Mary was at the centre of a seven mile radius circle which took in almost every member of her family, although her parents had by now been deceased for a few years. My enthusiasm, enhanced by the experience I had received in Wales, was fully restored. I was ready to go.

On the Monday morning, William Sales introduced me into the Area, following which the Staff Manager introduced me to the various members of my staff: Jack Hawley, Area Production Manager; Charles Dickens, his Operational Deputy; Ben Barraclough, Mechanical Engineer; Colin Rudge, Area Mechanisation Engineer.

Amongst those introduced was George Duffield, the first secretary I had had when taking over the Colliery Manager's post at Rotherham Main colliery some fifteen years earlier. Now Assistant Staff Manager, during the intervening years he had developed to be a person of stature and ability. My new secretary was to be Judy Smith who had for a number of years been with Sir John Barbirolli (resident conductor of the Hallé orchestra) and had travelled abroad both with Sir John and Lady Barbirolli and with the Hallé Orchestra extensively.

The Area was apparently running smoothly although there were one or two trouble spots one needed to be aware of at Denaby and Cadeby Collieries. As in former situations I spent many days visiting the collieries, undertaking inspections underground, and talking to the NACODs and NUM Branch Committees.

It was always my view that one requires to have a thorough background knowledge of every colliery for which one has responsibilities. Sooner or later, as final local arbitrator on disputes; or in addressing the leaders and men about re-organisation; or in giving effect to necessary change, you have to meet them. They know their local situations in great depth and the men can spot ambiguities and vacillations instantly, with immediate loss of face and respect to the speaker, usually expressed in the men's vernacular:

'That bugger's no bloody idea what he's talking about,'

not a very helpful situation amongst a group of men when trying to persuade them to accept change against the doubts or wishes firmly embedded in their minds.

Prior to nationalisation, the Manvers Main Colliery owners undertook an all embracing collieries integration and reconstruction scheme which, at the time, was quite substantial. I remember as a young student attending a mining engineers technical meeting at the Victoria Hotel, Sheffield, where G.C. Payne and E.J. Kimmins presented a paper on the concepts and working of the scheme. It embraced integrated underground haulage between a number of collieries and a central coal preparation plant which prepared coal qualities suitable for the housecoal, gas and steam generation markets.

It also had provision for the manufacture of steel producing coke, incorporating a new coke oven plant adjacent to both Barnborough and Manvers Main Collieries. The whole set up appeared to me to have been engineered to maximise proceeds from a number of collieries, in addition to enhancing the efficiency of a group of collieries as a whole.

The outlook, thought, design detail, and construction of the project at the time was a tribute to the design skill and combined efforts of some very capable people. However, there was little flexibility in the system to accommodate potential change and the coal preparation plant was very heavy on manpower, with upwards of three hundred men being required daily to run it. There were over twenty wagon loading tracks which involved extensive wagon handling arrangements and rail-track maintenance.

In 1957 I had flown out to the United States of America at my own expense and during the course of my visit to the Pittston Company's Moss Central coal preparation plant near Dante, Virginia, I inspected a plant which was handling some 18/20,000 tons per day (probably about twice the capacity of the Manvers Plant) with a staff of eighty men. By comparison, the coal preparation arrangements were much simpler. The bulk of the prepared coal went into steam generation. There were probably no more than four or five wagon loading out tracks to this plant. The value of this trip became apparent when I started to get grips with the problems prevailing, relative to accommodating the plant to the intensive change which had occurred over the years.

One of the greatest difficulties in dealing with circumstances of this nature is the impact of change, which can take numerous forms either in the market or in the methods of coal extraction or both. In the former case, the increased use of oil and natural gas, decline in steel production, expansion of electrical generation, growth of central heating, and the fall in house coal markets were the main factors. In the latter case, the development of coalface mechanisation had involved extensive technological change and had two effects: the increased production of fines (small coal) and the need for maximum underground flexible haulage and winding capacities.

Over the earlier years, change was comparatively slow but, with the successful development of mechanisation in all its forms, the rate of change rapidly increased. Unless the effects of change are timely assessed and accommodated, a point is reached when serious problems emerge and performance is seriously impaired or begins to decline. In this connection I have already mentioned in simple terms the reorganisation of the Elsecar Main Colliery coal winding situation. Here, with the integration of a number of collieries, the position was more intense.

Manvers Coal Preparation Plant

The Manvers reconstruction scheme was hit by these factors of change as had much of the Coal Industry, which I believe could not have been foreseen nor even imagined at the time of its conception. Something required to be done to offset the current adverse colliery situation. The coal-preparation plant needed to be re-fitted, reorganised, streamlined and got into better shape.

I discussed this with Ben Barraclough, the new Area Mechanical Engineer. We started out with a programme to this end and, over quite a period, we transferred some thirty men from underground to thoroughly clean up the plant from an accumulation of years of solidified coal and dirt spillage, consolidated sludge and neglect. After only a few weeks, I remember Ben informing me that more than 175 wagons of such material, estimated at over 2,000 tons, had been removed from the plant, but there was much more to be cleared before we could do much to modify the plant to meet the needs of the changed circumstances.

There was now no flexibility in the underground haulage system, since the logistics upon which former design obtained were no longer applicable. We were losing production at the pits linked into the integrated underground haulage arrangements; mine cars were often unavailable at the loading points where they were needed and it took wasteful time to switch them. Moreover, continuous power-loading at the coalface had changed the whole situation and had introduced a whole new set of parameters.

Action to resolve the situation was being followed in terms of both planning preparation and practical measure, with respect to modifying and updating the coal preparation plant and the provision of immediate schemes of relief to deal with the current difficulties. Unfortunately, however, the later situation was such that I wasn't able to pursue these matters completely to a successful conclusion, owing to matters beyond my control (to be discussed later). Until the whole setting had been thoroughly analysed and the problems sorted out, we embarked immediately on a system of surface and underground extended bunkerage as a temporary-cum-supportive measure.

A device I had seen in common use at most of the American pits was ideal in that it was relatively cheap and easily applied. It worked on the basis of coal being able to slide from a height down a helter-skelter chute and build up into a conically shaped stock heap. By a specially designed conveyor arrangement, this stock-pile could be reclaimed back into the conveyor system as required. Normal stocking capacity varied with the height of the cone which was naturally formed, usually about 750 tons, but this could be increased considerably with a bulldozer where space was available. The first 750 ton unit we linked into the Manvers colliery surface conveyor system.

PHOTOGRAPH 50 shows the coal stocking facilities installed at Manvers Main Colliery. The unit dealt with the output from the No.2 & No.3 coal winding shafts after passing through Bradford raw coal breakers. The Qualter Hall open-type spiral shown was designed to give a nominal capacity of 750 tons where the coal adopts a 35 degree angle of repose. To increase the capacity to over 2,000 tons, a small power shovel was used. Control was affected at three points, by the No.2 and No.3 banksmen and by the Bradford coal breaker attendant. The cost of the scheme at the time was £13,500. This equated to 750 tons at £18 per ton stocked. Operational costs for 750 tons stocked were estimated at 1.31d. per ton for reclamation.

PHOTOGRAPH 51 shows the Elsecar Main Colliery open spiral similar to that in Photograph 50, but the stockpile is over 4,000 tons. On one occasion, using a bulldozer, over 8,000 tons were spread and stocked.

A third unit of 4,000 tons was incorporated in the later Riddings (American) drift mine project. At Barnborough and other collieries underground, bunkerage systems were introduced as a further measure of immediate relief. The early relief these measures afforded to the overall situation was really something very special, particularly at Elsecar, Barnborough, and Manvers Main Collieries. Rotherham Main had been closed but Elsecar Main Colliery, my old pit, had remained stationary and in some ways had slipped back. My original replacement was still there after fifteen years and must obviously have been passed over a number of times for higher posts by the respective Area General Managers under whom he worked. On my second day after returning to the Area I made an underground visit to the Haigh Moor Seam at the colliery. I remember complaining about an exposed piece of machinery of serious potential risk. The overman, Harold Radley, who had given long and excellent service and who knew me well said,

'How long have you been away, Mr Round?'

'About sixteen years or so, Harold,' was my answer.

'You've not changed much,' he commented.

The colliery was definitely in need of new leadership and at the first opportunity I brought the Manager, G.C. Thorpe, into the Area and placed him well. He had given the Area faithful service, even though he hadn't been recognised or made any real personal progress over a long period of time.

We appointed Maurice Beedan as Colliery Manager, recruiting him from North Gawber Colliery in the North Barnsley Area. He was an excellent choice, broadly experienced, quite enthusiastic and with the sort of quiet but firm disposition I knew would be well received by both men and officials at the colliery. Within little more than a year he had restored the colliery's old spirit of adventurous achievement.

Jack Hawley and Charlie Dickens, I am sure, found me to be very different from my predecessor, being somewhat of a workaholic, with a fair degree of impatience in certain situations; the contrast must have been not a little disturbing at the start. However they soon got used to my approach. In Charlie Dickens I found many of my own characteristics. He too was greatly industrious and progressive.

My old colleague Colin Rudge had not changed a great deal during the sixteen year interval since last we had worked together. He was abreast of all that was taking place everywhere, still able to manoeuvre and get the things we wanted, be it information or equipment. His detailed knowledge of Area and colliery detail was of immense value to me on many occasions. Amongst the mining machine manufacturers he had contacts at the highest levels, a situation which helped us circumvent bureaucracy and red tape in difficult situations.

The tempo within the Area was building up. Things were happening. The contrast with the Anthracite which had demanded fantastic effort for very little apparent progress was indescribable.

Basically our problem pit was Silverwood Colliery, managed by Albert Tuke, working under his Agent Cliff Ratcliffe, formerly Manager of Earl Fitzwilliam's New Stubbin Colliery, during the time I was Manager of the sister pit Elsecar Main. Cliff had emerged from the coalface in much the same way as I had, but was a few years older than myself. It was to me a situation of excellent balance, but they had many labour difficulties. Unfortunately for me, after a few months, Cliff took early retirement and went to live on the south coast near Paignton.

The other difficult pits from the viewpoint of labour relations were Barnborough, Denaby and Cadeby Collieries. On one occasion, we had been subject to continuous pressure from the Cadeby men, while we were trying to effect changes and extend mechanisation. Finally I was asked to meet the Branch leaders along with my production staff. After a long meeting I requested facilities to address a weekend meeting of the men in the Conisborough Miners Welfare, and specified that it should be held between 10.00 a.m. and 12.00 noon with no drinks allowed into the hall. I had had experience of the effect of alcohol in such situations previously. Common sense and logic tend surreptitiously to leave the room on sensing the smell of beer. Although things were, from time to time, a little rough the men were quite reasonable. They listened and argued but came no little distance towards accepting what we wanted. Towards 11.45 a.m., they started to drift out of the room in small numbers, and although I could have usefully continued for at least for another thirty minutes, I brought the proceedings to a premature close by saying,

'If you lads are going for a drink, so am I,'

and finished, rolling up my plans and collecting my papers almost in the middle of an explanation.

Quite a humorous response rippled through the room. They accepted the situation. Had I been a drinking man the lads were in a hospitable mood - I wouldn't have survived sobriety. To have continued would have destroyed all the progress made for there always seems to be one soul amongst a group of men under the influence of a glass or two of beer who's capable of turning a pleasant situation into an adverse one.

Sir Humphrey Browne Visits Area

Coal face mechanisation had been well developed in the area, but maintenance of the equipment was a problem, despite the adoption of a fairly advanced inspection system. One of the major problems we had was with the introduction of the BJD magnamatic shearer, in which the mechanical haulage had been superseded by an hydraulic haulage, which had the advantage of smooth movement and the possibilities of higher production rates.

Our recording system indicated we were having wide variations of achievement between machines and with individual machine breakdowns. Machine performance varied from less than 1,000 yards to over 100,000 yards of productive travel before machine failure. Colin Rudge and Ben Barraclough were both pursuing the matter and it was receiving our attention.

Sir Humphrey Browne came down into the Area on one occasion for general discussions, as far as I could determine. Being subjected to quite an amount of questioning, I responded as I normally do, truthfully and without considering the effect truthful statements can have in chance situations. I suppose it's my make-up: I have never been able to change it, despite the fact that it has landed me in most embarrassing situations on more than one occasion. Political sagacity has always been a weakness, having been brought up in the belief that 'honesty of intent, expression and purpose' as a basic family premise was paramount. So in answer to the question, 'Are you having any trouble with your coalface equipment?' I answered,

'Yes, we are having difficulties with the new BJD magnamatic power loader,' and thereafter discussed the detail with him, indicating that the Area Mechanical and Mechanisation Engineers were dealing with the matter. It was a very pleasant meeting.

During the next afternoon I had a telephone call from Norman Siddall, who had previously been moved up to become Director-General of Production, W.V. Shepherd having gone on to the Board as Director of Production.

'Hey, what's all this you've been telling Sir Humphrey about the BJD magnamatic?'

I told him in detail all that had taken place during the meeting including the point he was raising.

'OK, I've had him in to see me.' he replied.

I don't know whether or not he had been embarrassed by Sir Humphrey Browne. I know he wouldn't have told me if that had been so, for he was more than capable of standing on his own two feet against anyone, irrespective of position or influence. However, a few days later he sent me a copy of a letter which he had received from Roy Landsdowne, the Managing Director of BJD (British Jeffrey Diamond), in response to matters he had raised with the company. Roy had formerly spent quite some time on the Board's Staff at Headquarters.

The import of the letter to me was couched in the 'old boy' network approach and suggested, in effect, that I didn't know what I was talking about - there were many of these machines working and 'no trouble was being experienced'. Arthur Yorke-Saville was the head of the British Jeffrey Diamond Company and also was on the American parent board of Jeffrey Diamond. I sent for Colin Rudge in seething anger.

'What do you think about this, Colin ?'

'The stupid clot!' was his immediate response. 'Will you leave this to me? I'll follow it up for you. I'll get Ben Barraclough and together we will go up to BJD's works in Wakefield.'

'Tell them, unless we get prompt action every bloody machine will be grounded,' was my final instruction.

Together with Ben Barraclough, Colin went to meet one of the top designers, Mr Pearson. I had met him on a number of occasions previously; he was a very capable and excellent type of person. They returned at about 5.50 p.m. I was alone in my office when they entered beaming.

'What's happened to you two? You look as though you've been caught up in the back end of a machine.' I observed.

'Front end, Boss. We were all right until we got to the Drambuie soufflé,' replied Colin in his disarming manner.

'Go away, the pair of you. I'll see you tomorrow morning; I have a meeting to attend at 7.00 p.m,' I retorted.

Next morning they came to see me, having spent all the previous day with the designers. Together they had come up with modifications in the form of internal unloading hydraulic pressure arrangements, which offset the risks of the sort of damage being experienced and offered good prospects for a complete solution to the problem. Colin told me he had spoken to Arthur Yorke-Saville, as instructed, and that I would be getting a telephone call requesting a meeting. Actually he rang me that morning and we fixed up a meeting in the afternoon. He was very concerned and wished to apologise for the whole event. 'Forget the apologies, Arthur, just get the machines working reliably. It will be good for both of us.'

A week or so later Roy Lansdowne, quite distressed, came to see me. During my Gedling and Bestwood days we had had many drinks together in the Wheatsheaf Hotel in Burton Joyce, where we had spent many happy hours discussing mechanisation and mining problems generally. 'Roy, what made you take the line you did? All you needed to have done on receiving Norman's letter was to lift up the phone and ask me what our problem was. We don't broadcast manufacturer's difficulties; we prefer to help them sort out their problems.'

'Charles, I behaved stupidly; it was an error of judgement,' he replied.

I then showed him the record of every machine's performance following its installation or re-installation, just to convince him our problem was an authentic one. 'Forget the whole situation, Roy, we are moving towards a permanent solution,' I advised him.

I have made the above points just to show how a simple truthful expression can have unexpected consequences. I couldn't nor wouldn't have done things differently had I known what the outcome was likely to be. That wouldn't have resolved the difficulties. Moreover it would only have been a matter of time before the problem surfaced over a much wider front and with even greater embarrassment, maybe to someone else.

Early in 1966 Irvin Spotte, President of the Pittston Company (the third largest deep mined coal operation in the United States of America), Dante, came over to see me. Irvin and I had been friends for some eight or nine years. I fixed up a visit for us both to the East Midland Division's South Derbyshire and Leicestershire Area's Donisthorpe Colliery, where we met Alan Walmsley, now the Area Production Manager working under William Unsworth, the Area's General Manager.

We were interested in a short coalface of some 60 yards in length, fully mechanised, which was achieving fabulous levels of coalface productivity (over 50 tons per manshift compared with a national average of about 5 or 6 tons). The mining layout was simple, nothing special about the equipment or organisation, but there it was, making a quite a useful contribution to the colliery's performance.

'What induced you to adopt this arrangement, Allan?' I asked him.

'It really adopted us. Our development had fallen behind. Output was suffering and something had to be done. We had the small parcel of easily accessible coal which limited the face length between 60 to 80 yards and in a matter of days had it opened up and the equipment was installed as you have seen. You know the rest.'

Situations of this kind often triggered off in my mind the extended possibilities they might have, e.g. profitable working under the Trent Valley at Gedling Colliery (Allan was familiar with this, for as a Group Manager for a period he had the colliery under his direction), working under densely populated areas to reduce potential mining subsidence damage.

Irvin and I were taken back to Allan's home for a meal before returning to Rotherham. Allan was a dedicated amateur radio enthusiast, had built his own transmitter/receiver station from surplus American war equipment and had just extended it to incorporate the transmission and recording of typewritten material which he demonstrated to us. He was naturally a person of great potential in whatever field absorbed his interests. On the way back, I told Irvin that within six months we would have a similar unit working within the Area. The system had great possibilities. 'Keep me informed,' was his observation, which I did.

Short Face 'in seam' Retreat Mining - Elsecar Main Colliery

Returning to Area, the detail of the visit was outlined to the production and mechanisation staff, following which we agreed to carry out trials. Colin Rudge found us the equipment without the need for standard dispensational procedures. The location for the trials presented no difficulty: Elsecar Main Colliery (Haigh Moor seam). Maurice Beedan was approached, and, with his natural enthusiasm and total commitment, there was no problem. It took but a few weeks for the Colliery Manager and Area staff to get the trial unit set up and into operation. It started up on 26 March 1966. We had roughly a month's problem with excessive roof and floor convergence, which we overcame before the unit settled down to a consistent performance. At its best, the face was producing about 3,400 tons per week at a face cost of around 2s. per ton and a face productivity of over 80 tons per manshift.

Since this unit established the feasibility of the first 10 tons per manshift mine, Riddings Colliery in the Barnsley Area, later to be described, the basic detail is worthy of short reference. Two basic concepts are involved in the achievement of high productive performance, 'in seam' and 'retreat' mining. The former requires all activities to be confined within the seam's thickness, whilst the latter requires the pre-drivage of coal roadways within the limits of the seam to the boundary of the panel size chosen. The coalface is formed at the panel boundary and worked back (retreated) to the point of origin of the coal-drivages. By doing this no dirt excavation is undertaken, a situation which avoids potential continuous mining restrictions or the requirements of cyclic performances which limits ultimate potential.

PLAN 2 shows Haigh Moor Seam Layout, Elsecar Main Colliery. The actual layout of this section of the mine in which the full gamut of coalface mechanisation in progressive successful applications was undertaken in the form of Buttocks, Meco Moores, Retreat Longwall and Shortface systems. The location of the last is clearly indicated on the plan.

PHOTOGRAPH 52 shows how the transfer conveyor loading unit which overlaps the gate conveyor has been applied within the seam's total height. Water tanks and control switchgear are all located along the LH side of the roadway.

PHOTOGRAPH 53 is a view along the face with the Anderton shearer loader cutting a web of about 27-30 inches and moving towards the tailgate. You may have noted that the heavy steel plough assembly shown in Photograph No 34 has been replaced; so has the cable carrier shown in Photograph No.36. The former has been replaced by the 'cowl' shown which is hinged to swing through 180 degrees for the return run-back through the face, the latter by special cable-carrying arrangements embodied within the conveyor waste side spill plates.

PHOTOGRAPH 54 Here the Anderton shearer loader is moving towards the viewer and is making its cutting run. The cutting drum, scrolled with its pick-boxes and replaceable picks, is to be seen in the background. The machine's 'chain haulage' arrangement is seen entering the front end of the machine, which carries a special sprocket indented with profiles to accommodate the chain links. The panzer is an updated model with a three chain flight assembly.

PHOTOGRAPH 55: this picture shows the tailgate which contains track and a small endless haulage for taking in equipment and materials when required and recovering the steel supports as they are withdrawn. These four illustrations indicate in a simple way what 'in seam' and 'retreat' mining is all about. The whole exercise, with its derived experience and findings, was recorded and later published in the mining publication *Colliery Engineering*, December 1966 issue.

Maurice Beedan was, like myself, a local lad and hailed from the small village of Jump about two miles from the pit. He managed the colliery with great skill, enterprise and industry, and the support he gave us was immeasurable. During his period of managership, a record weekly coal output in excess of 28,400 tons at an overall productivity of upwards of 5 tons per manshift was achieved – way beyond the Area and National averages at the time. In that simple but highly successful exercise, Maurice Beedan and the associated Area Staff, Charlie Dickens, Colin Rudge and others, prepared me a blueprint for what was later to be an even greater achievement.

During my early days in the area, I became involved at Silverwood Colliery following the retirement of Cliff Ratcliffe. The Barnsley seam was being worked but new mechanised layouts were being contemplated and examined and we were fairly well advanced with regard to working projections in the Haigh Moor seam. The location of the underground access drifts having been decided, the drivages were later initiated. Albert Tuke, the Manager, was doing a fine job but the pressure he was being subjected to was somewhat akin to that I had felt within the Anthracite and at one period it started to show. Colin Rudge, who had closer contact with him than I, had called in and said,

'Young Albert's getting overstressed. I've just come from Silverwood; he's been having it real rough.'

Recalling the kind treatment I had received in a similar situation I phoned him, discussed his problems and suggested he took a couple or three weeks off. I don't think at that point he had taken any holiday. However he agreed and he and Mrs Tuke together had a break.

SILVERWOOD COLLIERY UNDERGROUND LOCOMOTIVE ACCIDENT

Like myself, Albert could absorb and retain great detail and very little slipped his attention, a situation he profitably exploited when he opened up the early Haigh Moor seam mechanised units and organised some exceptionally high and consistent output and productivity performances. Albert made extremely good progress, ultimately becoming Area Director of the Castleford Area which took in the Selby Coalfield.

During the period we were together in the Area, he experienced a horrific underground manriding accident which occurred whilst men were being transported to their working places by locomotive and manriding carriages. Almost at the end of the transport run, a collision occurred with a stationary train. Men were trapped beneath and within the locomotive, in and between the distorted carriages, and others were thrown out along the roadway sides. A large number of men received serious injuries and, in the case of six men, these were fatal.

The report of the accident came through to me at about 8.30 a.m. whilst in the local Herringthorpe Valley doctor's surgery awaiting clearance for hospital treatment in the form of a minor operation. Being excused, in less than two hours I was at the scene of the accident and saw at first hand the carnage which had taken place. The colliery had a number of first aid teams which were amongst the best in the country, and they were fully in action when I came upon them, giving highly efficient treatment to the seriously injured. They were superb.

Further alleviating support continued to arrive. The injured were conveyed to the pit top, and those who had died were recovered from their trapped environments and taken to the surface. I recall getting back to the surface at about 4.15 p.m. where the Divisional Chairman William Sales was in attendance and waiting for me and to whom I reported the situation as I found it. He was being pressed by BBC reporters for a statement but directed me to undertake the task, which I did, but a statement as to the cause of the accident was then impossible to give.

The Mines Inspectorate and Albert Tuke were underground checking every facet of the Transport Regulations: examining maintenance records and signalling systems, etc. From my own observations and the subsequent reports I received, Albert Tuke handled that terrible ordeal with great personal control, courage, sensitivity and understanding. It was later determined that, without doubt, the calamity had resulted from a lapse occasioned by the locomotive driver.

REORGANISATION OF BOARD'S MANAGEMENT STRUCTURE

Towards the end of 1965 the National Coal Board were contemplating a Reorganisation of the Mining Industry on the grounds of specific changes which had taken place over the previous twenty years. For Governmental planning purposes during 1970 the NEDC (National Economic Development Council) produced a figure of 170-180 million tons for the 1970 output. The mining engineers on the Board's Headquarters staff postulated that such tonnage could be produced from 300 to 350 of the best pits and, from experience, an Area could effectively control twenty pits, giving an annual yield of 10 million tons. This bare-bones analysis indicated that seventeen Areas would meet the situation. The question of whether there should be an intermediate structure or not was academic in that the Board's view was 'No!'.

The 'Line and Staff' system of organisation was retained, except that Group Managers were delegated 'line authority'. The new organisational arrangements took the form at National and Area Levels as shown in the Revised Organisational Arrangements Chart which follows. However, during the placing interviews in January 1966 I personally found the manner in which I was interviewed was impersonal, clumsy and irksome, with the whole exercise lacking the degree of analysis it could well have produced.

The format followed discouraged any genuine, fair, intelligent explanation or discussion. One was expected to accept being told how and where one was to be placed, say thank you, and depart. The interviewing panel was Cyril Roberts, NCB Board Staff Director Member, Panel Chairman; C.G. Simpson, Director-General of Staff; and W.V Shepherd, Director General for Production.

'Mr Round, You know why you are here. In placing people for the post of Area Director we have to be guided by the experience they have,' was the Chairman's opening gambit.

In that simple observation he was seeking a convincing reason why I was to be displaced and was telling me I wasn't to continue in post as an Area Director. What was worse, it transferred the responsibility for non appointment to me in that I apparently failed to have the necessary qualifications.

'In that case, Mr Roberts, I believe if you examine the degree of experience I have had, it will match that of most of the people you interview, and indeed be substantially greater than that of many of your projected interviewees,' I replied.

He really was annoyed and it came through. W.V. Shepherd's looks were indeed cutting.

'There is the question of age to be considered,' was his next attempt.

Again, I felt that this man intended to prove the fault of omission was mine and mine alone.

'But Mr Roberts, I am now fifty-two years of age. You'll will no doubt be appointing people to the post both junior and senior to myself.'

There was no doubt in my mind about his annoyance. It came through to me quite strongly.

'The Board wish you to be given the post of Deputy Director (Mining) in the Barnsley Area,' he stated in a tone of finality.

'Charlie, you have the right specifications for doing that job because of the special situation which prevails there. You know the circumstances,' was the Director General of Production's immediate interjection. I honestly believe he was trying to diffuse the situation.

'The same specifications are equally capable of doing the top job too, Mr Shepherd,' I replied. 'However, my circumstances are such that I have no choice but to accept your offer. May I ask - if there are any vacancies for an Area Director at later stage will I be considered?'

Not a murmur: I knew then where I really stood and, having nothing further to lose, I observed, 'No doubt this interview will be adversely endorsed on my Personal Record File?'

'Oh no,' was C.G. Simpson's answer, which I'm afraid gave me little assurance, later, confirmed by the fact that there were three changes of the incumbent Area Director's post within a relatively short time - and the fact I was never again approached. I got up to leave. W.C. Shepherd came across, shook hands, and said,

'Don't go back and upset the Area.'

'It's too late. It's already been done,' was my parting remark.

The Chairman of the Panel refrained from any such courtesy.

During the interview my observations were addressed respectfully, quietly but firmly. I wasn't in any way uncivil to any members of the interviewing panel; however I think the failure of the Panel to answer my last question destroyed for me all credibility of their former statements and assurances as to fairness, including the undernoted explanation:

> On this chart are a small number of big jobs - Director, the two Deputies, Chief Mining Engineer and so on. The Board considered it necessary to get on quickly with earmarking people for these and it fell to me to get work moving on this. I sat down in November (1965) with a blank piece of paper and wrote the names of the Areas down the left-hand side and the names of the probable jobs across the top. Then I proceeded to fill in the blanks from my own knowledge of people available, and having been with the Board for 20 years I know quite a lot of people. Having filled in most of the blanks I then consulted with the Director General of Staff, Mr C.G. Simpson, who made various suggestions, and then with the Director General of Production, Mr W.V. Shepherd who added his quota. Shortly before the Airport Conference I placed my Chart before my Board Colleagues - none of whom I had consulted previously - and by and large my Chart survived.

> I have gone into a good bit of detail on this because, very naturally, some people who are disappointed with their prospects have seen all sorts of sinister influences underlying the selections - There have been none.

> We may have got some of them wrong - time will show - but this is one exercise in which I can say that kissing has not gone by favour.

Had the above explanation been given to me at time, I could have accepted my circumstances as being the 'luck of the draw' in view of the apparent 'football pool' nature of the exercise undertaken, rather than of 'lacking the required skills and abilities' which the failure of the Panel to address my final question to them would seem to have confirmed. This was hardly in keeping with the fact that I had been transferred back from West Wales to sort out the problem of the South Yorkshire Area, and had made an excellent start in the process.

It wasn't the failing to be appointed as Area Director of the enlarged South Yorkshire Area that caused me such strong disappointment, it was the destruction of my projected creative ideas and highly enthusiastic intentions which I wanted to see in operation, that really cut into my whole being.

However, as far as I was concerned, the whole situation was futile. I was to be replaced by Cliff Machen, the then Divisional Production Director whom I have always liked and respected, and whom I could have accepted cheerfully without question, as Cliff was a Senior Divisional displaced person of ability and experience, had they given me reasonable explanations.

The adversity of the situation to me was in the manner in which we were placed or misplaced and the ham-fisted way it was done, together with= the apparent lack of thought about the individual circumstance pertaining to each Area and the impact of 'changing horses mid stream'. You will recall my earlier analysis of the Manvers Coal Preparation Plant and the need to extend the underground haulage capacity throughout most of the collieries within the Area, amongst other things, to real beneficial effect. These things were in train in my own case, and there were forty-two such situations in all, amongst which no doubt some of the Area General Managers were similarly placed as myself, but lost the opportunity of employing their initiative and enterprise.

It wasn't easy for the transferred person to open up and use his skills, ability and experience in his new location, particularly in such situations where his immediate superior was devoid of ideas, lacking an imaginative, purposeful approach and the ability to direct the talents available to him, to the best advantage of the Area under his direction.

In a report the Director of Staff stated:

'In a handful of cases there was disappointment. In only two cases was there a complete failure to reach common ground and one of these has now to be sorted out.'

These two were the former Area General Managers of Alloa and North Barnsley. I knew them both and with regard to the latter I still had a part to play. The Board's Staff Director, from my experience, was way out with his assessment. Many good people were hurt and gravely unhappy, in my comparatively narrow orbit of activity, and in the circumstances unable to express their true feelings, a situation which must have extended to other Areas also. The Area General Manager of the Alloa Area apparently was the one sorted out at a later date, being reappointed as Area Director of a Scottish Area, whereas at the interview my enquiry about further consideration in that respect fell on stony ground.

Following our interviews William Sales, the Yorkshire Divisional Chairman, kindly transported a colleague and myself by car from the London Airport Conference Centre back to Doncaster. The atmosphere within the vehicle could more suitably have been transferred to a funeral hearse. As in past situations I was smarting - but so was my Divisional Chairman. Apparently he had been completely ignored, had but little information and had never been consulted relative to the exercise. The best thing about that journey back was calling at a nice hotel for a really good meal (the Chairman paid); the enjoyment of good food is to me both spiritually and physically uplifting for a short time.

When the final list of placing for the four top jobs in the enlarged Areas was issued, many formerly highly successful and responsible people found great difficulty in reconciling a number of the placings.

I have expressed the above as my personal opinions. I had them at the time but loyalty to the industry and those remaining whom I had belief in and respect called for me to accept each situation as it arose and do my best possible. However the exercise brought into perspective the fact that, despite the fortunate background, privileges, position and class some people may possess or feel they possess, they can fall far short of what we may imagine them to be when their actions are analysed. A little sincere consideration, recognition and expression of our former efforts would have helped many of us to overcome the disappointments of the situation. Morale fell in many Areas. In my own case I found that some people tended to be embarrassed in my presence; others whom I had met quite often, appeared to have evaporated.

'Boy - I really felt vulnerable, apprehensive and terribly alone,' but nevertheless as determined as ever in the pursuit of progress.

SIR HUMPHREY BROWNE'S EMERGENCY VISIT

The new Barnsley Area was to come into effect on 1 January 1967, so I was entitled to several months in which to continue my efforts in the old No.3 South Yorkshire Area as Area General Manager.

In April 1966, Sir Humphrey Browne, Deputy Chairman of the Board, rang me up.

'I would like to see you tomorrow morning about 11.00 a.m. and I will come up to Area.'

Naturally my mind slipped into top gear. So soon after such unpleasant experiences. Why?

Sir Humphrey duly arrived on time next morning and we settled down to coffee and biscuits. However, unusually for him in such circumstances, he addressed me by my Christian name.

'Charles, I've come to see if you could kindly grant the Board and myself a favour.'

'Of course I will, Sir Humphrey. What is it you want of me?' I responded.

With a broad smile, he said, 'Charles, I knew that was going to be your answer.' He continued 'You are no doubt aware that Mr Cecil Peake has left the National Coal Board. We would like you to take over the interim responsibilities for the North Barnsley Area until the North and South Barnsley Areas are merged.'

'When do you want me to start, Sir Humphrey? I'm ready any time.'

'Could you wind your affairs up here and start at the beginning of next month?' he requested, adding, 'I've told the Board on a number of occasions, Charles Round is a person of high integrity and is amongst the most able and loyal members of its Staff. He may be a little direct but that does not detract from his probity, industry and loyalty.'

It was quite rewarding for me to feel at least someone at Board level understood and accepted my makeup. I thanked him for his generous expression of personal support. That was the last time I met Sir Humphrey Browne for whom I had the greatest respect, although it wasn't his last communication to me. At no time was the interview referred to or discussed; there was no point, the affair was behind me and I had now something positive to tackle for a short time at least. As circumstances unfolded, it proved to be a most helpful event. I reflected many times on how I would have reacted had the request come through either the Board's Director for Staff or the Director General of Production at the time, but the answer always came out the same. I would have done exactly as I did, maybe a little less pleasantly, as they themselves had undermined for me what faith and respect I had for them.

Area General Manager
no.6 (North Barnsley Area)

At the beginning of the following month I walked into the NCB No.6 Area Offices. No one closely knew me but I was expected. The former Area General Manager's secretary conducted me to my office. She had prepared a list of the existing staff and, with her help, I spent all the day meeting them and talking to them. The morale of the Area was truly below rock-bottom. People were nervous and apprehensive, and the fact that their former Area General Manager had gone so suddenly without fore-notice or explanation had compounded the general situation. My last remarks to the Board's Director General of Production when leaving the interview were proving to be fully correct.

As with previous similar situations, I set about to get to know the Area and build up relationships, not knowing at the time that I was going to need both later. I spent a considerable amount of time underground, moving around with John Williams (Acting Area Production Manager) and the Group Managers in post. They really 'went to town' to fill me in with whatever I wanted. I found there were a good number of solid West Yorkshire personnel in the team together with some excellent planners.

THIN SEAM HAND WORKED MINING IN WEST YORKSHIRE

Basically the pits were in good shape. The small pits on the west side of the Area were doing fantastic things. Men were producing coal profitably, hand loading coal onto face conveyors in seam working conditions less than two feet in thickness. One morning, on picking up the morning production sheets, I noticed that at Emley Moor Colliery the faces were being manned with ten or twelve men in situations of face lengths which in other situations absorbed three to four times that number. I rang up the Group Manager and asked him if they were having intensive absenteeism.

'Why do you ask?'

'Because of the low manning levels you have on the faces,' was my answer.

'But that's all we need!' he shouted almost indignantly.

'Right, I'm changing into my pit clothes. We will visit the faces together,'

I needed to see this first hand, for the men were loading out up to 15 tons per man in a seam section of only two feet. Rennie Owen, the Group Manager, met me in the Emley Moor Pit yard adjacent to which was the Emley Moor TV transmitter mast towers some 1500 or 1800 feet into the sky.

It was a neat little colliery. The underground roadways were not excessively large, tailored over the years to meet the mining conditions and, being fairly shallow and with comparatively few subsidence stresses, were both stable and readily travelled. On reaching the coal face, I looked down on the gap formed by the extraction of the coal between the seam roof and the floor. It looked so small as to defy entry, yet to the right of me in that gap some 150 yards in length, ten men were shovelling coal onto the bottom belt which ran along the floor of the seam. Crouching down on my knees defied entry; only by twisting my lamp battery to the front of my stomach and taking an inverted lying posture was I able to get into the gap. By pushing and pulling with my feet and hands I was able to propel my body along the coal face, yet the men seemed as dexterous as cats in that environment. Moreover, as I passed each man he had already loaded out over 10 tons. It wasn't difficult to get trapped between the roof and floor, particularly if one tried to rise on one's knees. In that particular part of West Yorkshire there were a number of small pits such as Park Mill, Fanny, Shuttle Eye and Flockton all doing similar things. Those West Yorkshire men, not unlike the thin-seam men of Durham, were and are miners apart.

Incidentally during the following winter the TV mast became heavily iced and overloaded. It collapsed with a large bulk falling harmlessly into the colliery yard.

Shuttle Eye Colliery reserves were virtually depleted shortly after I got into the Area, but there was a small parcel of coal reserves in one section of the colliery take with a thickness of 1' 5" which the men pleaded with us to work, indicating to us other profitably worked out areas where the seam at 1' 7" in thickness had been extracted.

Woolley Colliery, dating back to 1869, was the major colliery in the Area. In 1966 its Colliery General Manager was one Calvin Round (no relation) more than ably supported by a youngster who showed at the time great potential, Trevor Massey, to whom I later gave a special assignment. Following this he made excellent progress within the Industry. Calvin Round was upgraded to Grimethorpe Colliery, a much more difficult pit. He wanted to take Trevor Massey with him but, for reasons of fairness, and indeed Woolley Colliery's interests, that wasn't possible. Calvin after but a few weeks in the post left us because of back troubles and was transferred by the Board's Staff Department to the Opencast Executive.

George Duncan, an importee from the Alloa Area, was appointed to the post of Colliery General Manager at Woolley Colliery. He had been well trained, had wide experience and a good personality. He fitted in extremely well and during the four years or more he was with us prior to his return to Scotland, did an excellent job. On his return to Scotland he was asked by his colleagues what was his opinion of the Yorkshireman?

'Well,' he replied, 'You know the Aberdonian? The Yorkshireman is the Aberdonian with the last vestiges of compassion drained from his system.'

The Branch Trade Union Leaders at the time were A. Phillips, President; E. Benn, Treasurer; G. Sunderland, Secretary; and A. Scargill, Delegate. The last was of little trouble to us in those days, the three top men being very strong and fair characters.

During my four years as Deputy Director (Mining) the colliery output was raised from 841,600 to 1,019,500 saleable tons with overall productivities of 42.8 and 52.9 cwts per manshift respectively. In this connection I would pay great tribute to another youngster, Gordon Sykes, a lad who played a tremendous part in that achievement. Quiet, unassuming and industrious, with a personal demeanour which would have fitted into any mining environment, he was amongst the best prospects I met during my career. I will refer to Gordon later, in connection with a National Coal Board Assignment we did together.

After a few weeks I really had a very broad and detailed knowledge of the North Barnsley Area including some of the closed collieries and, equally importantly, had built up some quite strong ties which were subsequently strengthened by the manner in which the two Areas were merged.

Indirectly, my sojourn of these few short weeks in the North Barnsley Area had extra rewards in that it afforded some early experience of the manner of person whom I was shortly to work for and with, the Area General Manager of the South Barnsley Area. The man was of very short stature and fully conscious of his size with no lack of the arrogance one often finds in short men.

We had, from the first day I walked into the North Barnsley Area, an excellent opportunity to further the Board's interests by quietly and gradually integrating the two areas without undue speed and the minimum of stress to all concerned. What I found really surprising was that, during the whole time I was there, the incumbent Area Director of the two merged Areas never invited me to join him for discussions but took a devious look at the North Barnsley situation by surreptitiously sending his Area Production Manager (with whom he had a special relationship) there without reference to myself, the then responsible authority.

I ran into this situation accidentally as follows. Shortly after moving in we had quite a serious situation at Hemsworth Colliery, involving spontaneous combustion. Because of the high risks of active fire likely to take place, on the crucial morning I went down the pit at 6.00 a.m. on the day in question. Whilst underground the Safety Engineer with me was called to the phone at about nine o'clock. The Area Production Manager from the South Barnsley Area was making enquiries from the Hemsworth Colliery Manager's Office as to the situation on behalf of his Area General Manager. Taking the phone over, I told him we were more interested in contributors, not outside reporters. If he wished to be appraised there was no objection to him dressing into his pit clothes, descending the mine and obtaining the information he was seeking first hand. When I got out to the surface he was still in the Manager's office. Addressing him in the strongest terms I could command at the time, I told him he could include in his report to his Area General Manager my reactions, which I have no doubt he did.

Had the Area Director shown the slightest interest in the North Barnsley Area and its affairs, prior to him taking it over, he could have been fully informed by either personal discussion or report. However, there was no immediate response to whatever surreptitious report his Agent submitted. What was more important was that we had controlled what could well have developed into a very dangerous and serious situation.

REVIEW OF NORTH AND SOUTH BARNSLEY AREAS PRIOR TO MERGER

Before discussing the merger of the North and South Barnsley Areas it is worthwhile to consider their relative backgrounds.

Subsequent to Vesting Date the North Barnsley Area was under the direction of Mr Harold Atkinson, a person of wide experience, business acumen and mining ability. I had formerly worked under him as a young workman employed underground at Wharncliffe Silkstone Colliery. My brother John and he, as musical instrumentalists, supported local bands, small orchestras, and ensembles at church and chapel ceremonial occasions. His area was consistently profitable and retained a highly volatile spirit. The whole of his staff held him with deep regard and respect. I recall that, on the occasion I was leaving Elsecar Main Colliery to join the East Midlands Division's Gedling Colliery, he sought me out during the annual dinner of the Midlands Institute of Mining Engineers to express a few words of encouragement. He gave me some tips and advice from his own experience, finally saying,

'Go down there, lad, and show them how you can get coal out.'

He retired prematurely, owing to a policy difference of opinion with the National Coal Board over the rate at which coalface mechanisation should be extended throughout his Area, particularly in a situation of thin seam mining for which at the time there was no suitable or really successful power loading technique.

The National Coal Board appointed C.V. Peake as his successor, with a mandate to carry out a policy of extending coalface mechanisation. Whatever the problems Cecil had to face and under what circumstances he faced them, I do not know, but for some reason or other the Area's performance declined. I have no doubt in my mind, however, that the later achievements at Woolley Colliery referred to earlier were favourably influenced by the efforts he applied in those early stages. It was at this colliery, under his direction, that a fully automated mining system was established: quite a 'moon-landing' feat in those early years.

The South Barnsley Area, at the start of nationalisation, was under the control of one Johnny Longden, formerly associated with Newton Chambers Collieries. Other colliery groups taken in by the Area included Old Silkstone Collieries and Wombwell Main Collieries. Johnny was probably the most unique mining engineer of his time, his claim to fame being that he was the only sane person in the National Coal Board and had a certificate to prove it. He could be funny and outrageous at times, but was a good practical pitman, physically strong and bullish, which together with his gruff voice could frighten those under him. He was subsequently replaced by his Area Production Manager who had qualified through Barnsley Technical College, in a similar manner to myself.

The Area had the philosophy that effective management could only operate on the basis of ruling by fear, which in the case of the incumbent Area General Manager was compounded by the arrogance of the small man, who was certainly no Johnny Longden. Unfortunately he attracted the prefix 'Little' which must have been irksome to him at times. When speaking to people in his office, he would stand on the elevated section of his fireplace, either to create an effect or because he was conscious of his situation. Both Johnny Longden and his replacement had strong connections with the then Board's Director General of Production, much in the same way as I had with Norman Siddall, then Area General

Manager of the East Midlands Bolsover Area. Norman in great contrast had the ability to stimulate people to want to apply their talents and unstinted efforts with great loyalty for and to him.

The new Area General Manager (South Barnsley) Area, had graduated by a route much shorter than my own, although I may have been appointed to the post earlier than he was. However, I was more fortunate than he, being moulded and conditioned by some very fine people and with the added advantage of having been exposed to a fabulous range of mining experience. The work was much harder and one had to accept a wider latitude of disturbance but it was exhilarating throughout.

His Area Production Manager, immediately prior to the merger of the two Areas, had been trained within the Newton Chambers Collieries organisation. Subsequent to him qualifying for his Colliery Manager's Certificate he spent some time in the Mines Inspectorate within the coalfields of Durham. During this period the experience he obtained was undoubtedly of great benefit to him, particularly in the later part of his career at Board Headquarters. He returned to the Barnsley Area to take up a post of Colliery Manager serving for a time at Thorpe Colliery. Further progress raised him to the post of Area Production Manager working to the Area General Manager, following which they established a very strong relationship, such that he was referred to by the Managers from time to time as 'AGM's Sidekick'. Apparently, when dealing with Group Managers and Managers in critical situations he would prefix his comments with, 'The Area General Manager is not going to like this,' or ' If the Area General Manager gets to know about this you know what will happen . . .'

The people to whom such remarks were addressed all knew the Area General Manager would have a full account of the particular situation before the end of the day. To me, such an approach lacks strength of purpose, if there is a failure to administer discipline where needed by referring it to be dealt with by a higher authority. However, it does project attention in such matters to the power of the Almighty. Thus the 'management by fear' philosophy, aided and abetted by the Area Production Manager, flourished. After a short period of time within the merged Areas, I found there was strong evidence of this.

Management by Fear: Concepts and Practice

GENERAL EXPERIENCE Throughout the whole of their lives people are naturally subject to fears which are roused by the basic difficulties of living in a modern society, involving unemployment and lack of understanding. It is, however, the directly induced fears built within the organisational policy of groups of people, such as early slavery, military discipline during conflict, and commercial and manufacturing practices, which tend to give rise to intolerable human situations.

The paternal nature of the Japanese enterprises and organisations appears to have the effect of sheltering the employees from the worst effects of exploitation by the rule of fear, substituting instead many forms of encouragement. The basic worldwide economic miracle achieved by the Japanese following the Second World War would testify to the system having been successful. No doubt such paternalistic management played a major role in the thinking and planning during an era of intensive change in which old, heavy, labour-intense industries were replaced by more modern automated ones.

During recent years, however, there are signs that this paternalism is tending to break down and move towards the more ruthless American 'hire and fire' non-protected approach, engrained within enterprises employing large numbers of people.

Newer fear-inducing concepts, which are currently tending to surface, embrace the removal annually of the worst performer in the management chain and, even more brutal, the dismissal of the lowest 10% of the performers within the management chain each year.

The shape of things to come, as being fashioned by IBM's New Jersey district office, is even more frightening, in that it appears to be based upon a 'spend your time out on the road' principle in which the executive is assigned to non-permanent workstations, each with its telephone, computer and plastic in-tray in the form of a 'cubicle' within a vast steel fortress, once used as an iron box warehouse. One day the executive's numbered cubicle could be 127, the next 51, an open discouragement from putting down roots or even thoughts of wanting to linger. The logic is both simple and harsh. With 350 desks for 800 employees IBM wants to make sure its executives are out on the road meeting clients and pushing for new business. Enthusiastic people need no such stimulus; they push themselves far more heavily than the unhappy person trying to beat the system.

For many years I have felt our hierarchies with their titles, frills and sumptuous working environments to be the awards of privilege rather than for accomplishment and solid achievement. They make but a negative contribution to future performances. Moreover, constructive creative ideas which would disturb (as most radical thought does) the equilibrium of the privileged are ruthlessly discouraged or crushed. Privilege, I feel, creates isolationism within organisations, impairing the potential performance of the whole.

I feel strongly that personal attempts to promote a 'management by fear' concept are fraught with grave injustices and involve degrees of ruthlessness which are unwarranted in the normal situation of man to managed relationships. Invariably one meets up with people who cannot be so motivated, which demands a far more intelligent approach to inspiration and motivation than the use of crude methods such as dismissal or psychological harassment. Compared with all the NCB Areas I had previously worked in, the atmosphere within the South Barnsley Area reeked of distrust and controlled expression. One Group Manager was a heavy chain smoker with breathing difficulties. He had been a contemporary of mine in our early days at Wombwell Technical College, and had an excellent record and a fine physique at the time. A colliery manager emanating from Hoyland Common, I'd known him for years, but his confidence had been greatly impaired. He was but a shadow of what I remembered. On one occasion, he told me of how he had been severely censured for the crime of allowing his wife to be conducted into the Barnsley Hospital by the colliery ambulance on a normal pit run, and that he had been made to make a payment of passenger cost.

Another colliery manager I understood was demoted apparently because he had called in for a short lunch time drink in his local. How one reconciles such a situation with the extended lunch sessions and afternoons on the golf course of the high executive is to me somewhat perplexing.

There was a former Deputy Area Production Manager (Operations) (reputed to have been imported into the Area at the behest of the Yorkshire Divisional Chairman) from the Durham Coalfield whom I knew well and later got to know even better. He had a former long and successful experience in the Derbyshire and South Leicestershire Area, and was regarded with great esteem by the mining engineers with and to whom he had worked in both situations. The Area General Manager and his Area Production Manager felt he had blocked the progress of the one they had wished to be appointed. He was subjected to intense pressure, following which he had a serious heart attack. His whole family felt stress to have been the basic cause.

Maybe such a situation might be dismissed as in the famous quotation of former USA President Harry S. Truman: 'If you can't stand the heat, keep out of the kitchen,' to which one might reply: 'If you cannot control the heat of the kitchen, boy, you've gotta problem.'

During my movements in and through some five Areas, I'd never met a similar situation. It could, and may well have been, total coincidence but things were different. Indirectly, however, I did have personal experience as to how things prevailed. At the time of the projected merger, whilst still in the North Barnsley Area, I had a BACM dispute running through the established procedures because of trouble with a colliery manager whom I had refused to endorse his annual salary review. He had taken it to the British Association of Colliery Management on the grounds that the general Christmas greeting sent to all the managers, of which he received a personally addressed copy, signified I was fully satisfied with his performance.

One Sunday lunch, I was called to the phone by a former Group Manager of this Area General Manager (then a permanent official of the National Association of Colliery Management). He shouted down the phone, 'I've just spoken to your Area Director!' He then lapsed into the crudest language, signifying my fate if I didn't settle the man's claim. My reply was simple and unmistakeable: 'Get stuffed!' as I slammed the phone down. I heard no more from him. Bullies tend to shrink from determined resistance.

I followed this particular matter through to arbitration where, in view of my new situation and the circumstances of getting the backing and goodwill of a new staff, we compromised at something like half the allocated review. In other circumstances I would not have compromised and would have taken the matter through to an imposed solution. The Area Director, although not involved, had apparently dug his oar into matters for which he had no concern. He never told me his former Group Manager had been in touch with him or vice versa. However, the event was helpful in that it alerted me to the nature of things I was going to have to deal with in the months ahead.

Area Reference

S.N.	–	Scottish North
S.S.	–	Scottish South
N.A.	–	Northumberland Area
N.D.	–	North Durham
S.D.	–	South Durham
N.Y.	–	North Yorkshire
D.A.	–	Doncaster Area
B.A.	–	Barnsley Area
S.Y.	–	South Yorkshire
N.W.	–	North Western
S.Y.	–	South Yorkshire
N D.	–	North Derbyshire
N.N.	–	North Nottinghamshire
S.N.	–	South Nottinghamshire
S.M.	–	South Midlands
S.A.	–	Staffordshire Area
E.W.	–	East Wales
W.W.	–	West Wales
K.C.	–	Kent Coalfield

The coalfields were split into 17 Areas plus the Kent Coalfield with 17 Area Directors and each Coalfield Manager in charge of four pits. Divisions and their staff were eliminated. Transfers of Divisional Staffs to Areas took place together with premature retirements.

Departmental Staff were retained at both National and Area Levels with further reinforcement to the latter following a special investigation undertaken by Dr Fleck of ICI. The upheaval was massive and many thousands of staff personal were individually interviewed and placed. Morale fell to a low ebb for quite a time following the exercise.

REVISED ORGANISATIONAL LINE STRUCTURE - 1967
NATIONAL COAL BOARD

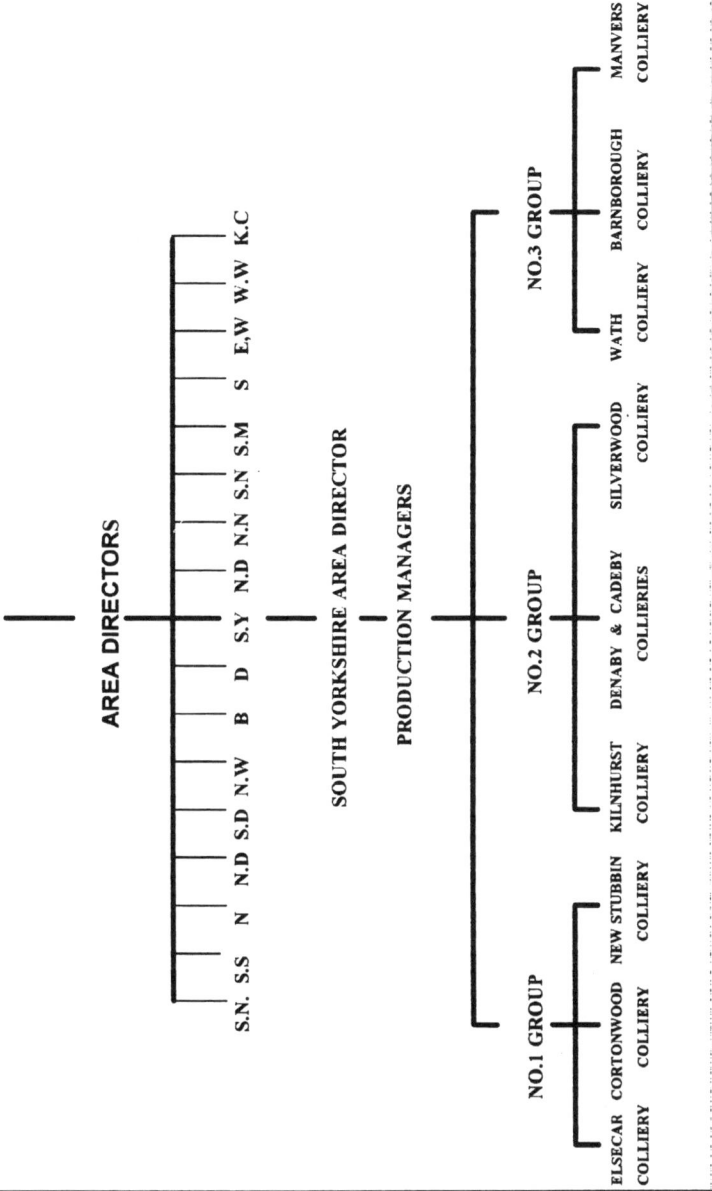

AREA DIRECTORS

S.N. S.S   N   N.D S.D N.W   B   D   S.Y   N.D N.N S.N S.M   S   E.W W.W K.C

SOUTH YORKSHIRE AREA DIRECTOR

PRODUCTION MANAGERS

NO.1 GROUP

ELSECAR COLLIERY   CORTONWOOD COLLIERY   NEW STUBBIN COLLIERY

NO.2 GROUP

KILNHURST COLLIERY   DENABY & CADEBY COLLIERIES   SILVERWOOD COLLIERY

NO.3 GROUP

WATH COLLIERY   BARNBOROUGH COLLIERY   MANVERS COLLIERY

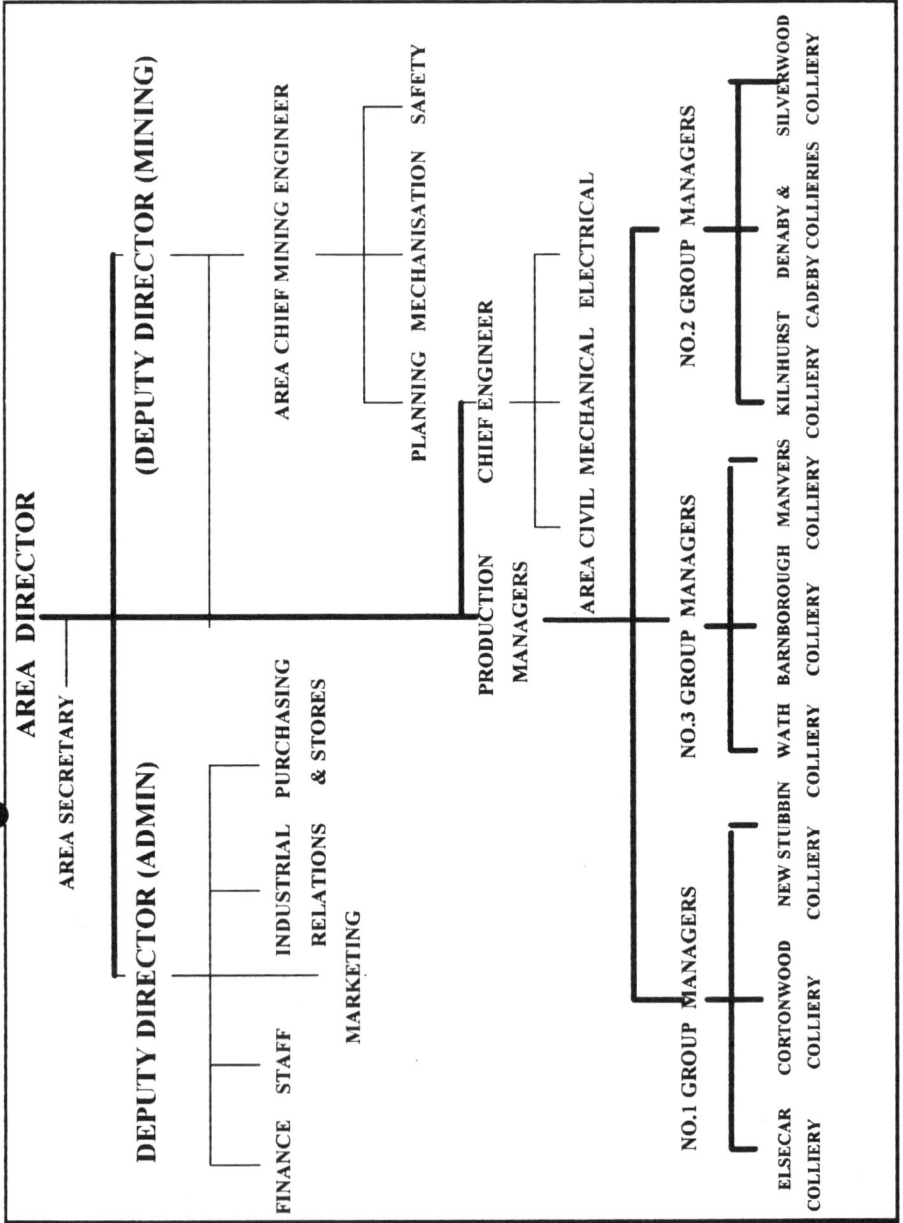

PHOTOGRAPH 44. ISOBEL RICHARDS (WPC3)
& DOROTHY ROUND AMMANFORD, W. WALES

WPC Richards was one of the very early recruits to the rank of Woman Police Constable. She served a great deal of her early career in Birmingham before returning home as a member of the Ammanford Constabulary. She was highly respected throughout Carmarthen and other places.

Here, after teaching Dorothy to drive over previous weeks, they are together on the occasion of Dorothy having taken and passed her car driving test in Llanelly. Dorothy would be eighteen at the time this photograph was taken. During her stay in Wales Dorothy was the subject of kind influences by a number of fine people both in police circles and generally. She started out with Lloyds Bank at Pontardawe, Swansea and remained with them during her career in Wales, following which she was transferred to their Sheffield Branch, Yorkshire.

PHOTOGRAPH 45. ABERNANT COLLIERY, SOUTH WESTERN DIVISION: MAIN
RETURN ROADWAY SHAFT PILLAR LOCATION

This photograph indicates the first stage in a sequence of developing damage to the pit-bottom roadways in the Red Vein seam at a depth of over 2,700 feet. The floor heaved excessively and broke into rough small material. This situation was common to all the roads being driven. 'Lateral thrust' (intense horizontal pressures) was generally in evidence with all drivage. Note the bottom LH corner of the photograph in which the feet of the steel arch lining have been pushed inwards.

At the time this photograph was taken instability was rampant within almost every roadway being driven in the seam. All were in or just outside the No. 1 and No. 2 pit bottom areas being affected. Such roadways, through which all the services and output flows, with a colliery of this projected size were expected to last from 50 to 100 years.

Various forms of lining were used to overcome the problem. Here we have steel arches set at close intervals and very heavily strutted with large diameter wooden struts.

PHOTOGRAPH 46. ABERNANT COLLIERY: UNDERGROUND ROADWAY DAMAGE

This is a second example of a pit bottom roadway which has been subjected to intense strata pressures. The first stage of floor heave has been followed by intense vertical and lateral pressures in that the steel support lining has been severely distorted and pushed to the right. As in the previous photograph, the strata between the damaged section and the repair section are badly contorted and fractured.

There were many such situations within the No. 1 and No. 2 shaft pit bottom areas which tied many men to roadway repair operations, the real problem being that it wasn't possible to obtain lasting repairs over a reasonable period of time. Pit bottom and main underground arterial roads are normally expected to remain stable for many years, particularly in the shaft areas involving the life of the mine, being dependant upon the volume of reserves to be worked and the speed of their extraction. The situation was particularly unfortunate in that the Peacock Seam characteristics and its quality represented the finest coal I have ever seen anywhere or at any stage in my career.

PHOTOGRAPH 47. ABERNANT COLLIERY: COMPOSITE WOOD-STEEL AND
CONCRETE BLOCK ROADWAY LINING

Following an inspection of difficult working areas within the Ruhr Coalfield, Germany, a circular steel lining, backed by timber and faced with special high stress resistant concrete blocks, was introduced in the No. 1 shaft side roadway. A reasonable degree of success was achieved but at great cost in terms of heavy labour requirement, time and money. The overall load on the resources and the problems of obtaining stable production were too great to handle at that stage, so we transferred working to the Red Vein seam some 500 yards higher up the shaft, where the situation was much easier.

Although conditions in the Red Vein seam were very good we had great difficulty in the construction of the advance roadheads. The men were not adjusting to the situation, so that the mechanised faces were not operating with the degree of success the situation warranted. To overcome this problem and introduce 'retreat mining' practice, steps were taken to procure an American type continuous miner which was delivered shortly after I was transferred back to the No. 3 (Rotherham) Area, South Yorkshire as its Area General Manager.

PHOTOGRAPH 48. AMMANFORD ANNUAL CARNIVAL: PRESENTATION OF
JUVENILE FANCY DRESS AWARD

On this particular occasion Mary and I were called to judge and award Fancy Dress Prizes
in the Juvenile Sections. The photograph shows the winner of the young ladies' competition. Maybe she will now be attending the Carnival with her own children.

The event is very popular and attracts a large number of people and celebrities. During
this Carnival I was privileged to get to know Sir Grismond Phillips, the Lord Lieutenant
Carmarthen. Later we received an invitation to his home following which he came along
to our Silver Wedding anniversary on 24 December 1963. He was formerly married to
Earl Fitzwilliam's oldest daughter. Princess Elizabeth and Princess Margaret as children
had great regard for him.

Dis. 1

CERTIFICATE of DISCHARGE

FOR A SEAMAN DISCHARGED BEFORE A
SUPERINTENDENT OR A CONSULAR OFFICER

DEPARTMENT OF TRANSPORT
AND POWER
No. **2515**

Name of Ship and Official Number, Port of Registry and Gross Tonnage	Horse Power	Description of Voyage or Employment
"USKBRIDGE" 183370. NEW PORT. 4097.97	273.2 1700 1530.	FORRIGN. CR.A.).

Name of Seaman	Year of Birth	Place of Birth
L. ROUND.	1914	YORKS.

Rank or Rating	No. of Cert. (if any)
SUPRR.NY PURSAR.	R1P.626691.

Date of Engagement	Place of Engagement	Copy of Report of Character*	
		For Ability	For General Conduct
22/6/62	TRoon.	IRELAND VERY GOOD R.14	IRELAND DECLINE TO REPORT R.14
Date of Discharge	Place of Discharge		
9/7/62	CORK.		

I **Certify** that the above particulars are correct and that the above named Seaman was discharged accordingly.

Dated this 9th day of July 1962.

MASTER

AUTHENTICATED BY

Signature of Superintendent or Consular Officer.

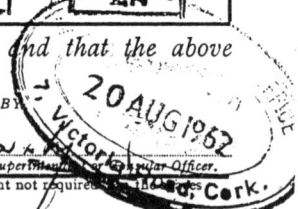

20 AUG 1962

* If the Seaman does not require a Certificate of his character, enter "Endorsement not required" in the space provided for the copy of the Report.

Signature of Seaman .... C Round ....

NOTE.—Any person who forges or fraudulently alters any Certificate or Report, or copy of a Report, or who makes use of any Certificate or Report, or copy of a Report, which is forged or altered or does not belong to him, shall for each such offence be deemed guilty of a misdemeanour, and may be fined or imprisoned.

N.B.—Should this Certificate come into the possession of any person to whom it does not belong, it should be handed to the Superintendent of the nearest Mercantile Marine Office, or be transmitted to the Department of Transport and Power, Kildare Street, Dublin.

A1660.D27805.6 Bks.7/60.WSMI. G27.

```
┌─────────────────────────────┐
│  s/s "USKBRIDGE"            │
│  NEWPORT, MON.             │
│  O.N. 183370  G.T. 3611     │
│  N.R. 1792  N.H.P. 2732     │
└─────────────────────────────┘
```

ABSTRACT FROM LOGBOOK (OFFICIAL) OF S/S USKBRIDGE.

At Sea. 27th June 1962.

    This is to certify that C.ROUND, Supy. did expose himself on

    deck (in bright sunshine) during working hours, on 26th June

    1962, and is now unable to carry out his duties.

        He hereby forfiets one days pay. .4 D.

                       George FIDDLER

                          Master.

CASABLANCA 30th June 1962.

    For absenting himself during working hours without permission

    for 2 days; C.ROUND, Supny. will be fined I days pay(Ist offen

    And will forfiet 2 Sundays at sea.

                      George FIDDLER

                         MASTER.

ACCOUNT OF WAGES

No. 50

Name of Ship and Official No.

s/s "USKBRIDGE"
NEWPORT, MON.
O.N. 183370 G.T. 4098
N.R. 2118 N.H.P. 2732

Name of Seaman	Rating	Reference No. in Agreement	Income Tax Code
+ C. Round	Suy	31	M with forms

Date Wages Begun	Date Wages Ceased	Total Period of Employment	
22/6/62	9/7/62	Months 31	Days 18

Dis. A No. 933436	Amount	Date 1st Payment	Interval
Insurance No. N-T Known	NIL	Allotment Note	

A. EARNINGS.

			£	s.	d.
Wages......1.8....Months @ £0....2.1=0....per month					7.2
........Days @ ....per day					
Promotion: Increase from........to........					
@ £........per month for........months....NIL..days					
Overtime: ....hours @ O.TIME...per hour					
N hours @ W.DRKE.D....per day					
W & ....months........ @ £........per month					
Bonus........days........days @ £........per month					
N.U.T. £1.16.184					

B. LEAVE

	Days				
Annual Leave Earned on Voyage....NIL					
Leave Due for Sundays at Sea....2					
Leave Taken During Voyage........					
Balance Due....NIL..Days @........per day					
Subsistence Allowance........Days @........per day					

	£	s.	d.
TOTAL			7.2

Less Distriling for........this........ per month
@ £........ per month

TOTAL EARNINGS

C. DEDUCTIONS

		£	s.	d.
Advance on Joining	... NIL			
Allotments	... NIL			
Pension Fund	... PRIVATE			
Social Insurance........weeks @........per week				
Union Contributions...FULLEB £0.10.per week				
Fines	1 days Pay			.4
Forfeitures	1 days Pay			.4
Income Tax (auit. Fen form) Rwd Printly				
Stores	NIL			
Wireless Messages NO COMMUNICATIONS				
Postage PAID PRIVATELY 0				
SHAOE PT.TIOUSLY				

N.A.T. INS. NIL

C.I.—Details of Cash Advances (in order of date)

Date	Port
	out...
	Cyal Irons otty...
	Cyal Irons oity Morphd
	out (Black Morphd

Advances Overleaf 8
Total 8

x C = out
x bal. of wages dep'osited in M.N. Total Deductions
x of... CLERK
FINAL BALANCE DUE ON DAY OF DISCHARGE 6.4

The above account of earnings and deductions, exclusive of allotments, is correct.

SIGNATURE OF MASTER........
SIGNATURE OF SEAMAN........

PHOTOGRAPH 49. LORD HALIFAX. MINISTER OF FUEL AND POWER VISITS
ABERNANT COLLIERY, SOUTH WESTERN DIVISION

L–R
Joe Harrison, Colliery General Manager
Alfred Kellett, NCB South Western Divisional Chairman
Lord Halifax, Minister of Fuel and Power
Sir E. H. Browne, NCB Deputy Chairman
Cyril Lee, SW Divisional Inspector of Mines.

Lord Halifax undertook a thorough inspection of all the surface plant and works in progress and followed up with an examination of the underground workings, showing a thorough interest in all aspects of construction, projected mining layouts, proposed mechanisation and the miners' reactions to the development situation.

As a person he was most inspiring to have met. Naturally he commanded great respect both for his courage and for his general concern for the workmen and their interests. He was probably one of the most responsible Ministers to have held the post in that he was genuinely concerned with the Coal Industry's future.

PHOTOGRAPH 50. SURFACE COAL STOCKING PROVISIONS, MANVERS MAIN
COLLIERY, NE YORKSHIRE DIVISION

Here the stocking facilities have been incorporated into both the No. 2 and the No. 3 winding shaft circuits. Normally upwards of 750 tons can be accommodated, but with the use of a small mechanical shovel stocking capacity was increased to over 2,000 tons. Generally it is a very cheap, easy and convenient form of surface coal stocking. Its contribution to improved colliery performance from the moment it was started was almost immediate, and helped greatly to alleviate problems from many sources.

There was quite a wide range of application at a number of other collieries within the Area which were being considered on the basis of priority such that the maximum Area production potential could be realised in the shortest time. As pit-head coal stocks subsequently escalated to unforeseen and crippling levels at almost every colliery the programme of extending this stocking system which had been planned and put into effect would have been highly contributory to the situation, although at the time there was no strong evidence that that was so.

PHOTOGRAPH 51. SURFACE COAL STOCKING PROVISIONS,
ELSECAR MAIN COLLIERY

The system shown in the photograph was designed for 1,000 tons capacity surface stocking but this was substantially increased by lateral spreading with a bulldozer. The actual stock shown is over 4,000 tons. During the week the Colliery raised its record tonnage of over 28,500 tons. We actually stocked over 8,000 tons by this means. As with the unit at Manvers Main Colliery, performance was boosted in no small measure by the application of these facilities.

The former timber stockyard can be seen immediately in front of the coal stockpile. However the quantity of mining timber currently stocked is but a fraction of that of former times. Stocks are invariably of steel roadway supports, together with mining plant and equipment.

Plan 2 – Haigh Moor Seam Workings, Elsecar Main Colliery, North-Eastern Division

PLAN 2 – HAIGH MOOR SEAM WORKINGS – ELSECAR MAIN COLLIERY

The reserves available to the Haigh Moor seam were quite limited in comparison to those which formerly related to the Parkgate, Thorncliffe and Silkstone Seams. The Parkgate, having been opened up before 1910, had continued working continuously for over 60 years before its final exhaustion.

The Plan indicates the Haigh Moor seam as having been extracted using a number of mechanised coalface mining techniques embracing the earlier 'retreat buttocks' described which did a great deal towards pre-training the men for the operation of the cyclic – Meco-Moore and the continuous Anderton Shearer Power Loaders mining systems which followed on in the early 1950s and 1960s.

The location of these developments is clearly shown, from which it will be seen that the 'Shortface Anderton Shearer Loader' applications extracted small parcels of reserves which in the circumstances prevailing could well have been left unworked.

An interesting feature shown on the plan is the Old Barnsley Water Level from which water was systematically drained by the Newcomen Pumping Installation for well over a century. Moreover, a further examination of the south-eastern extension of the seam's workings indicates that a useful parcel of reserves was abandoned by reason of a slight 'water inrush of 150 gallons per minute' within two exploratory headings. It was always my belief that this coal could have been safely and profitably mined by the judicious application of a simple borehole pumping station located and placed on the surface to best effect. This did not materialise because of later explained circumstances which were outside my control.

The two underground coal-stocking bunkers of 100 Cowlishaw Walker and 300 ton Butterley bunkers respectively were supplemented by an 'Open Spiral' surface bunker built to our specification by Qualter Hall's, Barnsley, a large local Mining Engineering Equipment Manufacturers.

This shows the 'working layout' of the seam and makes reference to the mining techniques adopted. The 'hand-worked buttocks' established an approach to the continuous mining technique. It proved to be a fine interim arrangement in preparation for what followed in terms of both cyclic and continuous power loading systems.

The first application was the Meco-Moore as formerly developed by the Bolsover Company, Chesterfield, Derbyshire at the time of Sir Eric Young (the first NCB National Production Director) with W.H. Sansom in a leading role. A substantial amount of the total seam reserves was worked by this method.

With the advent of the Anderton Shearer Loader into the Area as pioneered by Colin Rudge, the Area mechanisation engineer, the remaining reserves were worked out on the bases of 'in seam – retreat mining' which was most successfully practised. At the time of the 'short face retreat' trials the reserves were virtually worked out. However, we found an outlet as indicated on the plan which gave highly promising results and which paved the way to greater achievements.

Reference has earlier been made to the Old Newcomen Pumping System. Note the Old Barnsley Water Level which was drained for more than 140 years of continuous working, quite something by any standards of modern experience.

PHOTOGRAPH 52. MAINGATE COAL HEADING ROADHEAD, SHORT FACE
RETREAT MINING I: ELSECAR MAIN COLLIERY

The discharge end of the stage loader overlaps the gate 'belt conveyor' which permits delivery of the coal produced on the face to be loaded out for transit to the surface. As the coalface retreats the stage loader is moved back along the fixed anchor chain, with the belt conveyor shortened in convenient pre-determined increments. The support system is clearly indicated and consists of 5 inch diameter wooden props and 4 × 4 inch × 12 ft. rolled steel joists. The coal roadway sides in the active zone contribute to and simplify the heading support arrangements. The electric power cables and electrical switchgear control systems are also evident, whilst the piped water system at high pressure for dust suppression and firefighting purposes is to be seen running along the RH floor of the roadway. The material supply system is in the form of tracks over which suitable containers and bogeys are hauled by a rope system.

PHOTOGRAPH 53. SHEARING MACHINE CUTTING ALONG THE COAL FACE —
SHORTFACE RETREAT MINING II

This is a view along the face with the machine moving away from the viewer as it extracts a web of 30″–36″. The shearer is fitted with a 'cowl' behind the cutting drum. This can be rotated through 180 degrees so permitting bi-directional shearing. The amount of 'pick-up' spillage can be seen between the face and the panzer conveyor, the old box steel plough having been eliminated.

The powered supports on the right hand side of the picture are 4 leg units. The conveyor is pushed over into each new track by horizontal jacks, which also advance the individual supports. These are developments over the original experimental installation featured in PHOTOGRAPHS 33 TO 37 inclusive. A further but much later improvement involves the replacement of rope haulage by chain haulage which is much safer as wire rope breaking under load will 'fly and whip' with tremendous energy and great risk.

PHOTOGRAPH 54. SHEARING MACHINE IN ACTION –
SHORTFACE RETREAT MINING III

Here the shearer is cutting back across the face in the opposite direction. Part of the cutting drum with its scrolled line of picks can be seen in the background. The electric power cable and the dust suppression hose are to be seen at the front of the machine, the latter being most effective in creating a dust free atmosphere. Tracing through the various stages in Photographs 23 to 27 and Photographs 32 to 36 shows how we built progressive blocks of experience and integrated them in sequence so that the current breathtaking highly successful developments of the 1980s–90s emerged.

PHOTOGRAPH 55. MATERIAL SUPPLIES GATE – SHORTFACE RETREAT
MINING IV

This was the RH tailgate heading forming the supply road to the first short mechanised face. By reason of the roadway being driven in the solid, i.e. having coal sides, the support system is relatively simple and generally more lightly loaded. Wooden props and 4″ × 4″ steel joist was the standard system of roadway support provided. It was customary to reinforce the support in front of the retreating face by setting the central vertical posts. The coal at the foot of the RH supports is evidence of 'coal flaking' due to forward strata pressures over the roadway.

These coal headings were driven by a Lee Norse CM28 continuous miner at the rate of 80 to 100 yards per week. This was basically inadequate in that the comparatively high rate at which the coalface retreated left but little margin with which to maintain continuity of coalface production. The single rail-track and rope system took care of the general haulage handling requirements. This is an example of working within the confines of the seam.

9

Deputy Area Director (Mining)

The contrasts between the two areas were quite marked but the manner in which the two Areas were merged, in my opinion, fell far short of what could and should be expected from a competent manager motivated by a determination to act in accordance with the best interests of both Areas, and to maintain the responsibility of the position held and to his employers the National Coal Board. This was a situation which, to me, indicated the insular nature of the appointed Area Director and also the arrogance involved. The administrative merger took place almost immediately, but the real merger took very much longer.

Instead of initiating a policy of stretching a hand of welcome to new recruits into the Area, they were made to feel inferior interlopers by being discredited in many things alleged to have taken place in the North Barnsley Area. I had built up some strong ties so that many of the North Barnsley staff, planners, engineers, managers and group managers could open up their feelings about the way they were being received. Moreover, the atmosphere I had encountered was obvious to them also. My approach to them was to ride the situation. Bide your time, I told them, and ask as many awkward questions as possible about all you see in the south side of the newly combined area. It will let them see you are not as stupid as they may think you are and perhaps may cause some embarrassment.

Welcome to the Barnsley (Merged) Area

I have already indicated the lack of established liaison and contact which could easily have been developed and the surreptitious nature of the probes carried out by the Area Director's protegé during the time I was acting on the Board's behalf as Area General Manager for the North Barnsley Area. They were hardly the actions of a skilled, competent manager with imagination and the ability to utilise the experience and abilities of individual and potential members of his staff and colleagues.

The manner in which I was received into the Area fully reinforced the points already made. The first morning, on entering the Grimethorpe Area Headquarters Area Director's office, I felt a sense of concealed hostility. I certainly wasn't made really to feel welcome, though the Director stood on the elevated fire hearthstones making some pretence. I hadn't been in the office more than a few seconds when the Area Chief Mining Engineer (directly responsible at the time to the Area Director) and his sidekick, walked in unannounced. The Board had placed me between them and that was resented from both sides. There was no attempt to get down to any policies he had or had not determined about the merging of the two Areas, or to the liaison which he wanted to be established between the most Senior Member of his staff and Area Chief Mining Engineer in a situation in which potential conflict was bound to arise. The Board later realised this and, many months later, placed the Area Chief Mining Engineer under the direction of the Deputy Director (Mining).

The amount of time I spent with the Area Director during the first meeting probably lasted no more than ten minutes, during which nothing of import was raised and the Area Chief Mining Engineer made no comment. For the next fortnight this occurred every time I went in to see the Area Director (one had to pass through his Secretary's office to do so). The Area Chief Mining Engineer always followed me into his office merely to act, I believe, as a witness or in an attempt psychologically to put on pressure, for he never at any time raised matters concerning the running of the Area. I intended putting a stop to this nonsense, and on a suitable occasion, walking into the Secretary's office, I said to the Director's Secretary,

'Tell the Area Chief Mining Engineer I'm going in to see the Area Director.'

She blushed with some sort of pretence. I believe that previously she had been acting under instruction to summon him whenever I wanted to see the Area Director, as no matter what the time of day, the position was the same. As I went into his office he stood up from his desk and took his usual position on the elevated hearthstone.

'Oh,' I said, 'I've told your secretary to summon the Area Chief Mining Engineer.'

Promptly at the bidding in he came. As soon as he walked in I walked out, leaving them in a somewhat awkward situation. I had no further problem in that respect. Throughout my career I always tended to analyse such odd situations:

Why does the Director send for his colleague? Is he uncertain or nervous about meeting me alone, despite his apparent arrogant front? Do they feel that between them they can ensure I fall prey to their belief in the policy of fear?

I honestly feel that I was regarded as being in a vulnerable situation because of the interview experience referred to. Some people may feel that such thoughts border on paranoia: maybe, I don't know. They have always however helped me to work out approaches to a resolution of many such problems.

On my first day at the Grimethorpe Headquarters of the new Barnsley Offices, I was shown my office and literally left to find my own way from that point. About 10.30 a.m. my secretary brought me a coffee.

'Aren't you going to the Area Director's joint meeting of managers and group managers from both Areas? she asked.

'What meeting? Where?' I enquired.

During the short period I had had with the Area Director and his Area Chief Mining Engineer, no reference had been made to this joint meeting of all production staff. Undoubtedly notice had formerly been circulated by memorandum with myself deliberately left off the circulation list. Arrogance, stupidity or downright inadequacy. It was hardly the way to start out getting two Areas to merge effectively with an attempt to leave the most responsible and influential

No.2 man out in the cold. It registered, and indicated in those early hours that I was going to have a battle on my hands to carry out my duties to the Board and my statutory responsibilities to the State.

Finding out where the meeting was, although late, I walked into the room. Both the Area Director and his colleague were elevated on a raised platform. Both completely ignored my entrance, nor did the Area Director refer to my potential role within the Area or introduce me to the South Barnsley Colliery management for whom I had to accept future responsibilities for mining matters. He was holding forth on the achievements of the South Barnsley Area, pontificating on its successes, whilst his partner adopted a back-up role. There was no reference to any policies derived about the manner in which the two teams were to be integrated and a common purpose to be nurtured and developed to the benefit of all.

In that first fortnight, what respect I might have developed for them both stood little chance of emerging. I had to consider that the positions they held were far more important than their occupants. Thus I had to go along quietly, assessing each situation as it developed. It crossed my mind many times, and the thought is firmly fixed, that the Area Director could well have been encouraged by his mentor the former Director General of Production, now Board Member for Production, to attempt, in any way he could, to discredit my position and to make things intolerable. Subsequent events tended strongly to support and confirm that belief. Having worked with and observed him closely over some four or five years, I am satisfied that without such apparent support he would have backed away from any real conflict, as subsequent events proved. However one has to accept each situation as it is and patiently work out and apply strategies for dealing with them as favourable circumstances arise.

I wasn't alone in receiving the early stupid treatment dished out by the Director, who was also bemoaning the loss of his former Industrial Relations Officer and that he felt he had had imposed on him one whom he considered to be William Sales's (former Yorkshire Divisional Chairman) protegé Ian Farningham. Not unlike the Area Director, Ian was small but you didn't notice. Ian's demeanour and tremendous intellect gave him an air of natural confidence and pleasant expression which transcended the mere problem of size. He was undoubtedly a very capable man, much more so, in my opinion, than his predecessor Reg Thompson whom I knew extremely well through his parents, family, and marriage and his early training within the Yorkshire Headquarters of the National Union of Mineworkers.

DODWORTH COLLIERY

Since our relative paths were closely integrated it was inevitable that we should get together. How do we knock some sense in the Area Director's head? was our immediate consideration. One approach was to endeavour to deflate his ego whenever he became pompous: 'That will do for a start,' was our considered view.

'Why don't we tackle one of his old No.5 Area collieries?' was Ian's question.

'One that has difficult labour problems,' I suggested.

In this we were really doing our job and acting in the Area Director's and the Board's interests, although we doubted his ability to appreciate this. Together we went through the collieries. One stood out: Dodworth Colliery results were unsatisfactory. The NUM Branch President Jack Woffinden had given trouble for years as he had no respect for management, including the Area Director whom we understood to have retreated before him on a number of occasions.

The Area Director lived in Silkstone, the next village to Dodworth. He was sure to get all the gossip, and to help in that respect we opened the way for the Area Chief Mining Engineer to be present at all meetings and discussions. Lest anyone may think we were a pair of Machiavellian schemers, nothing is further from the truth. We were just trying to establish ourselves in accordance with the Board's placing so to do our jobs in an unfavourable and somewhat hostile environment.

We organised our first meeting with the Dodworth Colliery NUM Branch leaders. Representing the Board we had Dick Wilson, Colliery General Manager; the Group Manager; Group Planner; and the Area Chief Mining Engineer, Colin Rudge (Area Mechanisation Engineer) who had been moved over from the South Yorkshire Area for no logical reason. He was most unhappy about this, but I was more than glad to have him with me.

At the first meeting and at all subsequent discussions the NUM was represented by President, Secretary, Treasurer and Delegate. As with my Anthracite experience in a number of similar situations, Ian and I decided our best plan of action would be to record and act on all the NUM claims to the point of total elimination and exhaustion, then hammer away for better standards of performance. We were both aware that this would be rough on local management for a time and that we would be accused of siding with the men, but that would pass.

We opened up the first weekly meeting by reviewing the colliery plans and analysing machine performances. A careful recording of each meeting's activities with covering minutes was followed, with delegated actioned responsibilities. Claims and counterclaims were in abundance. Jack Woffinden could well have felt our role to be somewhat akin to that of Portia's in Shakespeare's *Merchant of Venice*. Management smarted. As the weeks went by, things began to change. NUM claims of any consequence dramatically fell; the onus was now falling upon the men. The management now understood the strategy and responded magnificently. Colin Rudge's personal contribution, both in terms of his staff's monitoring the mechanised coal face equipment, and by the provision of equipment at short notice, was immeasurable. We followed this through without let up for many weeks and brought about quite a reasonable improvement in the situation. Moreover, it encouraged the local management. Dick Wilson was greatly appreciative of our efforts. Being a very strong character his popularity with the Area Director and his colleague was little better than my own. The Director's reaction at the beginning was that of petulant scoff, much as we expected. His protegé kept him appraised of events as we anticipated. Incidentally, during all the meetings I ran, and there were quite a number, I cannot recall at any time the Area Chief Mining Engineer offering anything of a constructive nature or an expression of personal views. He was in a very difficult position.

Throughout this exercise, I was building up my knowledge of the South Barnsley Area and using the experience as appropriate. The overall situation within the Area was way behind those of the East Midlands Areas and indeed South Yorkshire. Barrow Colliery was the best. It was quite a good pit and had been very well run; however it certainly wasn't an Elsecar, Gedling, or Newstead.

DEARNE VALLEY COLLIERY

In my opinion the Area was somewhat inbred and lacked the imagination and breadth of outlook of the East Midland Areas. I recall that one evening on my way home to Rotherham I called in at Dearne Valley Colliery, and stood and watched coal being rope-hauled out of the mine in exactly the same way as I had done as a boy some thirty-eight years previously. A young lad of about sixteen was attaching and detaching the trains of small tubs onto and off the haulage rope. We started chatting about his job, how many trains per shift etc. he handled. Before leaving him he was informed that many years ago I was doing exactly what he was doing now. I added,

'You are doing the job very well, Son!'

He turned to me with great pride and said,

'We could do better, but are always in trouble with the tubs coming off the tracks.'

During a discussion with the Area Director, capriciously I enquired,

'Does anyone know you have a little pit called Dearne Valley in your old Area?'

Peevishly, he replied that of course they did.

'Then why is it forty years behind the times?' was my response.

'It's not worth spending money on, that's why,' he countered.

'Surely it must have been so ten years or so ago,' I followed up.

A similar situation existed at the adjacent Ferrymoor Colliery, this being located almost alongside the Area Headquarters. The mining situation in the Shafton Seam was excellent, not unlike the Haigh Moor Seam at Elsecar Main Colliery. They would both have yielded excellent returns if there had been a reasonable degree of foresight, imagination and enterprise, as later provided and recounted herein. Little wonder the men at these two collieries had been sporadically uneasy and at times troublesome.

UNDERGROUND TRANSPORT REORGANISATION

The visit to the Dearne Valley Colliery triggered off within my mind an arrangement of conveyor transfer whereby the inefficient rope-haulage system could be replaced completely without the need for a programme of heavy capitalisation, thus creating more favourable circumstances for the successful application of coalface mechanisation being introduced.

In those earlier days, whilst acting Area General Manager in charge of the North Barnsley Area, I had spent quite some time underground at Monckton No.3 Colliery (which had been closed but a few weeks earlier) assessing the recoverable useable assets which could be made available for subsequent application and use within the merged new Barnsley Area. I found a trunk belt conveyor system some three miles in length, fully equipped with belt, control switchgear, etc., in excellent condition.

Transferring such a complete capital working asset had far more to offer in this particular situation than breaking it up and using the material as an easement of standard revenue charges at the collieries within the area. I called in George Eaton, the Area Electrical Engineer, informing him what I had in mind. He thought it a tremendous idea. I asked him to find out for me whether the electrical system at Dearne Valley Colliery could carry the additional load, and if not, what would be needed to make it so. He was enthusiastically favourable, but after two days he hadn't reported back. I called him to my office.

'Did you check the points I asked of you, George?'

Embarrassingly he said, 'Sorry, I wasn't able to; the Area Director and the Chief Mining Engineer told me not to waste my time on such stupid ideas.'

'OK, don't worry. Leave it with me, George,' was my reassurance.

I really gave this situation a great deal of thought and the result was quite revealing. I had discovered the Director's attitude towards the proposals. I was acting in the National Coal Board's interest in seeking to improve the efficiency of one of its working collieries, a basic element of the job I was doing. Unless the colliery was fitted up with good transport arrangements, the pending coalface mechanisation would have stood little chance of success. In this situation I felt I had encountered my first major challenge in the process of which I could find out the strength of any support the Area Director had (or thought he had) from his mentor, the Board's Member for Production. At the interview he had said I had been chosen specially for the job I was now doing, in that I had 'the right specification needs for the circumstances prevailing within the Barnsley Area,' quite a useful fall back if needed to meet any backlash. There was no point in discussing it with the Area Director. He had declared his position. His disciple held the same views and as such could be ignored. In fairness, I felt it better to go ahead on my own rather than risk it being vetoed, which would have made it necessary for me to raise the matter openly at the next Headquarters Accountability meeting. I was prepared, but reluctant, to do this.

I faced the challenge head on and arranged to have the trunk conveyor system, together with its control gear, dismantled out of salvage sequence and shipped over to Dearne Valley Colliery. The surveyors at the colliery, on my instruction through the Group Manager, had put through alignment directions and datums and marked out roadway side cuttings and the like. I asked Ken Sutcliffe (Area Chief Engineer), quite an experienced and capable man, to work out the surface and power modifications required to facilitate a successful changeover, which he did admirably. This was followed up by meeting the NUM Branch at Dearne Valley Colliery. The men were wildly enthusiastic, having for many years protested for some reconstruction to be undertaken. They put the pit's holidays back for three weeks to give us more time to prepare for the final turnover. In addition, they found and detailed men to undertake the changeover work. Their cooperation was magnificent.

The salvage men at the closed Monckton Colliery were met and they too gave us great support by working to a timetable of essential salvage deliveries which they met in full. The installation was undertaken but, because of time pressures, we undertook temporary modifications which proved disastrous when the colliery restarted following its holiday vacation. Concentrating all the colliery resources, we sorted out the difficulties but lost three or four days production. Things were duly sorted out, following which the colliery began to go places and made some encouraging strides in both mechanised production and productivities. The Area Director had during the whole operation said nothing to me. I didn't need to keep him informed, as naturally I wouldn't want him to be involved in any stupid ideas of mine. It didn't matter really that the Area Chief Engineer reported directly to the Area Director under the Board's new arrangements.

However Ken supported me and did an excellent job on that and indeed many other occasions, particularly later with the development of the Riddings Project. My first impressions of Ken were erroneous and indeed unfair – I am glad I was able to correct my views within a very short time of entering the Area for he did his job admirably, working more in the interest of the National Coal Board and less with conflicting personalities in much the same way as I have always done.

From that point on, the Director and his follower constituted no serious difficulty, maybe a nuisance from time to time, but no cause for real worry. He had failed to face up to my first real challenge; though maybe he didn't realise it was one. What was more important, he had shown weakness in failing to undertake discussion and to use his power of veto if he felt I was acting irresponsibly. Had our roles been reversed things would have been different. My approach would have been:

'So you want to put this valuable asset into Dearne Valley Colliery. Are you sure this is the most rewarding application? Look into all possible alternatives, go away, bring me back notes of the salient points of your findings and we will then discuss them.'

In doing so, I would have been non-deterrent and securing my position, but satisfied at the same time that I was trying to get the best deal from what spare assets I had. Following agreement to a line of action, I would then have given all the help and encouragement I and the Area's resources could have mustered. In point of fact neither the Area Director or the Area Chief Mining Engineer was aware that the fixed assets were available to anything like the degree of detail I possessed following my former Monckton Colliery visit.

MANUFACTURERS' REPRESENTATIVES

Throughout my career from the level of Colliery Manager to that of Area General Manager and above, I have always regarded manufacturers' representatives as 'pollinaters', feeding us with updated information and experience about their products, services and developments, keeping us abreast of the wider technology within their range of expertise. On many occasions I was stimulated by discussions with them. Bob Thorpe of Anderson Boyes Ltd. was a excellent example, particularly as his range of experience extended beyond the coalfields of the United Kingdom. There were many others of his ilk who appraised me of experimental detail in its early stages which gave rise to useful thought, development, and eventual application.

During the early American Lend Lease Program following the Second World War I recall an ex-Manager of, I believe, Kiveton Park Colliery, South Yorkshire, acting as a representative for the introduction of 'duckbill loaders' (American mining equipment embodied in the program). We spent many hours together in discussion during which he fired my enthusiasm to learn more and more about American mining methods.

One particular Company's senior representative, however, was very close to both the Area Director and the Area Chief Mining Engineer, and, caught up in the situation of adverse relationships, went out of his way to avoid calling in to see me. He had ready access to both the senior offices by special arrangement. His products, conveyors and conveyor structure, had a strong monopoly in the South Barnsley Area. The Area Purchasing and Stores Manager provided me with a list of competitive prices and I discovered an adverse differential of some £2.50 per yard, for which there was no special or particular advantage or purpose.

Circulating a memorandum to my Production Managers and Colliery Managers, giving details of the costs of alternative available structures, I advised them to convert to more economical purchases, which they did in no small measure. I fully expected and was prepared to be censured or overruled by the Area Director but nothing happened, a situation which surprised me greatly in the circumstances.

However, in less than twelve months, this particular representative was trying time and time again to fix up a meeting. For upwards of three years I wouldn't see him because of his lack of courtesy and the fact that his prices were not competitive. When I did see him, he brought his sales records which indicated his company had been hit very heavily. He was informed if he wanted to get back into the Area it was more than necessary to adjust his prices to competitive levels. Maybe past close relationships with his two former contacts were so strong it was felt the position of this particular supplier was totally secure.

CLIFFORD MACHIN – FORMER PRODUCTION DIRECTOR, YORKSHIRE DIVISION

During late February 1967 the South Yorkshire Area started breaking all its former production records. I was very happy for them and wrote to Cliff Machin to that effect. This was his reply,

No.3 Rotherham Area
Wath-Upon-Dearne
Rotherham
6th March 1967

Dear Charles
Thank you for your letter of the 1st of March The results we get here are due to
you and your efforts in the past.
My very best wishes to you for the future.
Yours sincerely
Cliff
(Clifford Machen)

At the time I received that letter I was pretty low in spirits and somewhat depressed. It did much for me and served to illustrate the difference between competent 'big men' compared with the 'little ones' with whom I was now working.

Cliff Machin, prior to joining the Yorkshire Divisional Board as Production Director, was Area General Manager (Carlton Area) which included Grimethorpe, Frickley, Carlton Main, Monckton Group, Ferrymoor, Dearne Valley, Houghton and other collieries. He was well respected throughout and beyond his Area boundaries.

During one of my trips to Charleston, West Virginia, I met one Joe Kuti, American salesman for Westphalia Lunen Gmbh, a German mining machine manufacturing company. I understood him to be Hungarian by birth and that he had been displaced by the Second World War, finally winding up working in the mines in Cliff's Area. He studied to obtain the Board of Trade Colliery Manager's Certificate, and Cliff Machen did much to help him. After he qualified he was brought onto the Area Staff and carried out a great deal of varied work for Cliff. Joe and his family (all of whom I met at his Charleston home) were very grateful for the help Cliff gave them and they spoke highly of his humanity and consideration for others, to which I can readily testify.

Shortly after he took over No 3 (Rotherham) Area as part of the newly combined Rotherham and Worksop Areas, he built the miners at Elsecar Main a new canteen, and, knowing my deep association with the colliery, called me over to open it. He made sure there was quite a gathering for the occasion including his own and members of Headquarters staff. The photographs which follow commemorate the occasion.

PHOTOGRAPH 56 shows a number of the Rotherham Area and Headquarters Staff present on the occasion of opening the new canteen which was provided as a mark of appreciation to the Elsecar Main Colliery mineworkers for their contribution and co-operation over very many years. They were amongst the best anywhere and ranked alongside the East Midlands men.

PHOTOGRAPH 57 features Maurice Beedan, the Colliery General Manager, and myself preparing to enter the new canteen. It was a fine gesture on Cliff's part and was very well received both by the NUM Branch leaders and the workmen.

On 6 July 1968 Irvin Spotte (President Pittston Coal Company, Dante, Va, USA) and his wife Isabel, visited Mary and me. We had all been up to Ayr in Scotland for discussions with the Mining Engineers there, including Albert Wheeler. Bert and Mrs Wheeler were two of the most charming, graceful and natural people Mary and I ever met throughout my service with the National Coal Board. During the run-down of the mining industry from the 1980s onwards he played quite a responsible part at the highest levels within the new and contracted organisation British Coal, prior to which he was Area Director of the Nottinghamshire Area.

Cliff Machen had previously arranged with me to call on our way back from Scotland at the Wentbridge Hotel, Wentbridge, near Pontefract, for us all to have dinner with Mrs Machen and himself. They were excellent hosts and treated us really well. Since the date involved was very close to 4 July, American Independence Day, Mrs Machen had contacted the American Embassy in London about the sort of special dinner served on such an occasion. She was advised that turkeys were served on Thanksgiving Day, but these were not available. However we had an excellent meal; the company was delightful and the atmosphere was superb. Then dessert came along on a huge flat silver salver upon which the American flag had been precisely reproduced in strawberries and cream, a most remarkable sight, one that got through to both Irvin and Isabel.

Unfortunately Cliff died in tragic circumstances whilst he was in post. I herewith express in public my thanks 'heavenward' to Cliff, for he to me was a fine, kind and generous person.

On one occasion Geoffrey Barber, one of my Group of Production Managers, called into my office for a discussion about work that was falling behind schedule at one of his collieries. We had been talking for about ten minutes when the Area Director stormed into the office and viciously laid into him for no apparent purpose, having no doubt been fed with some gossip or other from his network. Turning to him, quietly I said,

'Hey, address yourself to me. The lad has been carrying out my instructions.'

He turned on his heel and walked out of my office. I had no idea what he was incensed about - nor did I want to know. Maybe he wanted to impress me at the Production Manager's expense. Geoffrey Barber's reaction was,

'Thanks, Boss - I've worked for the Area Director for quite some time but this is the only time I have seen him put in his place so neatly and quietly.'

'You're wrong, I haven't put him in his place. We're working to Board Procedure. If you're to be chastised for failing to do your job, then that is my responsibility,' I added.

Like the 'Seniority Rule' in the Anthracite, 'Board Procedure' could be adapted to meet almost any situation.

Rockingham Dirt Tip - Tailings Lagoon Collapse

One of the greatest tragedies associated with the Coal Mining Industry undoubtedly occurred at Aberfan, Wales in which over four hundred children and teachers and adults were fatally enveloped and suffocated in a wall of thick sludge which swept down from the top of a mountain behind the school. This had resulted from the collapse of a 'tailings lagoon' located on a mountain side. Tailings are the residue of the coal cleaning processes and consist of a mixture of water and very fine shale. The practice employed in the disposal of 'tailings' was to form a lagoon by creating a four-sided embankment and thereafter to pump the tailings inside the four dirt walls. As the volume of tailings increases inside the lagoon the four dirt walls are stiffened and heightened, thus increasing its capacity. Over a long period of time, the water drains out of the mixture leaving behind a volume of somewhat consolidated compacted fine dirt. The Aberfan disaster captured the world's news headlines and drew sympathy from every quarter of the globe.

A comparative short time, maybe four or five weeks, after the Aberfan disaster, at about 5.00 p.m. I received a call informing me that leaks had been discovered in the northern wall of the Rockingham Colliery dirt tip lagoon and that steps were being taken to stiffen up the lagoon walls to prevent breakthrough. I called in at the colliery on my way home and trekked across the colliery yard and dirt tip to make a full inspection. There were three immediate points of weakness, the worst at the north end. Ken Sutcliffe, the Area Chief Engineer, was there and had organised equipment, lighting and men to carry out remedial measures on a large scale. It was necessary to work continuously through the night to contain the situation.

At six o'clock next morning I was on the tip. The battle of containment at the northern lagoon wall had been lost; the wall was breached and the tailings were flowing into a valley some 200 ft. or so below. At the beginning the flow was intermittent. A volume of tailings would flow through a few yards and temporarily stop, building up a large amount, following which it gushed at quite a fair speed into the valley below. In this case there was no danger to the public but we did inform the residents in their elevated properties on the opposite side of the valley to be alert.

Cracks and chasms were forming on each side of the breach. The containment wall was now being broken and swept away, with the rate of flow and volume of material carried increasing minute by minute. With the higher speed and volume, the tailings spread laterally to form a river of sludge. For anyone to have been caught up in its path would have meant a very short survival time.

A considerable amount of tailings were disposed into the valley through which the colliery rail traffic passed. Sidings were blocked and a train of about eight wagons was completely enveloped. The sight during the active periods was awesome and the situation became frightening as the formation of cracks developed about us. We had to withdraw helplessly; nothing further at all could be done at that stage.

I was instructed by the Board's Director for Production to provide a report of the whole situation. This was quite a surprise to me as I understood that the Director (Special Duties) on Headquarters Staff had previously undertaken the same commission and had already submitted a report which I wasn't shown. In view of the sensitivity of the report, I handed it over to the Area Director who no doubt discussed it with the Board's Director for Production who probably came to some arrangement. It was never discussed with me. Having carried out instructions, there was nothing further I could do.

The cause of this particular incident was, without doubt, the unauthorised discharge of raw pit dirt into the 'live lagoon'. The volume of dry pit dirt increased the pressure on the lagoon walls until the north wall was breached at its weakest point, this being followed by total collapse and discharge of the voluminous 'slurry' contents into the valley below.

Some two or so years after this occurrence, my American colleague Irvin C. Spotte, President of the Pittston Company, Dante, Virginia, had a similar disaster on the site of one of his coal preparation plants on an even greater scale, but fortunately loss of life was much less severe than at Aberfan. However the problem was very serious. The Pittston Company were subjected to strong litigation with the claims being pressed amounting to many millions of dollars. I was caught up in this situation. The lawyers representing the claimants were suggesting I had been advising the President on tip security and he had failed to respond. This was not so. All I had done was to send him press and magazine coverage of the Aberfan disaster. I was in no position to give the sort of information as alleged. However I met the Pittston Company's attorneys, Donovan's (the founder being the Donovan who was involved with the Office of Strategic Services during the Second World War) in the Rockefeller Centre, New York and explained the situation to them. Subsequently, I was subpoenaed to appear before three American attorneys representing the American Plaintiffs and the Pittston Company's attorneys in London. The Lady Counsel who conducted the hearing had a list of agreed prepared questions. These she strictly adhered to, preventing the three prosecuting American attorneys from going beyond the scope of the enquiry specified. The hearing lasted three to four hours. I believe the Pittston Company settled out of court, but I never heard of or saw the terms of settlement.

Board Accountability Meetings

Like most well run organisations the Board had a system of 'accountability' meetings with its Area Staffs for the control of its operations. Monthly or bi-monthly the Board Chairman, in concert with a senior member of the Head-quarters Production and Finance staff, together with Administrative assistance fully briefed, would come to Area and examine us on our performance, shortfalls, or whatever they felt to be relevant. My experience following the changed organisation was exclusive to Lord Robens. These meetings, like all examinations, could be rough, inspiring, and often embarrassing. Irrespective of the nature of the accountability meeting, I personally found His Lordship's analyses keen and fair, and invariably after a rough meeting he would try to leave us in fair morale. Two embarrassing meetings, however, I will never forget. One concerned myself, the other the Area Director.

I'll deal first with the former. It seemed obvious to me that during this particular meeting with the accountability team we were going to have it rough. The team came into the conference room; His Lordship, looking grim and determined, was carrying a sheaf of computer printouts. As we were arranged round the table I was sitting virtually opposite Lord Robens. There were no preliminaries. He opened up looking straight at me.

'I want to know why the machine running times are so low in this area. I quote this as an example. The Anderton shearer on 10's Unit, Dearne Valley, is running less than 25% of its time. Why?'

'I can't understand or reconcile those figures, Lord Robens,' was my response.

His demeanour really hardened. He looked round the table, back at me, then he turned again to my colleagues in turn and questioned every member of the Area Staff.

'Do you understand these figures?'

'Yes, Lord Robens,' (Area Staff Manager).

'Yes,' (Area Mining Engineer).

'Yes,' (Deputy Director (Admin).

'Yes,' (Industrial Relations Officer).

'Yes,' (Area Director).

Turning to me with a hard expression he then said,

'We pay you a handsome salary to understand these things. Obviously the wrong man carries the post.'

My stomach had twisted into knots.

'Lord Robens, if it is possible to understand these things I will do so, but the figures you have quoted are not accurate,' I stressfully protested, pouring oil on an already heated situation.

'Why do you say that?' he demanded to know.

'You have quoted a machine which is producing over a thousand tons a day. These men have the shortest working time of all collieries in the Area because of their travel distance, and the times you are referring to are not measured. They are not realistic. With such running times the machine output would have been negligible.'

'Who is responsible for the preparation of these figures?' he asked.

'The Method Study Department,' was my answer.

'Send for the Method Study Engineer.'

Les Dawson the Method Study Engineer came apace.

'Sit here,' instructed His Lordship, 'I'll go into this myself.'

For an upwards of a hour he examined Dawson - with myself assessing my fate with every answer Les gave him. His answer to the final question was crucial.

'Do you measure the actual running time of each machine on each shift?'

'Goodness, no, Lord Robens, we don't measure any. They are assumed. If we had to, we would require to recruit 100 or 150 men at least,' said Les, relieving all the stress from within my whole being.

In a more kindly tone Lord Robens turned to me and said,

'You are obviously more versed with the situation than any of us round this table. What are your views as to how we should express machine performance?'

I gave him my views, some of which were incorporated in a Headquarters revised procedure which appeared shortly after the meeting. It was an embarrassing experience for me at the time, one I wouldn't have liked ever to have repeated. However, I believe Lord Roben's respect for me following the experience was enhanced, and I repeat that I found His Lordship to be a fair person as the foregoing example illustrates.

The second situation occurred some time afterwards and involved the Area Director and Dodworth Colliery. This particular meeting had been going smoothly for quite some time and there was a nice atmosphere. We were dealing with Dodworth Colliery which had shown quite an improvement following the work-effort Ian Farningham, the local management and myself had expended. (Ian in the meantime had taken a high appointment with a Clyde shipyard company and the new man in post had been with us probably less than a year.) The Area Director quietly rose from his chair, went to his office and returned with a sheet of paper. He slipped it directly across to Lord Robens. His Lordship looked up and said,

'I suppose the man was sacked?'

No-one answered; he looked round the table and repeated the question,

'Was this man sacked?'

Silence. None of us knew what the paper he had been given referred to. He then rounded on the Industrial Relations Officer and subjected him to intense questioning which lasted more than thirty minutes. I felt intensely sorry for the man, having been in a similar position myself, but I couldn't help him. The Area Director's role was passive. He made no attempt to aid his colleague who, it was later understood, like the rest of us was totally ignorant about the matter raised.

We found out later that the paper which the Area Director had handed over to His Lordship was a notice about pit-shaft winding times, indicating that those men who were not punctual would not be allowed to enter the mine. The notice had been signed and posted on the pithead at Dodworth Colliery by the Colliery General Manager. The famous or infamous NUM President Jack Woffinden had written across the notice, 'Take No —— Notice.' It was an actionable situation.

Lord Robens suddenly stopped, realising that the man was either ignorant or protecting his Area Director in a situation of misguided loyalty. There was no reciprocal loyalty forthcoming.

'Ah! I now understand - I'll deal with the matter later.'

We continued the meeting and finished the business in hand. On getting up he simply said,

'Mr Director, shall we go into your office, please?'

Thirty minutes or so later the Director came into my office in a most pathetic state. He was dejected with signs of tears having run down his face. For the first and only time during the whole period we were together I felt pangs of pity towards him. After he had settled down and was talking rationally he told me Lord Robens had chastised him very strongly for his disloyalty to a member of his staff and failure to initiate action relative to the event he had raised. Apparently his girl in the next office was aware that something serious had taken place and she too was highly embarrassed as was reported back to me by my Secretary.

'What made you do such a stupid thing?' I asked.

'I wanted to show Lord Robens the sort of problems we were up against,' was his answer.

'But it is our job to deal with such problems, not the Chairman's,' I replied, continuing, 'Had you told us about the notice and your intent we would have told you to forget it, or come to your assistance in the sort of backlash you have faced.'

In fact we had a good platform to stand on. Jack Woffinden had a very strong following at the Yorkshire NUM Headquarters, the South Yorkshire Area was having trouble with its 'home coal' arrangements, and the Doncaster Area was involved with strikes. We could have truthfully told Lord Robens that to have acted precipitously could well have triggered off a coalfield strike. I am sure His Lordship would have understood and accepted such an explanation with safe-guards, as a prudent approach.

However fate can be very capricious and do some unexpected things. There are two sequels to the above account. Several months later Jack Woffinden stepped out of line again and this time he was sacked. Why, and on whose authority, I do not know, but the sticky situation landed on my lap. I could have ignored it and passed it back to Industrial Relations, but I didn't. It had to be dealt with. The Dodworth men went on strike for a few days and then went back to work leaving the Yorkshire NUM officials to try and sort out Woffinden's situation. Sid Schofield, the Yorkshire NUM Secretary, came to see me. I explained what Jack Woffinden had done previously with regard to a manager's notice and that he had been let off, despite the fact there was no way we could allow the man usurp the authority of the Colliery Manager who had onerous responsibilities under the 1954 Coal Act. This new offence could not be over-looked. We spent a long time discussing it. Sid, with his plea of . . . 'In the interests of peace within the Yorkshire Coalfield' . . . was virtually pleading for his reinstatement. Jack Woffinden was a man with tremendous pride: I had determined that during our periods of former contact. It struck me that we could try and bind him down to reasonable behaviour under his own signature, so I told Sid we would be prepared to reinstate the man, providing he would agree to signing an attestation that he would in future abide by the Coal Industry's conciliation and consultative machinery. 'You can put that to him, Sid,' was my offer.

My experience with the Anthracite 'Seniority Rule' had broken through again. Those Welsh miners had taught me a great deal whilst I was amongst them. However, as I had expected, Woffinden would not sign such a deposition. Failing at Area level the Yorkshire NUM Secretary went down to try his hand at National Headquarters.

One Friday afternoon Sid rang me from Hobart House. Would I agree to Woffinden's restoration on the grounds we had formerly discussed?

'Yes, I would.'

Sid wanted reinstatement without conditions. He said, 'I'm taking the matter elsewhere.'

'Sid, you have the Area Director in the Headquarters building. He has the final authority, talk to him,' I advised.

Jack Woffinden was reinstated, on what grounds and on whose authority I was never informed; neither did I attempt to find out. However, the story does not even end there. Shortly after I had left the Area, whilst working for the Board at Headquarters, I called in to see the Colliery General Manager at Dodworth Colliery for a chat and to see how things were going. About ten minutes after my arrival the office door was thrown open and in came Jack Woffinden, white as a sheet and manifest with uncontrollable rage. Looking at me he shouted,

'You —— bastard - I'll dance on your grave.'

'Thank you for the advance notice, Mr Woffinden, I will now arrange to be buried at sea,' was my quiet reply.

The new Colliery General Manager (my former colleague Dick Wilson had moved upwards) was uneasy and dumbstruck.

'What's all that about, Mr Round?' he asked.

'Just a slight difference of opinion which dates back a year or so,' I told him.

Sadly, Jack later suffered ill-health and on his retirement his two sons became predominant in the NUM Branch leadership. I understood they did a great deal to promote the colliery's interests and were highly cooperative with the management. Their father had some good points, and for these I respected him and still do. In all my dealings with him he was certainly a 'man' and to me proved to be 'a man of his word'. Moreover, he worked hard for the men and was doing his job as he obviously thought it should be done. Indeed he must have satisfied his men for many years or they would have got rid of him. Over a number of years he had been used to management retreating before him in such situations, much as seems to have been the case in the final act.

Sir Humphrey Browne Resigns

Prior to starting out for work one Monday morning, in the morning post I received a letter which gave me no small measure of sadness. It was from Sir Humphrey Browne and its contents came out of the legendary blue. There was nothing in the grape vine to suggest any possible reason for his sudden departure. Most of us felt he was totally wedded to the Coal Industry and most unlikely ever to leave it. However it was a standard letter couched as follows:

National Coal Board
Hobart House
Grosvenor Place
London S.W.1
Deputy Chairman's Office

Personal

My Dear Charles
I am writing to tell you that, for a combination of reasons, I have decided to leave the Board and take a post outside industry.
You will know how deeply I regret this and what a wrench it will be. But there

are compelling circumstances.

I shall always treasure my life in the industry and the friends which I have in it. I shall leave at the end of May - effectively earlier as I am taking a holiday towards the end of the month.

With best wishes for the future of the Board and the part you will play in the industry

I am dreadfully sorry to be leaving people like you.

Yours ever,
Humphrey Browne

The last sentence was handwritten as a personal aside to myself.

About 9.30 a.m. the Area Director came into my office with a copy of this circulated letter in quite a state of excitement.

'There's been something going on at Hobart House, read this,' passing me the letter. It was the standard copy with no personal references. I handed him the letter back without comment. He then started to conjecture about things that might have happened.

'Sir Humphrey is not the sort of person who would advertise his personal affairs. Whatever has given rise to this situation will not be made public by him. I think it to be a sad day for the Industry by his loss whatever the cause,' I told him with sincere feelings, never having disclosed this situation to anyone until then. Later I understood that Sir Humphrey had taken over the Chairmanship of Woodhall Duckhams Ltd, an industrial complex. He was another fine person I'm glad to have known.

Lord Robens - Colliery Inspections and Impressions Notes

Lord Robens was not the type of Chairman who was afraid to dirty his hands or descend to the level of the shop-floor. He made countless underground visits within every coalfield and Area. Following each session underground, the Area Director would receive a short note setting out his impressions. At times these notes could be most devastating. In other circumstances they could be tinged with humour as a means of focusing strongly on the message he wanted us to get. For an example, early on entering the Barnsley Area I got caught up in the affairs of the Grimethorpe Colliery Brass Band. They wanted me to become its President. I would have liked to have accepted but in the semi-hostile situation I was in that would have left me open to even greater adverse exposure. I supported them quietly in the background. Together with Jack MacKenning the National Welfare Officer, we procured the Band a full set of new orchestral concert pitched instruments, at a cost of some £5,000. Today they would probably cost some £75,000 to £100,000, including percussion instruments.

George Thompson had been its Musical Director for many years. He had taken the Band over when it was little more than a 'Second Section' Brass Band, way behind its rival Carlton Main-Frickley Colliery Band (situated some five or six miles away) at the time of its great achievements as a 'First Section' Musical

Combination. Over the years he had developed and built the Grimethorpe Colliery Band to be amongst the top ten bands within the UK. Under his direction the band won the major prizes both in Belle Vue, Manchester and the Albert Hall, London. However, George retired and the band engaged an instrumentalist from the orchestral world, Elgar Howarth, an excellent cornetist and trumpet player with a great talent for composition and musical arrangements. Elgar transformed the band from being one of the top bands in the brass band world to a place alongside John Foster's Black Dyke Mills Band, probably the most consistent top ranking band of all time.

As its new Musical Director Elgar Howarth opened up a route between the orchestral and brass band worlds, through which other orchestral men travelled to the benefit of the latter. His innovation in developing a modern percussion section to augment the brass band's basic configuration of brass instruments was historic and lasting. Under his direction the performance of the band was phenomenal, a situation which established the band's existence as a household name throughout the UK.

It was about this time we had an early visit to Grimethorpe Colliery from Lord Robens, accompanied by Norman Siddall. The visit went off quite well but our machine performances were not reaching the acceptable standards we were striving for, nor were we getting the levels the Board required. In his note of visit impressions to the Area Director he had a paragraph to the effect:

'Tell the Deputy Director (Mining) the coalface machines must be made to develop a range of achievements and successes which are consistent with those being currently obtained by the Grimethorpe Colliery Band.'

The message did not escape us. We worked at it consistently and effected some improvement. Maybe I should have sought out Elgar Howarth for a few tips on harmony and composition. We might have done much better. Whilst on this subject, it is convenient to recount the following incident and its consequences.

Returning to Area Headquarters from an underground visit at about 5.30 p.m. I found a written note on my desk. Lord Robens wanted to speak to me. However I had to wait until later as Hobart House had closed. Next morning I rang up Hobart House Headquarters. Lord Robens was away, but his secretary told me what was wanted.

Lord Robens and Dan Smith (Mr Newcastle) had been together at a dinner. Their discussions veered round to the BBC's 'Desert Island Discs' programme. Both Lord Robens and Dan Smith had previously been featured therein. In the latter's choice of programme he had featured a brass band playing the hymn tune 'Gresford', which had been written by Robert Saint in memory of the Gresford Colliery Explosion (22 September 1934) in which 262 miners had been killed. Apparently, after great effort he wasn't able to get a copy which he would dearly liked to have. Lord Robens told him, 'I've a lad in my organisation. If he can't get you one, then there is no such recording.'

My job was to meet this assignment. I rang all the record companies. There was no response. On contacting the BBC, I was told that the recording might well have been taken from a concert tape, but their library couldn't help in any way.

It so happened that the Grimethorpe Colliery Band were making a record for Richardson Music Publishers, (Inc:Wright & Round's) Gloucester, which was run by the brothers Beckingham. I think Frank was my main contact. I explained the situation to him and asked him to include the particular hymn tune 'Gresford' within the programme being recorded.

'It can't be done. The record company won't do that - there wouldn't be any demand and they would lose money.'

'In that case, tell your record people they are enjoying the luxury of having a first class band of some thirty brass instrumentalists without cost to themselves. If they won't make me my Hymn record, I will arrange to bill them the whole of the Band's wages throughout the recording's duration.'

I got my Hymn record, and George Thompson and I did all the work associated with its production other than the actual recording. The record was made, Lord Robens and Dan Smith got copies and more than 40,000 records were sold. Two more volumes were subsequently made and I later received some very nice letters from Australia and America from people who had found it inspiring. Incidentally the original Hymn recordings have recently been committed to compact disk.

Visit to American Coalfields Sanctioned

During early 1968 Lord Robens made a second visit to Grimethorpe Colliery and, as on previous occasions, was accompanied by Norman Siddall. This visit subsequently had very wide repercussions. Prior to the visit I had accepted an invitation to write a technical paper for presentation to the Midland Institute of Mining Engineers. The subject I had decided upon was 'Dirtless Mining'. Having in mind the American mining practice of working within the confines of the seam, I wanted to bring out the advantages of low cost, high productivity production common to the system. The American's mining engineers were and are experts in this field. I had travelled miles underground at several American mines in which the normal height was never more than 3' 6" with everything being tailored to accommodate this low height. Moreover, the men were ridden virtually the full distance to their working places.

Lord Robens was in the Colliery Manager's bathroom. I knocked at the door and was admitted.

'Have you everything you want, Lord Robens?'

He looked at me and laughed.

'Charles, what do you really want to see me about?'

I mentioned the foregoing and asked him if I could go to America for further background experience.

'Why not come with me to Germany? I shall be going within the next few months,' he suggested.

'Thank you, but Lord Robens, the Germans turn out more dirt than we do and their mining systems are similar to ours,' I responded.

'Let me think about it,' he said.

We went underground and, on arriving at the roadhead leading into the mechanised coalface unit, Norman and I crawled onto the coalface. The Area Director and Lord Robens were standing together at the roadhead when the former did an amazing thing. He turned down the light of his safety lamp and started to carry out tests for firedamp at the roadhead lip.

'What's the Area Director doing?' Norman asked me.

'The coalface deputy's job. He's carrying out tests for firedamp,' I replied.

In the dry, humorous way he had in such situations, he continued, 'You do employ face deputies, don't you?'

'Of course,' I replied.

'You can now save one,' he observed.

We travelled through the face. The machines at this time were doing reasonably well. Nothing contentious and no adverse event which often marred such situations plagued us and we pressed onto arrive at the surface without incident. Before leaving, Lord Robens, whilst I was talking to Norman, addressed us both.

'That proposed American trip. Let Norman fix it up for you.'

Norman was aware of what I'd done. He turned and said,

'When do you want to go?'

'As soon as you can fix it up,' I replied.

The Director never showed me His Lordship's 'visit impressions' note on that occasion, but I did learn later it contained a sentence to the effect: 'I expect more from my Area Directors than the ability to carry out a junior official's job.' I suppose if any of us act in a stupid manner in the presence of highly competent people it will be noticed.

It was about this time that at home one evening I had a phone call from Mrs R.G. Baker, my old Elsecar Main Colliery Manager's wife. Her husband, now retired, had suffered a stroke which had left him partially paralysed down one side of his body.

'Mr Round, Richard has asked if you could call in to see him. He's not too well but he wants to talk to you' she said.

I went down to see him immediately and arranged with Mrs Baker to tell me when she thought he would be overtaxed by further conversation. Being confined to bed he was really pleased to see me and wanted to know how things were progressing throughout the Area. He had come up with an idea which he thought might be helpful. We discussed what he had in mind, following which he said,

'When I get out of here I'll make you one to try.'

At that point his two sons Richard and Robin, both of whom were running a small factory somewhere in the Midlands, manufacturing parts for mobile caravans, came into the bedroom. Turning to his sons he said, 'I want you to meet my old colleague and friend Charles. He's the type of person who goes out and does what everyone says he can't.'

Very shortly afterwards I attended his funeral. Again I pay public tribute to another fine person to whom I owe a great deal. Fortunately there are more of them in evidence than the mean and petty types.

Norman duly made all the necessary arrangements for the American trip and, flying out to Pittsburgh, I met quite a number of the top people in the American Mining Industry. When I met the President of the Coal Operation, he said,

'I've met you before, in 1957 I think it was. You are the person who used his holidays and personal cash to come running round our pits. What a guy! was our general impression.'

The American mining engineers were good me. Apparently I had a reputation for being aggressive in the American sense. I got to know quite a number of them during the several visits I made over there.

After two weeks I returned home and during the flight back I gave a lot of thought as to the sort of report I would prepare about the trip. The usual practice was to give a general account of what had been seen, including anything new that had been discovered. In this instance, however, I did a lot of work as to how I could implement in a practical form within the Barnsley Area, the experience I had obtained in the States. I referred to this in my report to His Lordship. He had allowed me to go in the first place, so obviously it was to him I had the obligation to report back. His reply was short and direct:

'I did not send you out to America for the exclusive benefit of the Barnsley Area. Arrange with Mr Siddall to come down to Headquarters to state your views and intent.'

Norman had obviously had the same message. When I contacted him, he arranged a meeting. However, during the intervening period, I worked on and completed a Stage I Planning Submission (Norman ultimately accepted it as such) incorporating my views and my proposed intent. I took Johnny Williams (Group Manager) and the Area Chief Mining Engineer to the meeting, although in the latter case I debated long and hard on his inclusion, coming to the conclusion that not to have done so would have been petty. Moreover, it would serve to ensure that the Area Director was informed without the need for personal discussion. I had hoped his inclusion would attract support, but they weren't interested and actually did all in their power to discourage the project, together with the apparent indirect support of the Board's Member for Production.

Meeting the Director General of Production at Headquarters we went along together to the Conference Room. I found the place full with some twenty or thirty people present.

'Who are all these people?' I asked Norman.

'Director Generals of Finance, Purchasing and Stores, Industrial Relations, and various members of their staff. Some I don't know,' he informed me.

Norman took the Chair and invited me to speak. For upwards of a hour I spoke to the cosmopolitan gathering about where I'd been, what I had seen, whom I had met and the like. For the next three-quarters of an hour, I outlined how it was proposed to convert the experience gained into the formation and construction of a small colliery employing some 217 men and yielding 464,000 long tons per annum with an overall productivity of 10.0 tons per manshift (the national situation then being a little over 2 tons per manshift) and in line with American practice. A question and answer session followed, which was quite interesting to me. Summing up the meeting the Chairman said,

'We have listened to quite an absorbing discourse and some very progressive proposals, and will regard your presentation as an accepted Stage I Planning Submission. Go back, prepare the next Planning Stages II and III, and submit them as early as you can.' Norman courageously thus gave his blessing.

The tremendous import of what the Director General of Production had done on our (and indeed the Board's) behalf came home to me much later, when I learned that the Board's Member for Production had commissioned advice and support from a large firm of American consultant mining engineers, Paul Weir & Company, Chicago. This was with respect to converting Ellington Colliery, Northumberland Area into a 10 tons per manshift (later reduced to 6 tons per manshift) colliery by the introduction of modern high production American equipment. I am sure Norman realised himself the risks and politics of supporting such a venture in these circumstances. I considered it a tribute to his courage and foresight to have had his backing in these circumstances.

Ellington Colliery was distantly working (some four or five miles) under the North Sea. The mining conditions were excellent, the seam was 14ft. in thickness and had a good roof and level floor. Later when I inspected the proposed site the potential appeared to me fantastic. In the Shafton Seam at our proposed new mine (Riddings Colliery) comparative conditions were: seam thickness 5' 0" approximately, gradient mildly inclined, roof a fairly soft mudstone, the floor a reasonably strong fireclay. The conditions were similar to those of the Haigh Moor seam at Elsecar Main Colliery in which I had formerly gained a great deal of relevant experience and where we had already established a blueprint for what we were seeking to achieve.

Following the submission of the Planning Stage II, we received a list of sixteen embarrassing questions (referred to by members of Headquarters mining staff at the time as 'Shep's Blockbusters'). They were designed at best to embarrass the proposals, or at worst to nullify them. Both the Area Director and his colleague were, I believe, enjoying and indeed supporting in any way they could W.V. Shepherd's negative intervention. The questions were dealt with without any further comeback, and the way was now clear for pushing further ahead with the Stage III planning proposals.

The Stage II and indeed Stage III proposals ostensibly were the processing responsibility of the Area Chief Mining Engineer (who had no interest other than that of informant to the Area Director). They were both produced under my personal direction. Having had great experience in planning, I was able to take it in my stride within my normal duties.

Fourteen copies were prepared, which I took into the Area Director for signature. He sat on them quite some time with no action. I was, however, called down to Headquarters on a particular day, so the day before, I retrieved the unsigned Stage Planning II Submissions, and took them along to the Area Chief Mining Engineer's office where he and his colleague were together. I walked in unannounced, placed them in front of the Director, and said, 'I'm taking these down to Headquarters tomorrow signed or unsigned; it's your decision.'

'I haven't got a pen,' was his childish response.

His colleague, with highly commendable alacrity, produced a pen. He was most adept in such situations. The documents were signed and duly deposited at Headquarters. A month or so later, Norman phoned me.

'Charlie, come down here for two o'clock tomorrow afternoon. Jack Weir [Mining Consultant Engineer] is over from the States. I'd like you to discuss Riddings New Mine with him - it's all been arranged.'

Walking into the Headquarters conference room the following day I found that Norman had called in John Adcock (Headquarters Mechanisation Engineer), Peter Rees (Mining Staff) and others. 2.00 p.m. 2.30, 3.00, and 3.30 passed but no Jack Weir (Paul Weir & Co: Chicago). Shortly after four o'clock, in walked W.V. Shepherd with Jack Weir, stating, 'Jack hasn't much time; he has to catch the 4.30 p.m. train from Kings Cross to Newcastle, (on a scheduled visit to Ellington Colliery).' He then went round the table and introduced Jack Weir to all those present. I was the last. Turning to Jack Weir, he said with a supercilious grin on his face,

'I'd like to introduce you to the conductor of the Grimethorpe Colliery Brass Band.'

I was seething with anger and contemptuous of the man who caused it. Jack Weir immediately left to catch his train to Newcastle and to Ellington Colliery, relative to their own project. The Director of Production had, I believe deliberately created a delay to prevent the meeting taking place: a full day wasted. As he left the Conference Room I followed him into the office across the corridor. Turning on him angrily I said,

'Shepherd, if ever you pull a stunt like that again I'll bloody well go for you in public, regardless of the consequences.'

'I'll not have you swearing in front of a lady,' he responded.

'My apologies, Miss,' I said, turning to the lady. Turning back on the objectionable character he had shown himself to be and looking him squarely in the face, with a very heated expression I said,

'You bloody bastard,' and walked out of the office.

Regretfully and shamefully I had fallen to the level of Jack Woffinden. Although the provocation was strong I make no excuses. I ought to have risen above such treatment. His references to the Grimethorpe Colliery Band, I firmly believe, could only have been fed to him by the Area Director who, I now more strongly believed, reported everything back to him in order to sustain his support and imaginary strength.

THE BOARD APPROVES THE RIDDINGS PROJECT

The submitted Planning Stage was accepted. We were now at the final hurdle, the preparation of Planning Stage III, which was completed very quickly. In clearing Capital Projects the Board had changed its former system. It now required the Area Director to present such Projects personally. In this case, the Area Director had no choice but to take me along with him as he didn't have the depth of knowledge with which to present the scheme. In any case, he could hardly have gone against his mentor W.V. Shepherd who, in my opinion, was so obviously against it.

We met the full Board. Lord Robens opened up by referring to the project being considered and asked the Area Director to outline the situation. He excused himself by saying, not untruthfully, that I had worked out the detail and would outline the project. I summarised in detail all that was involved but even at this stage W.V. Shepherd showed his petulance, and persisted with the types of question designed to embarrass what was being proposed. Thankfully I had Norman Siddall on my right-hand side, who came in with great support at the least sign of faltering on my part.

After a period of questioning from other members of the Board Lord Robens summed up the situation by stating,

'We have heard a full account of what is undoubtedly a most progressive and profitable scheme, one which you have all examined. I feel we should give it our support. Does anyone feel otherwise? Thank you, gentlemen, the project is approved at a cost of approximately £1,000,000.'

The Board's Member for Production's reactions were unmistakable and left me in no doubt that any failure on my part would invoke dire consequences to myself. That two hour train journey between Kings Cross and Doncaster was fraught with the most frightening of thoughts, not of the scheme about which I was fully confident, but of having indirectly involved Norman and His Lordship with commitment.

'Not the time to get cold feet boy, go out there and show them,' were my conclusions before alighting at Platform 2 at Doncaster Railway Station.

A working party under my Chairmanship was formed to execute the project and to monitor progress. In addition to the Area Staff, we had assistance from Headquarters personnel. Between January 1968 and March 1969 when the projected scheme was completed, twenty meetings were held. The appointment of Colliery Manager was handled with great care. We scoured the country and held a number of interviews. Finally we appointed Trevor Massey who had more than proved his worthiness for such a post at Woolley Colliery. Trevor did an excellent job, although he was subjected by myself to really intense pressure exacerbated by the nature of things already described. Such pressures he carried extremely well, without whinging or complaint. I had made an excellent choice in his appointment, subsequent to which he made remarkable progress in his career.

Working party meetings were held on Friday afternoons, which gave me the weekends free to contemplate, adjust and modify in the light of progressive factual detail. They were a source of irritability to the Area Director. We were making excellent progress and I had managed to establish an overall level of enthusiasm which had accepted success of the scheme as beyond question, with constructive relevant ideas from the members flowing freely. The Area Chief Mining Engineer ventured no involvement, either because of inability, or of being caught with his personal loyalties to the Area Director being paramount.

Area Director Publicly Threatens Board Discipline

We were approaching completion of the scheme and, during a normal working party meeting at about 4.55 p.m. one Friday afternoon, the Area Director burst into the meeting in front of some ten or twelve of its members (unfortunately no members of Headquarters Staff were present on that occasion), completely out of control. He advanced towards me at the head of the table waving some papers, pausing to pick up a large bolt eight inches in length (one we had been discussing on account of repeated failure). He came towards me shouting in almost incandescent rage.

'You won't get away with this, I'm going to report you to the Board,' he kept repeating, spluttering and wild-eyed. The situation was so ludicrous that I didn't in the least get angry at the time.

'Here is the phone – ring the Chairman right now,' was my immediate reponse.

The other members round the table looked on in disbelief, I noticed that he had a Capital Project in his hand. Instinctively feeling what it might have referred to, quite coolly I turned to him and said, handing the phone to him, 'Make your call now. I am not prepared to subvert Capital procedure for you or anyone else. This meeting is almost over. We will discuss the situation, whatever it is, in your office.'

He slunk out of the room, a very silly and stupid man. After he had gone out Johnny Williams the Production Manager said, 'Gaffer, how you kept your cool I'll never know. What a stupid exhibition,' (the general opinion of almost all the people round the table at the time).

The meeting lasted for a further ten minutes. I went straight to his office. He'd run home. The apparent cause of his outburst might have been my refusal to sign a Capital Project format subject to further explanation and supporting information. I had held up a Barrow Colliery project scheme for about £150,000–£200,000 which was supposed to directly save some 36 men, but having thoroughly examined it there was no such saving. I had sent it back for detailed proof before I was prepared to authorise it at my level. I wasn't able to verify this and he may have been attempting to show his authority in front of my staff. However, despite my former quiet bearing, afterwards I was annoyed. Over the weekend I prepared a report of the incident which I intended to send to the Board Chairman, but decided to seek the advice of Norman Siddall before taking action. On the Monday morning, the Director met me at the entrance to my door, totally dishevelled as if having been deprived of sleep.

'If I eat humble pie will you forget the whole incident?' he virtually pleaded. 'How can I or those other men round the table forget such a demonstration?'

'I want neither you, nor your humble pie,' was my observation.

I had neither trust nor respect for the man, knowing that had our roles been reversed, he would really have enjoyed crushing me. He disappeared for the rest of the day; no one knew where. Duly I fixed up with Norman Siddall to meet him the following Tuesday afternoon at Hobart House. When I was shown into his office his first words were,

'What have you been doing to your Area Director? He was down here to see Shepherd at about four o'clock yesterday, and looked as though he hadn't slept for weeks and had been pulled through a thick hedgerow. I met Shepherd on the stairs. He said, "We have to do something about Round." "What's the matter?" I asked him, "Has be been upsetting your little friend?"'

The Board's Member for Production apparently made no further comment. Norman then read the report I had prepared and said,

'Forget it, Charlie, it won't help - you'll be regarded as the one spreading the conflict.'

'OK, Norman - tear it up,' I said, which he did.

It was in such situations that I appreciated having him to turn to. That was the end of the incident except that, for the next three years or so, life became very much easier. If any members of the staff went to the Director on matters relevant to myself, he would say, 'Have you discussed this with Mr Round? See him first.'

Concerning the Submitted Project I feel a short account at this stage is worthy of inclusion in view of its historical interest and the phenomenal success which it had.

THE RIDDINGS DRIFT MINE

The Riddings Colliery Project was scheduled to become a producing colliery during 1970/71. It actually went into service as a revenue working unit on 1 April 1970. It was designed on the basis of producing 464,000 or 519,680 (long or metric tons respectively) per annum, with a total labour force of 217 men. The total cost of production (inclusive of interest and depreciation) was £2 5s. 2d. per ton with a cost per therm of 2.2d.

PLAN 3 shows detail of the surface provisions made, including 4,000 ton stock-piling facilities, diverted dirt handling provisions, road transport coal-loading arrangements, material stockyards etc, Manriding facilities conveyed the men in the early stages from the embarkation station near the lamproom to the bottom of the access surface drift. A cable belt was installed as the main coal transport system. (During my career at Gedling Colliery, we did much to help with the early development of this high capacity long distance conveyor system.)

PLAN 4 shows the underground layout, based entirely on 'in seam' and 'retreat mining', adopting small 75 yard fully mechanised faces. Drivage of the coal headings was undertaken by three Lee Norse C28 American continuous miners. These machines averaged some 300 yards of cumulative drivage per week, which left but little margin over the required rate of drivage, in that the production panels were being extracted at the rate of some 200 yds per week. Although I gave quite a lot of thought to the problem, unfortunately it was not until I had retired from the Board that I came up with a practical proposition subsequently patented, but it was never tried.

FIGURE 10 shows The 'Round' Roadhead Hydraulic Support System. This diagrammatic sketch indicates the format and principle of operation of the system developed in 1972. The concept behind the thinking was to try and mechanise roadhead support operations in step with the continuous miner, with the permanent supports being erected at a distance from the face of the coal heading safely within the restrictive 4 foot interval support distances operative. Circumstances were such that I was unable to pursue the situation to a successful conclusion.

The mining situation closely resembled that of the Haigh Moor Seam at

Elsecar Main Colliery such that Photographs 35 to 38 are virtually pre-replicas of what was undertaken within the Riddings Drift Mine. The explanations relative thereto hold good for this situation also.

Messrs. Dowtys produced a new support for us (their 6 × 8 rigid base supports together with their 'Mcco In-seam Conveyor System') for use in the coal heading drivages. Anderson-Mavor provided us with the American Lee Norse CM28 continuous miner (built in the UK) for the coal road drivages, whilst our old friend Bob Thorpe provided us with their latest 'state of the art' Anderton shearer-loader at the time.

These companies and their representatives entered into the spirit of the challenge we had set up with faith and great support. The Colliery's perform-ance, in terms of machine fulfilment for the first year of operation, was 24 machine strips per shift, 62 strips per day, 268 strips per week and an average of 70 tons per face manshift, with 9.7 tons per manshift overall giving a total output in excess of 470,000 (526,400 metric) saleable tons. The following year was even better with an output of some 500,000 (560,000 metric) saleable tons and an overall productivity of 11.7 long tons per manshift. During its first two years Riddings made a profit in excess of £2,000,000 after depreciation and interest.

The following two photographs were taken on the occasion of the official opening of Riddings Colliery on 8 August 1970.

PHOTOGRAPH 58 was taken after a visit underground to the coalface installa-tion.

PHOTOGRAPH 59: Lord Robens talks to Mary Round. They had met previously on different occasions. Here he was telling her about the Riddings Scheme, how it developed and what had been achieved, thanking her for her patience and enquiring about our daughter Dorothy's progress, having formerly met her at one of the Ammanford Technical College speech days, where she was officiating on behalf of the guests.

Several weeks after the Riddings Project was established I received a letter from Professor Dick Velzeboer from the Mining and Minerals Department at Delft University, Holland. Would I go over and give a lecture to his students on the Riddings Project? The faculty had about thirty Mining and Minerals students who all spoke excellent English. The lecture went down extremely well and some very pertinent questions were asked. They had studied and obtained quite a good knowledge of the project before I had spoken to them. Shortly after returning I saw W.V. Shepherd at some meeting or other.

'I've had a glowing account of your antics in Europe,' was his only comment.

Things were moving and progressing throughout the Area. Colin Rudge (Area Mechanisation Engineer), like myself, was never really happy working within the atmosphere the Area Director created. We found relief in doing and creating things which were positive and progressive. The fact that we were never chal-lenged allowed us get things done and the Area benefited greatly. Leaving South Yorkshire was a great disappointment to Colin, who had invested a great deal of past effort in its development and, at the time I left, we were both engaged in activities with exciting and exhilarating prospects. However, in the Barnsley

Area he was of great support to me. His sense of humour eased many a difficult situation. His substantial contribution to all that has been recounted I gratefully acknowledge.

What was achieved in the Riddings Project was truly remarkable and well before its time, since a further twenty years or so were to pass before any such overall productivity performances were to be achieved in the Selby Coalfield with mining equipment that was vastly superior to that which was available to us during that pioneering effort.

What of the Board's Member for Production and my Area Director? Were they happy with the project's success? We never heard anything from either of them, nor expected to, since from the concept to the completed execution of the scheme they didn't want to know, took no interest, and never made any constructive comment to me whatsoever. But I almost forgot: the Area Director made the final contribution. He organised the opening of Riddings Colliery on the 8 August 1970 without any reference to myself. Mary and I were quite surprised to find ourselves on the attendance list but he placed us both at the extreme left-hand of the long top table. Maybe he obtained a greater sense of achievement in organising this event than any of us did with the concept, construction, development and success of the project and the heavy industrious application it entailed. However I didn't mind really. With the completion and success of the project, I was relieved of great stress built up over the previous eighteen months.

Letters of Recognition and Personal Encouragement

In a Foreword to a technical paper on 'The Riddings Drift Project' written for the mining magazine *Mines & Quarries*, Norman Siddall, the Director General of Production (currently Sir Norman Siddall), summed up the situation neatly as follows:

> Many problems which face the industry and some of their answers are kaleidoscoped in the Riddings project and clearly identified in this excellent chronicle of events, in which there has been the courage to record failures as well as the successes over the period of design and construction. The diligent approach and painstaking detail with which the project was carried out is exemplified by the fact that it was completed ahead of schedule, within estimates of cost, and is achieving results in excess of those forecast.

In a private letter to me dated 21 June 1972 Lord Robens stated:

> *Riddings Drift was not the easiest project in the world but I am always pleased we pursued it to a successful conclusion. The scoffers came to scoff, but stayed to cheer.*

In a private letter to me dated 14 January 1972 Sir Derek J.Ezra (now Lord Ezra) said:

> *Your vast experience of mining, particularly in the Yorkshire Coalfield over the years has been a great help to the Board, and your personal contribution to the Riddings Project was an outstanding example of management achievement.*

In a private letter dated 11 January 1972 Wilfred Miron (formerly East Midlands Divisional Chairman) stated:

I should like to thank you for all you have done in the Industry, and in particular in the East Midlands Division in the days before you went down to South Wales as well as subsequently. You were always a tower of strength and one upon whom I could rely for skill, devotion to duty and good humour.

In a private letter dated 10 November 1970 Mr Alfred Kellett (Former Chairman of the South Western Division) wrote:

Dear Charles,

I shall always remember the energy with which you tackled the almost impossible problems of West Wales, keeping cheerful and optimistic despite all.

The miners of West Wales have a lot to thank you for and Abernant and Treforgan should remain live monuments to your efforts for many years.

I hope that in the home climate of Yorkshire you have found full satisfaction for your work despite the frustrations and disappointments caused by the reorganisation.

Yours sincerely

A.H. Kellett

The number of letters received from Norman Siddall throughout almost every stage of my career to me were invaluable. Some were funny, others more serious and advisory, but most importantly they were sustaining in their effect. I recall the first letter I had from him on the occasion of pressing him impatiently in two directions. This I thereafter referred to as:

The Parable of Ferdinand

The old bull Ferdinand and a young bull were grazing in a field separated from a herd of cows by a large hedge.

'Come on, Ferdinand, let's jump over the hedge and each enjoy the company of one of those cows,' yelled the young bull.

'Hold on,' said Ferdinand, 'Let's wait while they open the gate and enjoy them all.'

Quite recently during a TV chat show I heard the above story a second time some forty years afterwards. Maybe the parable has had some effect, I don't know. Such judgements must come from others. When one is consumed by a natural enthusiasm to develop and achieve, working for people of the above high and generous character becomes a privilege, for they bring things out in one's makeup one never thought existed.

Lynemouth and Ellington Colliery Investigations

On a number of occasions I have mentioned the uncertainty of fate. Here follows another classical example. The Riddings Project now behind me, out of the blue I was asked by the Board to undertake an investigation of Lynemouth and

Ellington Collieries in Northumberland and was given the choice of one of three youngsters to assist me. Gordon Sykes at Woolley Colliery stood out amongst them - I selected him and what a fine choice it was. He was very intelligent and unassuming, had an experienced background which would be quite useful, and was most industrious.

You may recall that at the time of W.V. Shepherd's reference to me as 'Conductor of Grimethorpe Colliery Band' Jack Weir caught his train up to Newcastle and from there to Ellington Colliery, relative to W.V. Shepherd's project for raising the colliery's overall productivity to 6 tons manshift. His visit took in the American installation which was in its early stages of development.

Thinking the situation through was most perplexing. Here was I, after an interval of eighteen months, being asked to determine why the performance of Ellington Colliery had declined in a situation where all the advice and experience of one of the world's largest mining consultancy organisations was on tap; where W.V. Shepherd, the Board's Member for Production, had initiated a major project and had maintained a direct interest in the colliery through its Area Director. I really had serious problems. To go in and do the job properly would call for a lot of analytical digging and a presentation of the true facts as I (or anyone for that matter) interpreted them to be. There would be embarrassment at some level which in the circumstances stretched right up to the Board.

To undertake a superficial investigation and present a political innocuous report could avoid such embarrassment, but to do this would constitute deceit to the Board, to myself, and indeed to the Ellington miners involved. Such an approach would not provide a basis for a resolution of the colliery's difficulties.

It was not a question of choice. I had to go in and do the job properly without fear or favour, as this is in my makeup and the only way I know, despite the problems it created. I'll deal only with Ellington Colliery, the situation at Lynemouth Colliery being very much the same. The seam at both collieries was about 14' 0" in thickness and the mine workings were some four to five miles out under the North Sea. Young Gordon Sykes and I started by thoroughly investigating the mining circumstances. One could discount the situation of mining layout or the systems relating thereto, as the colliery had been highly profitable within the logistic parameters involved. I felt the basic problems were to be found in the supporting services: transport, ventilation, organisational arrangement, working conditions, labour practices and supervisory arrangements. These we subjected to intense 'in depth' investigation and application from Monday to Friday working from 8.30 a.m. to 5.30 p.m. each day, Young Gordon was a treasure. One could set him off on a line of analysis or action and he'd follow it through quietly, with great thoroughness, and what he produced was truly accurate. That was obvious to me after making a few early checks. It saved a tremendous amount of my time, as I could fully rely on him. Like myself, he also had the ability of winning people over into accepting a part in whatever was being undertaken.

Whilst we were engaged in our undertaking, one morning I called in to see the Colliery General Manager, Tom Smith. He had a problem. Water had broken into his newly developed Yard Seam, and had flooded the coalface so that the whole of the new district was out of action. People were calling in from Area Headquarters enquiring about the situation and then running back to their offices.

'Tom, isn't any one going down to assess the situation?' I asked.

'You've seen what's taken place. I'm going down myself within the next thirty minutes,' he replied.

'Right, Tom, I'll come with you,' I told him.

'You will? It's not your problem,' he replied, quite astonished.

'It is. You're in trouble, your pit's in trouble; we've got to find out why,' I responded.

Tom was an ex Newcastle policeman well over 6ft. tall and of great physique. Off we went underground. From the manriding road access to the new Yard Seam the distance was pretty short. At that particular end of the coalface water had broken in and was flowing through quite steadily. The equipment along the face was lying under water. However, all power to the district had been cut off so there was no danger from that aspect.

As we went further into the district the depth of water had risen up to my waist. Some fifty yards further on it had risen to just below my nipples. Beginning to get concerned, I enquired,

'Tom, when do we need to get out of here? I'm getting worried.'

'You're quite safe unless we come across some flat-fish,' was Tom's limit of concern.

'The water's coming up to my bloody chin, Tom.'

'Don't worry, Mr Round, stick with me, I'll keep your head above the water,' was his assurance.

As we moved into shallower depths and through the worst, he turned and said,

'You know, Mr Round, if you really intend to carry out many inspections under these conditions you must get a pure flannel suit; no matter how wet you get you won't get cold.'

Flannel suits for many years were the common working apparel for shaftsinkers and drifters working in wet conditions.

We established the location where the water had broken in. It appeared, from the cursory inspection made, that were there was a strong possibility that the district would need to be sealed off. That night, the two of us, together with his Group Manager, Bill Davidson, went out for a drink to a remote pub way out in the wilds somewhere. What a night we had! Tom recounted his police experiences in Newcastle, in particular his attendance at football matches. Next morning, I called in to see him before going underground.

'Hey, what did you two sods ply me with last night? I must have had at least six whiskies.'

'You had fourteen doubles; we didn't want you to get a cold after your morning trip,' he said.

(Fourteen doubles would have been my last drink ever - Tom's embellishment.)

'No wonder I slept well; I must have been almost out.'

'You were,' he said with a twinkle his eye.

The moral here is: 'If you go paddling in water up to the armpits in the morning, don't go out drinking with a couple of burly Northumbrians at night. You might not get back.'

The impression I got from this experience was how lightly the Area staff that morning seemed to view what I would have considered a serious incident, and that their apparent lack of support of the Colliery General Manager with difficulties might well have been more than it appeared to be.

Having thoroughly checked through the standard of underground services and communications at both pits we turned our attention to machine performances, checking back records. We discovered that at Lynemouth a number of strange practices operated, whereby some machine crews were being subsidised by work undertaken on the night shift, which gave the impression the machine performances were higher than those which actually obtained. The problem here is that when the circumstances which permit such subsidies are no longer present or possible, the true level of performance gives the impression that the men are no longer working as they did, the basis of a possible excuse, in relation to a drop in colliery performance.

We had numerous discussions with the NUM leaders (who were quite helpful) at both pits, one of whom persuaded me to attend the Northumbrian Miner's Gala which took place one weekend when I didn't go home. I enjoyed the occasion but it wasn't quite in the same league as the one I saw in Durham.

Before completing our task, I asked to be taken to the W.V. Shepherd's/Paul Weir & Co. American Project Section, in which the modern American equipment had been installed. I found the equipment was operating under gravely disadvantageous circumstances, such as water, sludgy floor conditions, and delays in the material supplies. Coal clearance in no way matched the potential of the equipment and working time was less than it might have been.

The basic mining situation with 14ft. of coal, a good roof (roof-bolting having been successfully practised on quite a wide scale) and a reasonable level floor really did have great scope for achievement under normal circumstances and I can quite understand the basis of choice for the projected venture. However, I felt that an inadequate analysis of all the circumstances in relation to such a project prevailed and there was no apparent real enthusiasm at all levels pushing it along.

Lord Robens in his Foreword to 'The Riddings Drift Project' made the following statement:

> There is no substitute for careful preparation of management, men and machines. Only by ensuring that every 'i' has been dotted and every 't' crossed before operations begin, can any team really reach the potential levels of productivity of any system. Riddings Drift is a living example of this type of planning skill and will take its place in the top league of productivity performances.

I never discussed with Tom or anyone the American Project, but my private thoughts were:

As conductor of the Grimethorpe Colliery Brass Band I had chosen music the Band could play, and had all the bandsmen with their music in place keeping in strict tempo with the requirements of what was being played. This was a situation which was not to me remotely in evidence with the Board's Member for Production's American project situation.

Having finished our reports Gordon and I first went and discussed them with the Colliery General Manager and the Production Manager. They accepted them as being factual and representative of the circumstances prevailing. The reports made no personal criticisms, nor expressed conjecture. We merely stated such facts as 'supplies took from 24 to 48 hours to reach their destination; hot and humid conditions drained energy from one's body irrespective of effort; men's working times were low in many situations; serious coal clearances were in evidence; a large amount of coal was being lost through spillage etc.'

At the end of the reports we made appropriate recommendations. I realised our reports would give rise to embarrassment, but in the interests of the collieries and their employees that was to me inconsequential. Because of this I sought a meeting with the Area Director which would have been helpful to him in that any explanations he may have had could have been incorporated in the reports. Moreover, any advice I may have had could well have eased the position. He said he would arrange a meeting after he had examined our proposed submissions, but, despite several requests, he failed to arrange one. I had a similar experience with another Area Director whose excuse was the amount of detail produced. I often wondered if he, and others like him, were aware of the voluminous detail provided in their respective daily press, *The Times*, the *Financial Times*, the *Telegraph*, the *Independent*, etc., or other signs of executive prowess usually prominently displayed in strategic places on their desk tops for visitors to see, and whether such information as contained therein had more to contribute to their operations than the vital information they were failing to obtain, relative to the vital and efficient discharge of their responsibilities?

There is little point in managers being pushed for production at the price of overstretching communications, services and the overlooking of essential developments. All systems grind to a halt under such policies once they have absorbed the applied momentum. After submitting my report to the Board, I received a copy of a whinging letter which the Area Director had written to the Director for Production. The man obviously lacked the decency, or hadn't the courage, to send me a copy, despite the fact that the work we had put in was all on his behalf.

His former excuse to the Board that the men had reduced their effort, the basic point of complaint, had been destroyed by the facts now to be laid before the Board. The real point was one of neglect in allowing the pits to deteriorate in the manner they had. I didn't blame the managers; the fault wasn't theirs. It lay at higher levels within the Area organisation. In his letter he virtually suggested elements of fabrication. However, I had the basic detail fully prepared to prove the situation. I could understand him writing to W.V. Shepherd as he had a direct connection with the colliery through his project. Whether he sent a copy of his correspondence to Norman Siddall I don't know, although I doubt it.

Prior to being called into the Board Meeting we were placed together in a small room. Even there the man wasn't prepared to discuss the situation before we were called. He did issue what I felt to be a thinly veiled threat, to the effect that matters weren't finished with. I don't know what he was trying to imply and I never did find out. Maybe he sent another letter, the contents of which were never made known to me. It is just as easy secretly to send two attempted personally adverse communications as one. However, as circumstances later indicate, my future opportunities for a return to my former post as Area Director had been blighted. Maybe this situation partially contributed.

We went into the Board Meeting. The Board weren't after blood but were only concerned to have the situation rectified. I thought they were extremely tolerant, fair and decent in the circumstances.

I have no doubt that thereafter through the Board's Areas, I was represented unfavourably amongst the Area Directors. If that was so, then that must be the price one must pay for trying to be factual and honest. In fact, I recall at a later date meeting his successor; I believe I had previously met him in Scotland in connection with a visit to a Webster Miner installation. This is what he told me:

'When I first read your report I thought what a bastard, so I went to see for myself. I couldn't fault a paragraph. How right you were.'

Under his direction work was initiated. It took several months to sort matters out following which Ellington Colliery did some fantastic things and made substantial profits (so I was later told). Lynemouth Colliery, the more difficult of the two, was subsequently closed.

It would not have surprised me to learn that either or both of the managers involved had been subjected to heavy criticism, maybe censure, following the event. Neither man showed or indicated any disloyalty to their Area Director or to anyone else, and in a similar situation I would not have wanted them to have acted any differently than the manner in which they had conducted themselves throughout our investigations. But then I never had a personal 'ego' problem and I always subordinated personal matters as secondary to the interests of the job I was doing at the time no matter at what level I was acting.

However I thought the treatment received from the Area Director to be but a poor reward and acknowledgement for the tremendous effort young Gordon and I had applied on his behalf and the fact we had pin-pointed the basic problems involved. Our rewards did come, however, in the form of the letter below:

National Coal Board
Hobart House
Grosvenor Place
London S.W.1

C. Round, Esq., Deputy Director (Mining)
Barnsley Area

27th August 1970

Dear Mr Round
Performance Improvement Team Enquiry into Lynemouth and Ellington Collieries

The Steering Committee on Organisation and Policy Planning have considered the report of the Performance Improvement Team on Lynemouth and Ellington collieries, and have accepted the recommendations of the team.

The Committee have asked me to express to you their thanks for your contribution to the work of the team, and their congratulations on the excellence of the report. I am sending a similar letter to Mr Sykes.

Yours sincerely

(sgnd) Martin Shelton

M.S Shelton.

I was also rewarded by watching the excellent progress young Gordon Sykes subsequently started to make in his career, to which I hope I had been able to contribute, and by the later knowledge of Ellington Colliery's excellent recovery under its new Director, which more than justified our endeavours in face of the childish petulance which followed.

At the end of February 1994 on TV I watched the last four ponies being withdrawn from that colliery, thus completing the total closure of the mine upwards of twenty-four years after we had submitted our detailed report which opened up a new lease of life at the time.

Lord Robens' Invitation to Join Headquarters Staff

Later Lord Robens invited Gorden Sykes, myself and others who had carried out similar tasks in other Areas, to a dinner in London. At the dinner table he invited me to sit to the right of him. Throughout the dinner he referred to the things we all had done and thanked us individually for the parts we had played. Towards the end of the dinner he turned to me quietly and said,

'Charles, how would you like to come down and work for us?'

'Whatever you want me to do, Lord Robens, I'll gladly do the best of my ability.' I had no idea or inkling of what he had in mind. When you have developed true respect and faith in a person you naturally trust without needing verification. I was never badly let down.

'We'd like you to do the sort of thing you have just done on a National scale, to go in and help Directors sort out any special difficulties they may have.'

'That will be fine, I will be happy to do as you suggest, Your Lordship,' signified my acceptance of what later became described as the post of Director (Special Duties) located in the Board's Doncaster Headquarters. The new post had to be cleared administratively through the Board Member for Production when I subsequently met him in his office. He was quite pleasant and made no reference to anything in particular or generally for that matter. His final words were,

'Keep your eye on Kent,' which I erroneously took to mean 'Get stuck in!' for the Kent Coalfield had been a source of trouble for many years.

I wound up my affairs in the Barnsley Area and left in a manner even worse than my entry. I just walked out. The Director avoided me, although he was in his office. Reflecting on this, I'm glad it happened that way. Anything he might have said I would have considered with little faith: such were and are the honest feelings I had and have. I was glad of being spared the ordeal, although it was a new but strange experience.

Subsequently the Area Director was transferred (in my opinion by his mentor) to fill the third vacancy in the South Yorkshire Area following my earlier transfer, in a matter of some four years or so. Cliff Machen had died rather suddenly, and his replacement John Booth prematurely retired from the National Coal Board after a comparatively short time in office.

Johnny Williams, the Production Manager, and others rang me up to say the Area Chief Mining Engineer had been round to all the staff cracking the Area whip towards boosting the presentation funds relative to the Area Director's forthcoming departure.

Mike Eaton, who had previously taken over my post, was elevated to the Director's chair. On one occasion someone told him I was in the Area, and he asked me to call in his office before leaving. My former excellent secretary Mrs Jean Lewis, a highly capable and fine lady who worked extremely hard (she always regarded me as a 'workaholic' but on no occasion did she fail to keep up with the pressures involved on my behalf), who was now working for him, took me into his office. I sat down at his desk and found it strange to find no one standing on the elevated hearthstones. He sat opposite.

'Thanks for calling, Charlie, I've been wanting to meet you for quite some time.'

'Why, Mike?' I asked.

'I wanted to thank you for the way you integrated the two areas, and the manner in which you built up the combined Area so that I inherited a first class set-up.'

Mike was totally different in character to his predecessor. He had a quiet charm, a natural sense of decency and a generous spirit, not unlike that of Cliff Machen upon whom I often felt he had modelled himself.

During Margaret Thatcher's confrontation with Arthur Scargill in the miners' strike of the 1980s he virtually operated in the role of a Government spokesman. How he fared in association with the politicians is something I have to ask when I next see him. I'm still in the business of acquiring new experiences, although I would prefer them not to be too exacting for an octogenarian.

With the succession of vacancies over a comparative short period after I had been moved into the Barnsley Area, the fact that I was never approached to undertake a return to the Rotherham Area (particularly following the success of the Riddings American Project and the acknowledged Ellington/Lynemouth investigations) confirmed to me the adverse personal nature of the interview with the Board's Staff Director and Board Deputy Chairman referred to and indeed indicated the inadvisability of expressing forthright views to authoritarianism. Narrow-mindedness exists at all levels within an organisation. However compromising one's natural character for personal aggrandisement and self interest, in a situation where true motivation was always to achieve the best one could within and on behalf of the organisation being served, would have exacted heavy personal price with no guarantee of success. It might well have been that secret misrepresentation by the Area Directors of both the Barnsley and Northumberland Areas, in the manner described, adversely contributed to this situation, one is never able to determine such matters.

Disappointments arising out of such situations are best overcome by seeking new outlets for endeavour, irrespective of the difficulties obtaining within a monopolised undertaking. Cheerfully accepting things as they are, not as one would like them to be, and moving along with a positive approach avoids the cancerous growth of inner bitterness which can, and no doubt does, destroy a person from within, particularly as there is so much pleasure and satisfaction to be obtained in creative endeavour, whatever the form it takes. Such was my experience with the Riddings Colliery American Project, which in other circumstances would never have seen the light of day.

The Kent Coalfield

Settling in at the Board's Doncaster Headquarters took but little time and, in the absence of something positive to do, I decided to spend as much time in Kent as possible. Obviously, I could hardly keep my eye on Kent if I didn't know what the Kent situation was in terms of its collieries and circumstances. Booking in at a small Dover Hotel for a week at a time and returning home to Mary and Dorothy in Rotherham on the Friday night for the weekends, I made a start with a degree of background knowledge I needed to check for myself.

JIM STONE: FORMER AREA DIRECTOR DONCASTER AREA

Jim Stone who at one period of his career was Area Director of the Doncaster Area had undertaken on behalf of the Board an investigation (similar to the one we had done at Ellington and Lynemouth) into the situation prevailing within the Kent Coalfield. His report was quite a critical indictment of the Management and the situation within the coalfield in which the three basic collieries were Bettshanger, Snowden and Tilmanstone. As in my own case, his report could not avoid embarrassment to the coalfield management such that coalfield criticism took the form of personal gossip as to the manner the assignment was undertaken, with allegations of drinking sessions and his general behaviour during the execution of his commission. Irrespective of these diversions the basic structure of his report was sound.

My first approach was to determine the mining situation prevailing at each of the collieries. With this in mind, I spent a substantial amount of time underground thoroughly examining all parts of each mine. I went to much greater depths than Jim Stone. However my inspections confirmed to a great extent his findings and opinions. The Manager in charge was running the Area in much the same way as operated within the Barnsley Area, with an atmosphere of suspicion. People were afraid to talk, and ideas appeared to be blocked, yet all the collieries warranted positive action.

The first meeting I had with the Coalfield Manager was to the effect that he considered Jim Stone had misrepresented the true situation prevailing and the manner in which he carried out his assignment left a great deal to be desired. He gave no indication of what he considered the true situation to be, and offered no ideas or suggestions as to any positive line of action designed to effect changes in the situation prevailing. Maybe he didn't know it but I knew Jim far better than he did - Jim was a good practical pitman and certainly wasn't short of ideas. That had been well established throughout his career. True, Jim had somewhat of a buccaneering approach, but he was able to get things done

Within a very short time I determined an opinion to the effect that there were no inspiring qualities in the Manager and that his Mining Engineer who, I believe, was quite competent, capable, and promising, was being cast in a similar role to that of the Area Chief Mining Engineer in the South Barnsley Area.

CONFLICT WITH COALFIELD MANAGER

Betteshanger Colliery's conditions were quite similar to those of some of the Yorkshire collieries. My opinion differed strongly from that of the Coalfield Manager. However, one of the power loading faces had a problem of randomly met hard brassy nodules in the seam and was having difficulties in that respect. Both Headquarters Authorities and the coalfield were pushing to introduce a new power loader format, the 'Webster miner'; the prototype was under trial at Castlebridge Colliery in Scotland. At the best it was unproven and at the worst it was unsuitable for this particular situation, being purely an experiment in a set of circumstances which could not afford it. Visiting the Scottish Webster miner I felt the machine, which hardly moved during the period I was there, was no substitute, or likely to be so, for the current equipment developed at that time. Events proved this to be so for it never really broke through as an established production machine. The AB Machine I recommended was an updated version of the Anderton shearer loader, more powerful and reliable than those in former use. It was installed and effected quite an improvement in the prevailing situation.

Following my inspections I produced a report together with recommendations for all three collieries. I took the report to the Coalfield Manager, but as in a previous situation, I couldn't get him to enter discussions. At about this point Lord Robens had retired and had taken the post of Chairman of Vickers Ltd. Sir Derek Ezra was newly in post. Kent always had been W.V. Shepherd's (now Board Deputy Chairman) province. Politics were in strong evidence in a number of ways. Phil Weekes was now Director General of Production; Peter Rees was involved in National Planning together with Douglas Simpson and others. There was a strong integrated Welsh (including the Coalfield Manager) contingent involved in Kent, one way or another, and this had been so for quite some time. Moreover, as stated, the Board's former Director for Production had been associated therewith for many years.

About a month or so after trying to enter discussion with the Coalfield Manager, without reference to the report already submitted, I received a copy of one he had subsequently commissioned his Mining Engineer to prepare, in which there was strong evidence of identity or coincidence. I couldn't reconcile what the man was trying to do. I couldn't understand why he wasn't prepared to get down to a purposeful examination of basic possibilities which had been outlined in the report submitted. I wrote to him to that effect and pointed out the apparent plagiarised aspect of the report. He protested to the Director General of Production who called a meeting between us which did little to resolve the position.

Had not the circumstances changed at the 'top' I would gladly have let the situation develop. It could well have reinforced Jim Stone's earlier opinions and observations and may have focused attention on what I felt to be an inadequacy at the highest management level with regard to the problems of the coalfield. I never heard anything more of what transpired afterwards; I made no enquiries.

The Deputy Chairman Visits Kent – I Seek Early Retirement

However, so far as Kent was concerned a climax developed later, with the visit to Kent of the Deputy Chairman who was to have a meeting with the management and the NUM. I went to the Coalfield Headquarters and on seeing me he asked,

'Who told you to come down here?'

'You did,' I replied.

'I did no such thing. You were told not to come,' he retorted.

'You told me to keep my eye on Kent and that is what I am doing. I received neither communication nor instruction not to be present,' I protested.

'All right - you sit in but you don't say a thing,' was his instruction.

'As you say, but I want to see you before you leave here,' was my response.

The meeting, I felt, was fairly innocuous with nothing of real import involved following which I took him into a little office I had. Respectfully I said to him,

'I've had enough - I wish to be considered for early retirement.'

He looked at me, astonished.

'I'm not asking you to go,' was his response.

'Neither are you making it possible for me to stay - I'm serious,' was my reply,

'Go back, think over the situation and then come down to my office in a month's time,' was his suggestion.

Although I would love to have stayed and given whatever support I could to Sir Derek Ezra (whom, on the pleasing experience I had accompanying him during an underground visit to Betteshanger Colliery, would have been happy to work for) and thereafter to Sir Norman Siddall in their later capacities as Board Chairmen, it was not to be. I went down after a month's interval and confirmed my decision, having come to the conclusion that I had no future in a situation embracing adverse high level influence which constantly indicated a degree of apparent hostility and animosity I couldn't reconcile. I have no doubt I could have fought the situation and won through but there would have been little point, I'd lost the motivation of years and the pleasure and happiness I had in trying to do a good job.

Premature Retirement at the Age of 58

Sadly, at the age of fifty-eight I retired and left the National Coal Board. The Headquarters staff under Phil Weekes now (Director General of Production) together with other members of the Headquarters staff, called me in for a quiet drink and were most kind and generous to me during the last couple of hours I spent with the National Coal Board.

I have no complaint as to the treatment received from W.V. Shepherd, or the Barnsley Area Director, or indeed others. 'Mining is a man's profession', in which he has to deal with human as well as natural situations.

The former I found somewhat of an enigma in that he was most variable, but capable of practical expressions of real kindness. During the time I was working under Don Severn in No.4 (Huthwaite) Area as Area Production Manager, Tom Adams of Dowty Hydraulics whilst driving his car had a fatal heart attack. We had over the years worked a great deal together. I wrote to his widow who lived on the outskirts of Chesterfield overlooking the Wingerworth coke oven plant expressing my condolences. She wrote back and asked if I could go down to see her. This I did and found that she was in a real distressful situation. She and Tom had adopted two children who were both at public schools. She really had problems and was in need of work. Ascertaining that she had been fully trained as a nurse and formerly a matron in a large hospital, I explained by correspondence the full situation to W.V. Shepherd who at the time was Area General Manager of the Bolsover Area. Quietly and without fuss he took steps to have the lady refreshed and updated at one of the Chesterfield hospitals, following which he made her Nursing Sister at one of his collieries (Grassmoor Colliery, near Chesterfield I understood it to have been). On finally meeting him as Deputy Chairman of the Board to finalise my retirement situation he went to great lengths to ensure I had calculated my pension rights correctly, sending for the Board's actuary to determine this. He explained that other people in a similar situation had mistakenly determined their situation and had suffered accordingly.

As Barnsley Area Director, like others at the time when the Board reorganised its management structure in 1967, he was virtually overnight transferred from a balanced situation into a maelstrom of disturbed management elements at all levels within his new enlarged Area organisation. His new Second-in-Command had much wider experience and had been trained under a succession of first class mining engineers and highly skilled managers, having worked his way by sheer hard work and ability, to a level more than equal to his own before they came together. Quite probably, in his mind, he felt that here was a threat to his position, undoubtedly an experience new to him in the changed circumstances. Unnecessarily, I believe, he subjected himself and others to needless stress and pressure. With a more enlightened approach, imagination and creative ability he would have had alongside him unquestionable support, loyalty as freely and generously given as to others placed in a similar positions. Working with him was quite a new experience, certainly not a rewarding one from any aspect of relationship or added positive experience of value.

The Area Director had his problems but, to me, he represented but a very faint shadow of the excellent people by whom formerly I was directed throughout my career at all levels. In those early days of contact, by the very nature of these things outlined, he destroyed for me the personal respect his post warranted.

However, that apparently is the way of things in this modern age, with blue and white collar workers now being subjected to the sort of treatment formerly carried by the workman on the shop-floor or by colliery employees.

Both were my superiors in position, doing their jobs and being motivated in a manner known best to themselves; more importantly, the negative approach they subjected me to had spin-off benefits. I had to adopt and develop a positive counter approach which I believe added strength to my character. In short, I lived a great life and had a wonderful career standing on my own two feet with the unlimited support of Mary my great partner.

I have no regrets, nor do I expend useless thought in my imagination of how things might have been. Providence was good to me, Although I put into my career considerably more than I received in comparison with others, such is of little consequence when judged against the happiness, achievements and satisfaction obtained throughout a long and worthwhile career.

Mary, John, Dorothy and Harry my gardener of some twenty-five years, in the circumstances took my decision to retire prematurely as the best thing I could do in the circumstances. They were concerned, realising that in those past few months I was most unsettled and terribly unhappy.

Twelve months previously I had bought a house in the Domesday Book village of Wickersley, Rotherham, a delightful place in which to live. Lord Robens was most kind. He offered financial help, should I need it, and allowed me to have the benefits of the Board's standard agreements in such situations even though I was moving into my own property. I hadn't asked him - he volunteered his assistance. Maybe when he left the Board he took part of me with him without either him or me knowing it.

Mary was most happy. She had almost all of the remaining members of her family and mine within a radius of some seven miles and at a later stage was absorbed in her youngest grandchild, Dorothy's son Paul.

Dorothy had met and married an excellent boy, John Fordham, in July 1970. He had lost both his parents several years earlier and was living alone in rented accommodation. He had been trained and had qualified as an accountant and was working for a group of companies in Sheffield. He subsequently obtained a post with Butterworth Jones & Partners, Accountants, Taunton, Somerset.

My son John, at this time in his early thirties, over in the Irish Republic had established the small company 'Industrial Detergents' against tremendous pressure and competition from his former employers Diversey Ltd, and was beginning to expand. He, together with a colleague, Edward Robinson, formerly Sales Director from the same company, had built up their company from virtual obscurity into one of the leading Irish companies in that field. Starting out with their manufacturing processes housed in an old pigsty at Shankill on the outskirts of Bray, the business flourished to the point of having to be transferred, with the help of the Irish Government, to new large factory premises situated on the Bray Industrial Estate in Wicklow. Since then it has been further enlarged on different occasions to meet the needs of extended growth.

Under his own steam he really had developed in ability, experience and business acumen and, not unlike myself, was no stranger to enterprise and hard work. His children were growing up and he and Sue his wife, with whom he shared a common interest in golf, had built a strong family background.

However, even in these favourable circumstances with no serious health problems I was very depressed and unhappy. I couldn't settle down to my former hobbies: music, recording, brass bands, etc., I was completely lost. Without something real and purposeful to do with an end product arising, I had no fulfilment. This was an experience I didn't like. Thankfully it lasted only a short time and once overcome has not re-emerged.

PHOTOGRAPH 56. OPENING OF CANTEEN AT ELSECAR MAIN COLLIERY

Front Row, L–R: Geoffrey Thorpe, Former Manager; Clifford Machen, Area Director; Gilbert Walsh, NUM President; Jack Burton, Deputy Director Mining; Charles Dickens, Area Chief Mining Engineer; G. C. Payne, Former Area General Manager.

George Eaton, Barnsley Electrical Engineer, stands immediately behind Gilbert Walsh.

PHOTOGRAPH 57. MAURICE BEEDAN:
GENERAL MANAGER, ELSECAR MAIN COLLIERY

Maurice Beedan proved to be an excellent acquisition following his appointment as Colliery General Manager at Elsecar Main Colliery. He was received quite well and was given excellent support by both workmen and officials alike. His quiet, unassuming manner, stintless industry and application together with his natural enthusiasm opened up a new lease of dynamic life for the colliery.

New developments could be given to him with the greatest of confidence knowing that all possibilities offered would be extracted to the full. Under his leadership the colliery established a record weekly performance of 28,500 tons with a productivity of over 5 tons per manshift.

During the Second World War he served and had distinguished service as a pilot in the Royal Air Force, undertaking a wide variety of important duties.

PHOTOGRAPH 58. OFFICIAL OPENING OF THE RIDDINGS DRIFT MINE

Prior to the opening of the Colliery on 8 August 1970, an underground inspection was carried out with Lord Robens, Sir Norman Siddall, Trevor Massey and others present. Trevor Massey, Colliery Manager at the time, is on the extreme far left. Lord Robens, Chairman of the National Coal Board, is in the back centre (headpiece of cap lamps in place) whilst Sir Norman Siddall, Board Member for Production at the time, is at the extreme front right.

During the lunch which followed the official opening of the colliery, Lord Robens outlined to the guest and media present the circumstances relating to the colliery's conception and construction, indicating that the value of the experience gained extended over a much wider area within the Board's organisation.

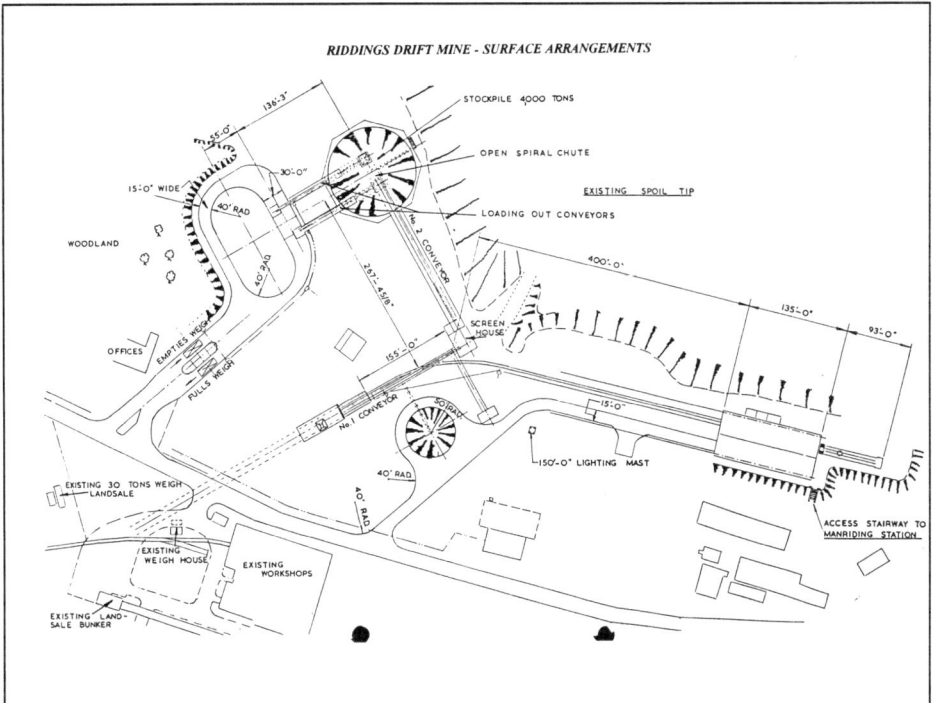

PLAN 3. RIDINGS DRIFT MINE: SURFACE ARRANGEMENTS

Every effort was made to keep the surface arrangements (as with other aspects of the scheme) as simple as possible. Lamproom, maintainance and bathing services were shared with the adjacent South Kirby Colliery. The coal was road transported to Grimethorpe coal preparation plant three miles or so distant. Stocking and loading out facilities had a capacity of 40,000 tons (roughly two days' output). A private firm of haulage contractors was employed, being cheaper than undertaking the work with the internal Area transport organisation.

Simple screening provisions covered the removal of foreign materials and dirt, the latter being diverted into a small bunker of about 200 tons. The small amount of dirt produced was disposed on the adjacent spoil heap by lorry and buldozer.

Material stocks were disposed along and on either side of the manriding track and within the prefabricated building covered manriding station. Fork lift trucks and small mobile cranes kept the manpower to a minimum.

Coal feedback facilities were provided directly over the Main Drift (No.1 Conveyor). Used in emergencies they proved to be quite a useful asset.

PLAN 4. RIDDINGS DRIFT MINE: UNDERGROUND MINING LAYOUT

PLAN 4. RIDDINGS (AMERICAN PROJECT) DRIFT MINE

The layout of this new mine was on the basis of an annual tonnage of 500,000 long tons per annum with a total manpower limitation of approximately 200 men and an associatied overall productivity of 10 tons per manshift. Designing a new mine to meet such tight parameters really provoked a great deal of detailed thought, effort and later appreciation.

In accordance with the American approach to the design of new enterprise, currently referred to as the 'KISS concept' (Keep It Simple Stupid) we analysed each operation and activity with care and in great detail. Each projected activity was examined on the basis of: is it necessary? can it be eliminated? does it need the degree of labour being projected? and is there a mechanised substitute? Having fixed a limitation to be used, the effect was to direct a great deal of detailed thought towards an attempt at cutting down the total projected labour content of the total volume of work to be undertaken in raising the tonnage stated: somewhat the reverse approach to what formerly took place.

The final layout is shown in Plan 3, together with the adjacent colliery boundaries and general geological fault systems. Despite the excellent results obtained I was never satisfied with the weekly perfomance in opening up the reserves with coal headings. We had three Lee Norse CM28 continous miners averaging about 300 yards of road drivage per week; there was no margin particularly so that in a situation in which the 'short retreat faces' were being extracted at the rate of 200 yards per week, it was extremely tight and any maintainance trouble with the miner had the effect of producing 'gaps' in the continuity of the coalface operations and loss of potential performance.

Our problem was that of freeing the CM28 from the restrictions of coal clearance by direct loading onto the gate conveyor belt via a bridging media and the support installation which involved setting 4 × 4 H section joist and wooden props. These provisions did not compare with the American approach which involved roof bolting, and the use of pick-up loaders and shuttlecars, the efficiency and reliability of which greatly outshone our efforts. Because of this I spent a great deal of time trying to work out a partially mechanised system of support setting.

It was about 1972 or 3 when I came up with a possible answer: Figure 10, which was based on the principles shown in Figure 12A. Although patents were taken out on the original design in America, Germany, South Africa, Hungary and Poland I was never able to get it off the ground, I felt greatly confident it would have made an excellent contribution to the rates of coal drivage being achieved at the time.

At the present time this problem has been greatly eased by the adoption of super long faces of over 300 metres (350 yards) in length, with panel face traverse distances of 3000 yards, so the need for the formerly required high drivage rates is not so intense, even with the phenomenal coal face performances being currently achieved with modern fully mechanised systems. Reference back to Figure 8 shows one such example.

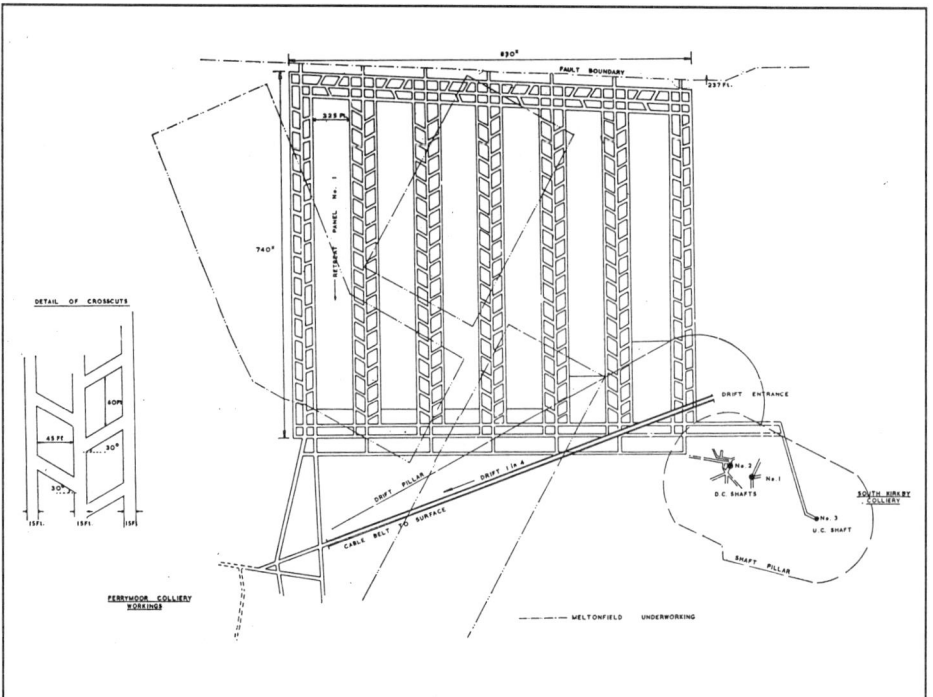

PLAN 4A. ORIGINAL TEST SECTION: WORKING PANELS

This plan represents the NW section of the seam reserves which served to prove the projected system so to establish the experience relative to both coal heading requirements and the associated extractive speed of panel extraction. Further experience envisaged was that of changeovers from an extracted worked out panel to the commencement of its successor. Eight experimental panels were worked in this part of the mine from which was determined the road-heading drivage rates needed to be increased to match the rate at which the individual panels were retreat-extracted at the rate of over 200 yards of retreat per week.

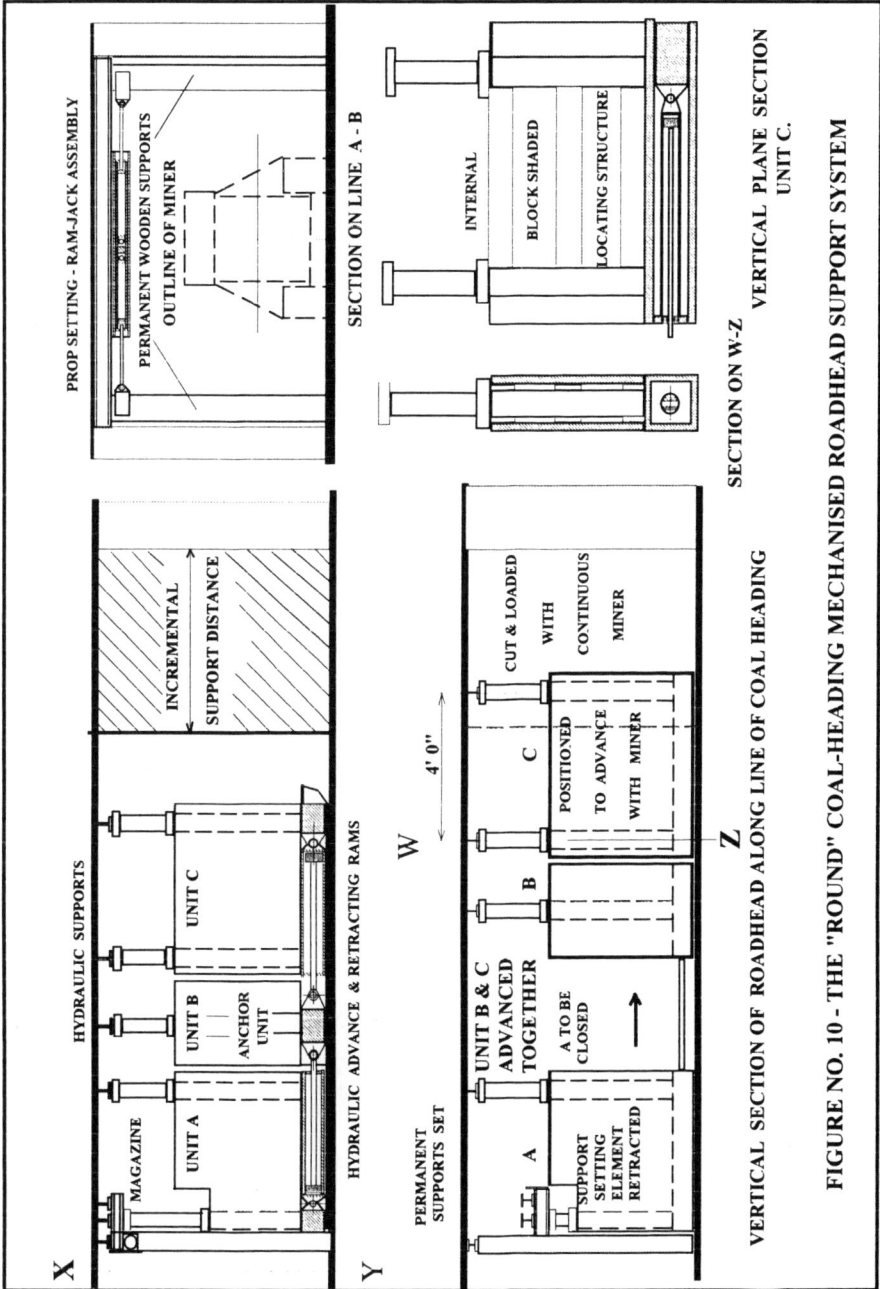

FIGURE NO. 10 - THE "ROUND" COAL-HEADING MECHANISED ROADHEAD SUPPORT SYSTEM

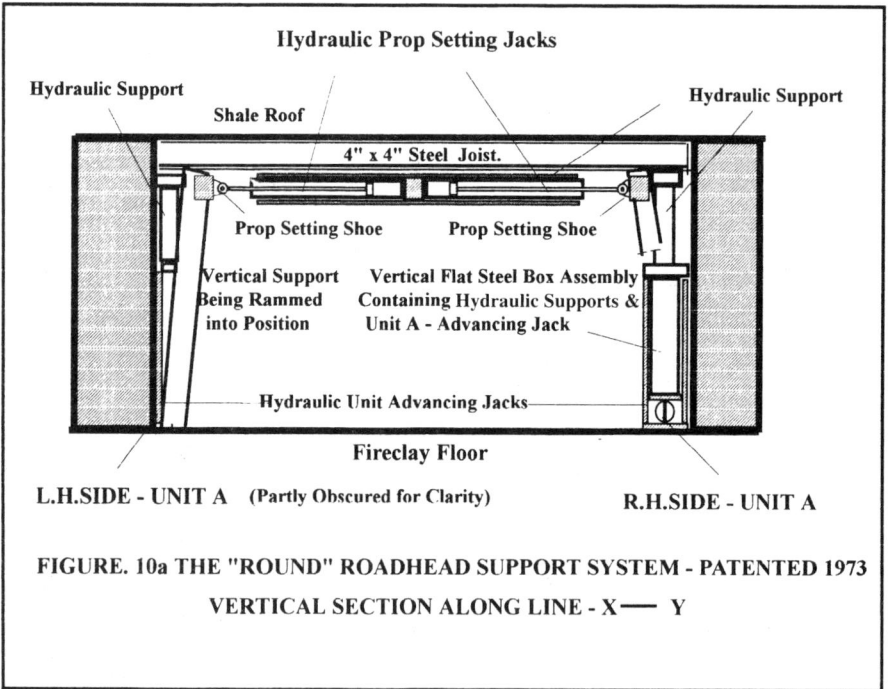

Hydraulic Prop Setting Jacks

Hydraulic Support

Hydraulic Support

Shale Roof

4" x 4" Steel Joist.

Prop Setting Shoe   Prop Setting Shoe

Vertical Support   Vertical Flat Steel Box Assembly
Being Rammed   Containing Hydraulic Supports &
into Position   Unit A - Advancing Jack

———— Hydraulic Unit Advancing Jacks————

Fireclay Floor

L.H.SIDE - UNIT A   (Partly Obscured for Clarity)   **R.H.SIDE - UNIT A**

**FIGURE. 10a THE "ROUND" ROADHEAD SUPPORT SYSTEM - PATENTED 1973
VERTICAL SECTION ALONG LINE - X—— Y**

PHOTOGRAPH 59. LORD ROBENS TALKS TO MARY ROUND; OPENING OF
RIDDINGS (AMERICAN PROJECT) DRIFT MINE 8 AUGUST 1970

Having been placed almost obscurely at the end of the 'top table' Lord Robens had missed us so he sought Mary out for a short chat. She had met both Lady Robens and His Lordship on previous occasions. They always made her feel comfortable and easy. During his speech he traced the history of the Riddings Project from the 'bathroom incident' referred to in the narrative, to its successful completion. On this occasion he was telling Mary of the nature of achievements involved and thanking her for the obvious patience and tolerance she must have exercised during the execution of the project.

Dorothy he had met at Ammanford College during his Presidency at a Speech Day when she was given the responsibility of looking after the main guest. During the chat he wanted to know of her progress.

After some 15 months of intense effort and application against great odds with the highly successful accomplishments now behind me, on this particular day all the accumulated stress was released which left me free to enjoy the occasion to the full.

10

Early Retirement
and Consultancy Work

Fate and Providence, both having been exceedingly kind and generous to me
in countless ways, hadn't deserted me. An old friend, Tony Burke of Burke-
shire Engineering Co. Ltd, whose small company had done an excellent job for
us in the abrication of steel and its surface erection during the construction of
Riddings Colliery, asked me to join him with the company he had just sold out
to, John Folkes HEFO, a group of companies situated in the Midlands.

Tony started out as a young appproved electrician at Kiveton Park Colliery,
following which he graduated to become an electrician. At the time Kiveton
Park Colliery was involved with the introduction of American lend lease mining
equipment, following the Second World War. Tony went on an early Sheffield
mechanisation training course involved with shortwall machines, roof bolting,
duckbill loaders, boring equipment etc. His father, at the time a colliery overman,
was actually involved with this equipment underground.

Later, leaving the mines, he joined Bailey's Jones and Bailey's, a subsidiary of
Guest Keen & Nettlefold's, as a salesman for roof bolting equipment which was
being introduced as a new method of roof support underground. Four years
later, he started up a steel fabrication company in Maltby which he later moved
to a new factory site in Carlton-in-Lindrick. After this he sold out and it became
part of the John Folkes HEFO Midland Group.

Some twenty-two years ago on the Dukeries Industrial Estate in Worksop,
Notts., he built up a highly flourishing business, Clayland's (Electrical & Engineering Distributors) Ltd together with a Plant Hire Company. Tony, also as
evidenced by his remarkable success, was a natural salesman with a good business
brain and managerial ability.

LORD ROBENS – CHAIRMAN OF VICKERS LIMITED

During the time I spent with him, the John Folkes HEFO parent company in
a certain situation wanted an introduction to the Engineering Director of Vickers
Ltd. Millbank, London. I wrote to Lord Robens, now the Chairman of the
company, asking if I could meet him. There was no problem. A meeting was
arranged and I went down to see him. As soon as I mentioned my name at
reception the young lady beamed. She made a telephone call and with great
courtesy conducted me to the elevator and told me which floor I was to reach.
I walked out of the lift virtually into the arms of Lord Robens. He had come
out to meet me and conduct me to his office himself.

I have often contrasted this welcome with my experience with the then Deputy

Chairman's and Jack Weir's in connection with our Riddings project, the difference being that Lord Robens to me had that 'human touch and understanding' not readily met at such high levels of administration.

Putting his arm over my shoulder he said in a very kindly voice,

'Young man, I want a word with you. Why did you leave the Coal Board?'

'I am sure you have a good idea yourself, Lord Robens, I really was unhappy.'

By this time we were sitting together on a very large settee.

'I have, Charles. You no doubt upset some people with what you undertook and achieved.'

Coffee and biscuits were brought in and he continued,

'Charles, had I still been in post at the time you would not have been allowed to leave. You had much to offer the Industry. All that has been lost.'

We never discussed personalities or situations. That was neither appropriate nor warranted. Such matters lay in the past. I told him of my friendship with Harold McEneny, the Director of Engineering at Vickers Ltd in Barrow in Furness, then in post.

'Yes, Charles, I know. We were having dinner together on one occasion and you formed the topic of conversation for quite some time.'

'Tell me about yourself, Lord Robens, and how you find satisfaction from your new job?'

In those days the public in general, personally or through the media, were aware of the fluency His Lordship had, no matter what the subject or situation. In this personal setting that fluency was tinged with both excitement and great infectious enthusiasm. I revelled in the occasion. He told me of the things Vickers were doing, the great strength he found in his Board members, his need and ability to travel abroad, how they stood up to worldwide competition. It truly was fascinating.

Over ninety minutes had passed before I asked him about the contacts I was seeking. He asked Mr J. Hendin, the Chairman of the engineering group, if he could kindly come into his office. Introducing me to him as one of his old colleagues he explained the nature of my request. There was no difficulty about establishing the proposed contact and this was duly made.

In an early part of these memoirs I mentioned, in a lesser situation, that in the absence of transport between Rotherham and Rawmarsh I could have flown the distance. In this case the distance between Millbank House, London and 4 Tanfield Way, Wickersley, Rotherham, Yorkshire, was very much greater but I could have taken it in under the same conditions, such was the stimulating effect of my morning with his Lordship.

Shortly after this meeting His Lordship directed Dunlop Engineering Ltd, Coventry, towards seeking my services. They were engaged in looking into the possibilities of designing a new radical format of underground vehicular traffic for the transport of men and materials and wanted background information. I prepared them a report not only of the circumstances involved but also of the basic problems and constrictions which needed careful consideration. The report was well received by the Head of the Section involved.

During the period I had with Tony Burke he restored my well being and resolve. I express my gratitude both to him and Joan his wife for their interest and kindness.

PHOTOGRAPH 60 was taken during an offcial dinner occasion. Tony and I wound up our affairs with John Folkes HEFO Ltd about the same time.

Mining Subsidence – Property Damage Claims

Unlike C.V. Peake, I never, officially or otherwise, advertised my services as a Mining Consultant nor seriously considered ever becoming one, but periodically someone or other was in trouble and needed help, be it a colliery manager deep in trouble, or a student seeking advice or information. I was never able to refuse helping people I thought to be in genuine trouble. If I felt their situation involved injustice I fought like a tiger, tooth and claw.

One such case involved Mrs Wright and her oldest son Stephen Wright of Temple Normanton, Chesterfield. It involved the Bolsover Area of the National Coal Board. Over quite a number of years Mrs Wright and her husband had built up a thriving egg-producing small holding within a locality which was subject to intense deep mined and opencast activity. Her property was severely damaged by mining subsidence. The Board denied liability, but the Wrights fought a protracted battle over a period of some four years or more, a situation which affected both their health and their income. However, they ultimately won their case at real cost to themselves. Shortly afterwards Mrs Wright's husband died, leaving her to complete bringing up their four boys.

Because of business and other limitations imposed by the circumstances, at a much later period, after obtaining certified assurance that it would be safe to build their new domestic property adjacent to the old without the risk of future subsidence damage, she engaged an established reputable firm of architects to undertake the design with particular reference to the environment and their former experience. Her youngest son was in an advanced training stage for a career as an architect and had designed a heavily reinforced concrete raft upon which the new house was to be constructed by a highly reputable Chesterfield firm of builders.

The new house was duly built. After having lived in the property for upwards of two years without incident or problem, at about mid-day one morning the house was subjected to an intense instant shock which coincided with heavy blasting operations at an adjacent opencast executive site roughly a mile distant. All internal house clocks were stopped at this precise time.

What had happened was to me consistent with the explosive shock waves travelling through the strata hitting the house foundations (i.e. the reinforced raft) which had the effect of shaking the property and inducing heavy vibrations through the whole of its structure. In less than twenty-four hours, random cracks and fractures appeared throughout the whole building. In simple terms: imagine a stiff jelly on a plate, with the edge of the plate given a sharp heavy blow at one or a series of adjacent points. The effect upon the jelly is not particularly hard to visualise, although the elasticity of jelly is totally different from that of a rigid structure.

A property damage claim was submitted by Mrs Wright to the Housing Manager of the NCB Bolsover Area. The Acting Area Housing Manager accepted liability in the form of a letter to Mrs Wright on behalf of the NCB's Opencast Executive, on the basis of the explosive shock wave, but no immediate action was initiated in a situation of continuing deterioration.

After a few weeks Mrs Wright had been reduced to a very depressed state. She had been unable to obtain action and find support with which to press her claim, and her solicitor was in no way able to help. She heard of me by chance, through her security adviser who had months previously installed my domestic alarm system. Ringing me up she explained her situation.

'I'm sorry, Mrs Wright, I really can't help you - I've retired.'

Her distress came over the telephone line unmistakably; her voice was filled with deep emotion.

'What am I to do, Mr Round? What am I to do?'

'I'll come down and look at the problem. I can't promise I will handle it, but if I can help you in other ways I will.'

Within twenty-four hours I was inspecting the property. Mrs Wright and son Stephen were present. It was undoubtedly a genuine situation and their plight got through to me. I agreed to take the case.

In view of the fact that the Board, through its Acting Housing Manager, had by letter certified acceptance of liability I thought the situation could be cleared quickly, with little cost to Mrs Wright. I arranged a private meeting with the Acting Housing Manager who was somewhat new; I understood he'd been recruited from a Local Authority. Leaving him, I felt things should proceed without difficulty.

Further weeks went by with no action. However, in the meantime a new Housing Manager had been appointed. In a situation of this nature I had a good idea of what was taking place. The new Board official and his advisers were looking for a way in which they might break the claim, despite the previous certified acceptance of liability. The Opencast Executive represented by the Bolsover Area Housing Manager were concerned about the claim's possible effect of creating a precedent in a location in which they were extensively active and where explosives were being used to break the overburden.

I fixed up a date for a meeting with the new Housing Manager and did something that was not normal in such situations, I took the client and her son, her architect and her builder along with me. The Housing Manager wasn't at all pleased and muttered something to that effect, but he himself had brought with him his Deputy, the Area Civil Engineer, two others and a minute taker, which I thought to be extravagant from the Board's control of cost viewpoint.

The action I had taken produced a balanced meeting, a situation which gave me freedom of action and a reinforcement of the prevailing facts. Stephen, Mrs Wright's son, a school teacher, took minutes for our purpose. In such situations the taking and circulation of agreed minutes did not usually take place. In the

Housing Manager's opening statement he conveyed to me quite clearly that they were out to break the claim.

'We are not denying the claim, but we must be sure what the cause is. I have brought the Area Civil Engineer who has had a great deal of experience in such matters. We have worked together many times in situations of this kind.'

He then introduced the Area Civil Engineer who opened up in what we all considered a pompous, patronising manner.

'In view of the property's comparative newness the problem of "lateral thrust" is a matter of vital importance. It cannot be discounted and could well be the cause of the property's damage.'

Time and time again he stressed this aspect of 'lateral thrust' (in simple terms natural forces acting horizontally all round a building's foundations and supplementing the vertical gravity loadings) but he never produced to us any confirmatory evidence or examples at any time to support his assumptions. You may recall a reference to this point was made in connection with Abernant Colliery and the associated photographs showed the effects of such phenomena at a depth of some 3,000 feet. In this situation, at depths of a few feet, there could be no comparison of the relative forces involved in situations of this kind.

Here he was, pontificating on the effect of 'lateral thrust' at little more than five feet below the surface subsoil in a situation in which the property had been stable and secure for more than two years. If one was to assume the foundations had stored with maintained stability any such lateral forces (much as the lining of underground mine roadways does) for at least for two years, one could not discount the release of such forces triggered off by the destructive shock waves arising from the opencast blasting operations, supplementing them with a greater intensity of damage to the property.

It was in essence: 'We're denying the claim and contend the house was badly constructed.' I had formerly had experience in a number of areas embracing many hundreds of houses, including many new ones, for example Calverton and Cotgrave, but had never before come across anything remotely similar to what he was rigidly suggesting. The only situation to my personal knowledge in which a house had collapsed in a matter of hours was an occurrence one weekend on one of the Nottingham Boulevards when a single property collapsed completely during the night hours, the houses on either side suffering no damage. We traced the cause to be sudden strata movement associated with a main geological fault which ran directly beneath the property and which had been triggered off by Radford Colliery workings.

We accommodated the man time and time again with diggings and tests, all of which failed to produce whatever he was looking for. He never made clear to us what that was, if indeed he knew himself. My patience was running out and things were getting irksome between us. In one letter to me he virtually suggested how I should conduct my clients' case. In my reply I indicated that from my experience to date he would be the last person from whom I would wish to seek advice. This badly upset his dignity for which my client was later penalised.

Finally, to save face, they wanted to nominate a firm of consultants to investigate the whole situation. I said I would agree on the condition that we would be given the opportunity to study whatever report was produced. They wouldn't agree – we were at an impasse. To break this impasse, I informed the Housing Manager I would take no further part in negotiating directly with them, and would work only to instructions from my client and her solicitor. I had got as far as I could, with obstinate and highly opinionated people determined to break a valid and legitimate damage claim over which I had no control. This letter was twisted in interpretation to mean that I'd withdrawn from the case and had thus forfeited some four years or more heavy work that had been put in, even though I still continued to work with my client and her solicitor.

The solicitor agreed to the consultants who carried out a survey, at what cost to the National Coal Board I don't know. The outcome was that the claim was upheld and my client got her house put into good order. It has remained so over quite a few years now, with the discontinuance of opencast blasting and total cessation of mining operations: so much for the mystique of 'lateral thrust' and 'implied bad property construction'.

They paid the solicitors' fees but refused mine (under the Board's arrangements they were entitled to do this irrespective of the circumstances although it was very rarely done). Whether the client's architect and builder submitted a fee to the client I do not know. I would be greatly surprised if they did, as both were most honourable men; they were seeking to defend their professional competence and reputations which had been brought into question by this man's suggestions and actions. I presented my modest fees to Mrs Wright and Stephen. When they told me my fees had been denied I said,

'In view of that, Mrs Wright, give me half the fees, I'm extremely sorry you have been treated like this. Having being fully informed by copies of all correspondence and having been in attendance at all meetings throughout you know what we were all subjected to.'

Both Mrs Wright and her son Stephen were highly indignant.

'We will do no such thing, Mr Round. But for your skill and tenacity those people would have stripped us of all we have. We're deeply grateful and will never forget all you have done for us.'

I'm glad I was directed by Fate to give them support in their difficulties. Mrs Wright was a delightful lady and had worked extremely hard (having lost her husband at a crucial age), over many years bringing up her four sons. Two became teachers, one an architect and the other, Christopher Wright, is a leading art historian on seventeenth/eighteenth century European art, whose services are sought by a wide range of American art dealers and connoisseurs. At one time he was Assistant to the Russian spy Anthony Blunt for some twelve years at the Courtauld Art Institute. Sadly, Mrs Wright did not live more than four or five years after having her property fully restored. She died suddenly in the presence of her vicar. I know what stress she had suffered during the situation described, and that was her second experience.

Within the experiences I have encountered, the price of arrogant and pompous little men's dignity has come high, unfortunately in this case to the innocent, and indeed to the National Coal Board, more particularly so as mining is certainly not a 'kid glove' occupation.

Following this subsidence case I was pressed to undertake three others in the Barnsley Area. I won't go into detail with all of them but one in particular was unique. During my career as Deputy Director (Mining) Barnsley Area, late one Friday afternoon George Hewitt, a surveyor whom I knew very well from his early Elsecar Main Colliery training days, had come into my office.

'We've got a property damage problem at Darfield Colliery. It's a very old building, at the moment I am not sure whether or not it is listed. A large coping block of stone has broken away and fallen to the ground and cracks have appeared in the walls.'

Together we had been to inspect the building, Middlewood Hall. It was a magnificent building, the outside of which evidenced to no small effect mining subsidence activity. We had checked Darfield Colliery plans in all seams and determined the fact that one of the underground coalfaces had approached beyond a distance sufficient to give rise to concern as to potential subsidence damage. The problem was undoubtedly ours, but there was little we could do about it at that point other than to stop the workings so as to minimise potential damage. Such a decision was vested at a higher level. The coalface continued and was worked to its boundary, intensifying the early damage we had examined.

It was upwards of two years after this incident that the owner, Arthur Wainwright, asked me to act for him in a subsidence claim which had been denied. The National Coal Board had a problem, as the damage was extensive throughout the building and in the adjacent stables and outhouses. To correct the situation called for heavy expenditure. This situation was compounded by the fact that there was also an appreciable element of depreciation in the form of wear and tear over many years. The Housing Manager seized on this element to deny all liability.

Protracted negotiations were leading nowhere. In the meantime Arthur Wainwright and his wife Pat had been reduced to living virtually only in a large kitchen and damaged lounge. Pat, a no-nonsense Yorkshire lass, had become desperate. Filling a bottle of contaminated dirty water from the kitchen tap, she sought an interview with the Housing Manager at the Grimethorpe Headquarters, taking the sample with her. Her own description of the event to me later was:

'I walked in to a smarmy Good morning, Mrs Wainwright, please sit down, what can I do for you, approach. Holding the bottle of dirty water in front of the man's face, I said, "You can do something about this. How would you like your wife to be subject to such treatment?" I poured a quantity of the water over his desk and then I left.'

Getting to know Pat over a period of more than seven years, I have no reason to doubt the truth of the lady's statement. She could when pressed be a formidable lady.

After more than three years irksome battling this situation took quite an unexpected turn. The National Coal Board made an offer of around £26,000.

It was below what I would have expected and certainly wasn't acceptable to the owner. The total cost of total repairs I would have thought to have been over £100,000.

In the old village of Worsborough, a few miles away, lived a very rich entrepreneur whom we understood to have made a considerable amount of money buying old buses, stripping them down, reconditioning the engines and exporting the finished product to China and other Far Eastern countries. He was operating a thriving business in Birdwell near Barnsley. He became interested in the property, bought it and converted it into flats, knocking down old outbuildings of little use to the estate. To my mind he was most shrewd; the property, Middlewood Hall, was located in a beautiful environment and I believed he would receive a fair return on his investment. As things later developed, in the favourable housing market at the time, I have no doubt he did. Arthur settled with the National Coal Board, bought a new house and moved away.

In dealing with these two cases I became acquainted with a fine old couple, Mr and Mrs Alfred Elmhirst of Hound Hill Farm, Worsborough, near Barnsley. He could trace his family history way back before the sixteenth century and the very large farmhouse he lived in together with the farm lands they owned were also traceable back to those early years; neither land nor property had ever been out of the family line. Alfred Elmhirst, formerly a practising solicitor, was governor of the local schools' authority and my nephew Charles Round was headmaster of the Worsborough Bank End Primary School so had frequent contact with him. Together they successfully petitioned the appropriate authorities towards the reconstitution of an old water mill in Worsborough Dale.

He had brushed up against the National Coal Board Area authorities and had protested through the local press about the manner in which some of its staff were behaving by exerting psychological pressures on old people with legitimate claims. He outlined one case he had taken up involving an old lady in her eighties. Alfred Elmhirst called in his architect to carry out an investigation such that the Board's liability in this case was established; he took the matter over for the old lady and the Board met the claim. Examples of this nature were quite rife and quite regularly there were many justified complaints in the local press.

On one occasion when I met Alfred we were standing adjacent to Hound Hill Farm's back entrance, discussing a subsidence problem of his own. He stopped quite suddenly and pointed to a tree.

'Charles, do you know what that is?'

'It's a tree, but I can't tell what sort it is from here.'

'No Charles – the significance. In 1645, during the Parliamentary Wars, my ancestor at the time was one Richard Elmhirst who was alleged to have Royalist sympathies. He had set up fortifications to Hound Hill Farm in common locally with two others, one at Wentworth Woodhouse [Earl Fitzwilliam's Seat], the second at Wentworth Castle. There were other additional locations in Yorkshire at the time,' Alfred explained.

The place was overrun by the soldiers of the Parliamentary armies. They were in the process of constructing a scaffold when Thomas Fairfax, Commander of the Parliamentary Armies, charged to reduce all forts to rubble, rode into the farmyard with his retinue.

'What are your men doing?' he demanded of the Officer-in-Charge.

'They are building a scaffold with which to hang the traitor Elmhirst,' was the reply.

'Stay your hands – let us not be hasty. I have a kindness for this man,' Fairfax instructed.

Apparently Thomas Fairfax and the ancestor knew each other quite well and had transacted business in York on many occasions. Civil charges were subsequently brought against the latter on the grounds of 'Supporting the Royalists'. Richard Elmhirst raised the plea that his fortifications were for the protection of his wife and workers together with their respective families against marauders from both sides, the 'forts' having been built exclusively for that purpose.

Although he was formally charged as a 'Royalist Sympathiser' and found to be not guilty, he was fined £450 on the basis of assumed guilt. Subsequently he was fined a further £100 but was allowed to retain his farmland and properties.

The farmhouse is steeped in history which is felt immediately one passes through the back door, and possesses a fine library of ancient and other volumes and works of all kinds.

PHOTOGRAPH 61 shows the back of the farmhouse, taken from within the farmyard. It is indeed a beautiful building well located within a rural environment. The location of the proposed execution of the farmer Richard Elmhurst was to the right of the building behind the trees.

PHOTOGRAPH 62 shows part of the fortress work and old pillbox running alongside an old B roadway. A substantial amount of the fortress has been reduced over the years and the stones shipped away to other locations.

The Waddilove Committee
on Mining Subsidence

With the development and extension of coalface mechanisation throughout the country, mining subsidence damage really became a problem. Following one particular year's Audit Review, I understand the Board's attention was drawn to a situation in which such costs had reached over £32,000,000 during that year. I have no doubt the circumstances had been exploited by many bogus damage claimants, but in accordance with the experience outlined. I also believe quite a number of defenceless bone-fide claimants suffered extensively. However there were situations in which the Board's officers virtually assumed the role of judge and jury in the administration and execution of claims. It became a matter of national concern and was raised on the floor of the Houses of Parliament a number of times. The government of the day set up a working party referred to as 'The Waddilove Committee'. I submitted written evidence to the working party and was later called in person before an early sitting to give further and extended confirmation. The Waddilove Committee's Report was subsequently published and certain of its recommendations put into effect by the Board.

Overseas and Miscellaneous Assignments

SPANISH ASSIGNMENT

During the second half of October 1972 on behalf of Fletcher Sutcliffe & Wilds Ltd I undertook a very interesting assignment in Spain. It had two facets:

I. MEETINGS IN MADRID

I) EMPRESA NACIONAL DE ELECTRICIDAD

This concerned the possibility of a £6 million loan for the purpose of mechanising their two mines situated in Turuel, and involved the Export Credit Guarantee Department, which amongst other things required a mining report concerning the mine and the feasibility of the mechanisation proposals.

The meetings took place in the Electricity Authority's main offices in Madrid.

Señor F. Prades, Dr Ingenieuer de Minas (Petrolmina SA FSW Spanish Agents) and I met the electrical company's Mining Engineer (Petrolmina SA Agents) Señor L. Gerezuela, Dr Ingenieuer de Minas Empress Nacional de Electricidad, for discussions which proved most interesting. The company operated two mines, Innominada and Oportisn, in Andorra, Turuel, with a combined annual output of less than 1 million tons.

Their reconstructional proposals embodied a surface drift equipped with heavy duty conveyors. Complete reconstruction of the two mines embodied outputs of 600,000 and 500,000 tons per mine respectively. The seam pitched at about 18 degrees and was 15 metres in thickness and was being worked in five successive lifts of three metres each.

II) MINAS FERROCARRIL DE UTRILLAS SA

Attending were Señor Jose Ramon Ceurero, Company Chairman; Señor Francisco Angula Barquin, Chairman Petrolmina SA; Senor Fernando Pradee Sanchez, Managing Director, Petrolmina SA.

The Company was engaged in a policy of expansion and modernisation. A new electricity generating station had been built by Siemens Gmbh on the outskirts of Utrillas near the mines in readiness to meet the expanded output from the two mines Innominada, Andorra, Turuel, and Minas Ferrocarril de Utrilla. It was understood at the time that the chairman of the electricity authority involved with the new generating station had taken over an agency for Russian built mining equipment the type used being much cheaper than its British or

European counterparts. Experience with the Anderson Boyes mining machinery and equipment had been very good.

A new shaft capable of raising coal at the rate of 500 tons per hour had been strategically located to exploit the expansion of the workings scheduled to be fully mechanised with ranging drum shearer loaders and powered supports.

III) BRITISH EMBASSY, MADRID

Mr Richard Burr, Commercial Officer at the British Embassy, Calle Fernando el Santos 14, Madrid 4, was most helpful with regard to possible alternative Spanish Agents relative to British Companies setting up trade links with Spain, and for introductions and advice accordingly. He introduced me to a number of people amongst whom was a mining friend of his, Señor Juan Pablo Floran of Sociedad Minera Penarroya SA. We had quite useful discussions.

2. MINE VISITS

I) MINE INNOMINADA

The coal seam being worked was 15 metres (49' 3'') in thickness and pitching about 18 degrees.

The floor of the seam was a fairly hard shale; the roof consisted of gritty water-bearing sand. The coal was worked in five separate 'lifts' or 'slices' of approximately 10 feet, each starting at the top and descending down to the floor. The mining engineers had developed a special technique using chicken wire mesh carpeted along the 140 metres length of the coal face, secured at the ends and sides to all the previously laid chicken wire mesh carpet. After the coal was extracted the strata broke down on top of the carpet and consolidated under pressure. This then formed the roof of the next descending slice. The technique which made this possible had been well thought out and was effectively and efficiently applied. The application of the next step using ranging drum shearer-loaders was a simple one compared with what they had mastered so well.

FIGURE 11 is a sketch plan with sectional elevators and shows the layman what the above is all about. It shows part of the colliery workings about 1965. There is no reason to doubt that, with modern bidirectional shearer-loading and powered support, lifts of 5 metres could be taken with a prodigious surge in production performance. Coal transport consisted of belt conveyors of 27.5 and 31.5 inches width running at a speed of 250 feet per minute. With the proposed mechanization belt widths of 36 and 42 inches with speeds of 500 per minute should be contemplated. The colliery output at the time of my visit was some 200,000 tons per annum.

3. Other Mines Visited

i) Minas Ferrocarril de Utrillas - Turuel

ii) Minas de Villabona, Villabona - Fluorspar Mine

iii) Hunosa State Controlled Mines a. Zone de Aller-Hunosa Oviedo. b. Garcia Conde

Mine Garcia Conde is worthy of a few short paragraphs as I found it fascinating, having had experience in the Anthracite workings at an inclination of 45 degrees. Here the workings were inclined at 78 degrees. Moreover, the coal was being extracted by a Russian built power loading machine suspended from a winch rope, the winch being located at the upper level. This coalface was 180 metres in length with the seam 1.20 metres in thickness. The roof (hanging wall) was a weak type of shale, the floor (foot wall) being a fairly strong fireclay. Support to the workings was effected by six rows of wooden struts placed horizontally with bars set to the hanging wall. Working steeply inclined seams was subject to special provisions. There is a requirement to solid fill the void left by the extraction of the coal, retain the dirt within the void by boards and chicken wire, and to restrict void filling to a distance equal to six rows of supports from the working face.

FIGURES 12 & 12A STEEPLY INCLINED WORKINGS are elevational sketches showing the mining layout adopted. Since the shearer-loader is unidirectional a stable hole is prepared at both the upper and the lower levels.

Travelling underground to the coalface, I recall walking along an upper level road for more than 1000 metres to the coalface, which took the form of a large hole in the floor, adjacent to which was a large winch firmly secured by steel bolts and vertical beams tightened to the roof of the roadway. Some 10 metres or so further along the road was a small tipper which discharged the dirt content of a string of small cars through a second hole in the floor. My Spanish colleague shone his light down the first hole and said, 'We go down there!' All I could see was what appeared to be an abyss with the timber struts having the appearance of the rungs of a vertical ladder descending into oblivion. For the first time in the whole of my mining experience I was initially scared. Unless I summoned up the courage to make the descent through the coal face I would have wasted not only my time but also the time my two Spanish compatriots had devoted to me, and they were both watching me intensely.

'Could you please go first and show me how one gets into the face?' I asked one of my colleagues, for that to me was the most tricky part of the exercise. I carefully watched him hang on to the side of the hole until his feet touched the first horizontal strut following which he lowered himself down hanging on to the strut and finding the next one with his feet. From thereon it was on like climbing down a ladder. I followed my guide with a very twisted stomach. However, once I was able to hang over a strut and find the next one below, after descending a few metres it wasn't too bad. The men did not appear to wear any form of safety harness whatever, which surprised me greatly. About the middle of the face at a vertical depth of 300 ft. perched on the horizontal struts with my Spanish colleagues, I asked them about the accident rate to those working in such a high-risk situation. Apparently it was quite good but, they added, occasionally if the men had drunk too much wine their safety was impaired. They added, 'If they fall we always know where to find them.' On moving from the coalface after a vertical descent of 600 ft. my Spanish colleagues came up to me smiling. One of them said, 'We never thought you make the drop. In fact, we had bets on the chance you wouldn't.'

'You would have won your bet on me climbing up a 200 metre chimney stack on the surface; I could not have done that,' I told them. Apparently they had many visitors but some of them failed to make the drop, and returned from the mine to the surface by the same route they had taken to go in.

SPANISH COAL PRODUCTION 1971 & 1992

Coal Classification	Number of Mines	Production Yearly	Percentage Total
1971 Lignite	51	2,740,344	19.07
Bituminous	55	8,853,804	61.63
Anthracite	93	2,772,811	19.30
Total Spain	199	14,366,959	100.00
1992 Lignite		18,530,000	55.10
Bituminous		8,650,000	25.72
Anthracite		6,450,000	19.18
Total Spain		33,630,000	100.00

The current total number of mines in the UK is probably less than 20, so hardly any mine is secure with the announcements of further closures taking place sporadically. As a comparison with the Spanish situation: for the 37 weeks ending 11.12.93 deep-mined output in the UK was approximately 32,550,000 metric tons.

I was quite interested in their anthracite production situation as compared with my experience in West Wales. The average sized colliery in Spain produced 29,815 tons per year. Our average sized anthracite colliery produced 124,900 tons yearly.

AMERICAN AND CANADIAN LITIGATION

At the beginning of 1974, my old colleagues Russell Bracegirdle and Allan Walmsley, former Manager and Undermanager of Clifton Colliery respectively, now Mining Engineer and Managing Director of Fletcher, Sutcliffe & Wild's,

Mine Machinery Manufacturers, Wakefield, Yorkshire, had litigation problems. Kaiser Steel Corporation, Sunnyside, Utah, USA and McIntyre Porcupine Ltd, Edmonton, Alberta, Canada, were suing for damages in excess of $3,100,000 and $4,700,000 respectively. Russell Bracegirdle rang me up, explained their situation, and asked if I would act on their Company's behalf.

'Of course I will. I'll be over to see you tomorrow and will get started immediately.'

The basis of the Kaiser Steel Corporation's case in simple terms was that Fletcher, Sutcliffe and Wilds Ltd allegedly, knowing the detailed mining circumstances, had supplied them with inadequately designed, unsuitable and unserviceable equipment. McIntyre Porcupine Ltd's case was that the powered supports supplied were seriously unstable and did not provide safe and secure conditions for the men working near or adjacent to them.

The depth of research and the detailed reports I produced with each of the two cases surprises me even today. We were fighting in uncharted areas. There was no way in which it was possible to design a set of powered supports such as to guarantee successful operation within unknown and unpredictable mining situations. I had always felt the onus fell to lie with the coal operator, a situation we had always accepted in the UK from the first attempt at experimenting with their use.

A meeting was to be held between the opposing attorneys and five of us from FSW went out to Salt Lake City. With our group was young Bill Morrell of the sales staff, formerly a young student under my charge during my Gedling days as Group Manager. His father, Bill Morrell, was Manager of Hucknall Colliery and both a friend and a colleague.

We met our American attorneys for a couple of days or so before the scheduled meeting took place. The leading attorney remarked,

'This is the first time in my career at law I have been given such well prepared detail. Usually I have to start with a few sheets of scribbled preparation.'

A line of action was agreed and the meeting with the opposition took place the following day. Kaiser Steel Corporation personnel came along with their attorneys and Cecil V. Peake as their mining consultant; we were similarly represented by one of the leading law firms in Utah, based in Salt Lake City.

Their claim of inadequately designed, unsuitable and unserviceable equipment could, amongst other things, be opposed by the inability of Kaiser Steel Corporation to specify precisely the range of mining conditions to which the supports would be subjected. They were of standard UK design which had proved themselves within the coalfields of Great Britain. As to suitability, it was for the user to determine and decide whether the specifications offered would meet the geological and physical parameters of their projected application.

Unserviceability of the supports arose from the abuse they received during operations, gross lack of maintenance and the use of explosives in blasting out support units which had become solid under intense strata loadings.

It was on this latter point that Cecil V. Peake and I crossed swords. After a discourse (Cecil could be quite fluent but not always convincing) on why the FSW supports had failed in the situation, he made great play on the fact that there had been recourse to the use of explosives in the liberation of chocks which had squeezed solid between roof and floor.

My response was that Cecil was fully aware that the practice of blasting out solid supports was prohibited, by National Coal Board edict, throughout Great Britain because of the high risks associated with methane explosions for general safety, and indeed for other reasons, i.e. the unpredictably of effect. Was he suggesting that the manufacturers should build into their designs provision to meet illegal and unsafe practices? In the early days of the development of powered supports a number of mining engineers, including myself, had under varying situations met the experience of solid supports, and had liberated them by pneumatic drills. The time factor was heavy, substantial production was lost and the work exacting. In some cases the installation was abandoned. The next set of supports installed took into consideration the problems associated with the previous set, so that ultimately the problem of the 'solid chock' was largely eliminated by design evolution based on experience. In my belief the particular difficulties of this specific installation was somewhere in that cycle.

However there was another common problem associated with the 'solid chock': that of inadequate maintenance, and there was strong evidence in this situation to support that to be an important contributory cause. Over many years I had found the American approach to maintenance was not so intense as ours. We had built up strong inspection and repair teams at all collieries and in addition had central maintenance shops strategically placed within all coalfields. The maintenance staffs in the United States were much smaller. It was the manufacturer who carried the spares and very often he who installed them at machine or plant breakdowns. This latter situation was not an efficient proposition; when dealing with powered supports, maintenance had to be organised and continuous. Two days later a settlement out of court was reached. FSW agreed to pay the plaintiffs $200,000.

During the weekend interval break between meetings we spent a short vacation in Park City, an old silver mining town later developed as a ski resort. At the top of the ski run at a height of some 10,000 feet in the Rockies the weather was glorious and the air like sparkling wine but the feet tended to be a little cold even though one felt warm in the sun. W.F. Morrell, R. Bucknall, M.V. Bricknall, the FSW Directors, and myself stood talking to a group of American ladies from Connecticut on top of the slope. One lady offered me her skis, saying,

'Have a go, it's not difficult.'

'Lady, have you ever seen me ride a bicycle?' I said.

'No, do you do something special?' was her reply.

'Let's put it this way: if you and I were on the steepest part of the run and you were 50 yards above me, you'd be in mortal danger. I dare not expose you to such a risk.'

She and the whole group all burst out laughing.

'Another rejection, the story of my life,' was her final comment.

I enjoyed the time I spent amongst the Mormon community; they had great consideration and took time out to talk to you.

The Smoky River project referred to the development of a new mine which involved the application of the longwall mining systems in association with pitching gradients of some 21 degrees. Prior to the purchase of equipment a feasibility study was undertaken based on reports from geologists Paul Weir & Co., Mining Consultants FSW and McIntyre Porcupines Mines Ltd. I did not visit the particular site, nor was I involved in the negotiations, although I had done a great amount of work in connection with it.

However the case was ultimately settled out of court at, I understood, a cost to FSW of $400,000. I received a kind letter of thanks from Booker's parent company of Fletcher Sutcliffe & Wilds, thanking me for the support given to them.

Several weeks after I returned home I received a package of fourteen LPs from the leading law attorney in Salt Lake City who had conducted the Fletcher, Sutcliffe & Wild's case. These featured the Mormon Tabernacle Choir and have given me hours of pleasure over the intervening years.

OTHER AMERICAN ASSIGNMENTS

From October 1973 onwards I undertook a number of assignments on the American scene, including a mining plan for a proposed new longwall mining system for Clinchfield Coal Company – Virginia Division, and a strata control and review of longwall mining Layout C for the Jewell Valley Corp, Virginia.

My first visit to the USA coalfields had been in 1957. Ron Hamilton, Chief Engineer (No.9 Area SW Division) and myself joined a charter flight from London to New York. I recall that the aircraft was a converted Super-Constellation of World War II and there was a flight time of twelve hours, with a touchdown at Gander, Newfoundland on the way out and another at Shannon, Eire on the way home. Whilst in New York we made contacts which facilitated visits to mines in both Pennsylvania and Virginia. At the time, we were primarily interested in the manning of coal preparation plants. The visit to me was fantastic. I met quite a number of Company Presidents and Vice Presidents, and one in particular was Irvin C. Spotte, Vice President of the Clinchfield Coal Company – Virginia Division.

We met under the following circumstances. Ron and I had spent a full morning looking round and assessing the situation at the Moss Central coal preparation plant, near Dante. When we adjourned to the company store about mid-day for a luncheon snack, standing alongside me was a gentleman I hadn't met or been introduced to. After a while I turned to him and said,

'Are you associated with the coal preparation plant we have just visited?'

'Yes, it comes under my jurisdiction as Vice President of the Division of the Clinchfield Coal Company,' he replied.

Having broken the ice we started chatting, during the course of which he asked,

'You are from England: are you familiar with the mechanised longwall mining system they are practising out there?'

'Yes,' I told him, 'I have been closely associated with its development from the start.'

'We are interested in trying out such a system at our Moss No.2 Mine in a seam of about 42 inches in thickness. Have you any views?' was his follow up.

'Could you get me some coveralls? I will go with you and have a look at what you have in mind,' I volunteered.

'Would you do that?' he eagerly responded.

The outcome was that he found me the attire, following which we travelled down the valley and entered Moss No.2 Mine about 4.30 in the afternoon.

We examined a number of operating American mining sections, afterwards visiting a number of de-pillared extracted areas. By this time we were on Christian name terms as though we had known each other for some considerable time, a situation I found refreshing as opposed to the stiff and starchy protocol approach often met in home relationships.

'What do you think of our chances, Charles?'

The seam was quite flat with about 300 yards of cover, with a good shale roof and firm fireclay floor. Compared with some of my successful home situations it looked quite promising. Leaving the mine about 8.00 p.m., we continued discussions at Irvin's Dante office, leaving there to travel to Bluefield. Virginia about 10.30 p.m. Thus I made a friendship which lasted over twenty-five years, during which he became President of the Pittston Coal Company. We met many times afterwards both in the UK and the USA. During his tenure both as Vice President and President, Irvin introduced a number of successful longwall installations. During one of his early efforts however he ran into great difficulty. A full set of hydraulic supports was destroyed. He sent me photographs – all support units were twisted and distorted way beyond anything I had ever experienced or previously seen. The strata loads in that particular situation were greatly in excess of what could have normally been predicted.

Irvin was and is an excellent mining engineer and one of the most industrious persons I have ever met. We both derived common benefits from our long association. I understand he is now a mining consultant living in Florida, and have no doubt he will be extremely busy. Whenever my mind wanders over the American scene, I gratefully recall fate engineering our earlier and lasting contacts.

LEE NORSE COMPANY

In 1974 I was invited to undertake a UK market survey on behalf of the Lee Norse mining machine manufacturers, relative to its equipment and products. The Company President at the time was Joe Clements (one of America's foremost football players of all time), whom I met in London to discuss my approach to the enterprise. After carrying out the assignment I submitted my report and Joe sent over Jack Marsh, his Product Manager, whom I introduced to National Coal Board management personnel, with whom he made a number of friends. Jack made several trips over during which we got together, maintaining a friendship over some twenty years until his recent untimely sad demise.

Four or five years ago, together with our daughter Dorothy, her husband John and our grandson Paul, we spent a few days with Jack and Jean, his charming lady, over in Scottsdale. Arizona, taking in the Grand Canyon. What an awesome and beautiful experience. The grandeur of it is firmly and permanently imprinted in my mind. Travelling from Scottsdale to the Grand Canyon to me was a historic experience, particularly the small but beautiful town of Sedona. My mind boggles at the countless billions of tons of eroded material apparently washed by the Colorado River from the Sonora Desert and Grand Canyon into the Gulf of Mexico over limitless aeons of time.

Baldwin Francis Electric Engineers, Sheffield, UK

Together with one of the Company's Staff I flew out to Milwaukee, Wisconsin to visit Cuttle-Hammers, a large electrical equipment manufacturing company, in connection with inter-related UK and USA market outlets. One of the Directors I had formerly met in the UK, and had conducted him on a visit underground in the highly mechanised Haigh Moor seam at Silverwood Colliery, Rotherham Area. He was interested in UK mining electrical equipment, particularly switchgear and electrical motors.

Following the visit he took us for lunch at a delightful restaurant outside Milwaukee which together with the Annunciation Greek Orthodox Church in the city had been designed by Frank Lloyd Wright, the celebrated American architect. Another site of great interest we took in was the Mitchell Park Horticultural Conservatory, which embodied three large spherical glasshouses. They were each some 360 feet or more in diameter. One contained desert plants, another 'American plants', the third having a large variety of plants and foliage possessing wide horticultural interest.

Before flying to New York by arrangement we called in at Charleston. West Virginia, to meet one Darwin Ensign, then in his early seventies. He was the former owner of an electrical equipment manufacturing company which he had sold out to a large electrical products manufacturing company. He was interested in a new consortium of his former staff who were setting up an enterprise for the manufacture of electrical mining equipment. We met at the airport and from the first moment an empathy developed between us which was sustained until his death several years later. The emerging company was looking for outlets and UK agencies and had established contact with Davis (Derby) Ltd, manufacturers of electrical signalling and their equipment. The Managing Director of Davis (Derby) Ltd, Charles Wallace, I knew quite well from his former days working for the National Coal Board. Moreover, he was distantly related by marriage to Mary.

We spent quite some time in discussion with Darwin and two of his former colleagues and then went on to visit the new factory, which gave me the opportunity to assess the whole situation, including capacity, products, and quality of products, upon which I later produced a report. Having three hours of idle time awaiting my flight to New York, Darwin insisted I went to his home and meet Mrs Janet Ensign, a fine lady who suffered stoically with arthritis in every joint of her body, despite which she was remarkably cheerful. Formerly she had been a professional vocalist and with my own love for music we had very much in common, so we soon became friends.

After I had returned home the report I had produced was voluntarily sent on to Charles Wallace, and had the following outcome. In a letter sent immediately with the new American Company Secretary, he wrote to the effect,

'I have known Charles Round for many years. If he says you're good, - that is satisfactory to me, I trust his word anywhere – we are prepared to link up with your Company.'

This they did to the mutual satisfaction of both companies for many years. Charles Wallace died prematurely, but we do still meet Jean his charming wife quite often.

Following the setting up of the joint relationships, Darwin, Janet and I met on several occasions both here and in the USA. On one occasion we managed to get them together with Jim Nash, Mrs Nash and their charming daughter Kay. The occasion was a mining exhibition in London, following which they toured the European capitals – their first visit overseas. Mary and I conducted them on a run to Windsor Castle, Runnymede, and other places of interest. See PHOTOGRAPH 63.

PHOTOGRAPH 60. TONY BURKE, OWNER AND MANAGING DIRECTOR (ELECTRICAL AND ENGINEERING DISTRIBUTORS), CLAYLAND'S ELECTRICAL AND ENGINEERING DISTRIBUTORS, DUKERIES INDUSTRIAL ESTATE, WORKSOP, NOTTS

Tony Burke started out as an apprentice electrician at Kiveton Park Colliery, located between Sheffield and Worksop. In the post Second World War days of American Lend Lease Mining Equipment, he took an interest in the early stages of the American mechanisation of pillar and stall mining activities, following which he joined Bailey's, Jones and Bailey's Ltd, as a travelling salesman. They at the time were pioneering the development of roof bolting as a form of underground roadway support which had been successfully established in the American mines.

Following a successful launching of a small engineering company which he sold out to a Midland group of diversified companies some 22 years ago, he formed his current enterprises which have prospered greatly under his able direction.

PHOTOGRAPH 61. HOUND HILL FARMHOUSE, WORSBOROUGH, BARNSLEY, YORKSHIRE

Early (1645) Royalist Fortress. Substantial fortress walls were built round the full extremities of the Hound Hill farmhouse, farmyard and outbuildings adjoining the spacious yard area prior to the Parliamentary Wars of 1645.

These were subsequently over-run and substantially destroyed by Baron Robert Fairfax's army. The stone being recovered was transferred and used elsewhere.

The farmhouse is set in a delightful and pleasing environment with country land on all sides. It is a beautiful historic building steeped with an atmosphere of its period and contains a very large library of ancient volumes and other works.

To the west of the road the area was rich in old mines running along the seam outcrops and also close to the larger collieries of Barrow, Barnsley Main, Silkstone and Dodworth, so the property had its problems with mining subsidence.

PHOTOGRAPH 62. HOUND HILL FARM, WORSBOROUGH, BARNSLEY,
YORKSHIRE.

During the Civil War between Charles I and the Parliamentary Armies which ended in the former's defeat at Naseby, Northants in 1645, a number of Royalist fortresses were constructed in parts of Yorkshire including three in South Yorkshire: Wentworth Wood-house, Wentworth Castle and Hound Hill Farm, Worsborough, Nr Barnsley. Baron Robert Fairfax, Commander of the Parliamentary Armies (1645–1650), was instructed to reduce these fortresses to rubble.

The photograph shows part of the Houndhill Farm fortifications including an early form of defensive 'pillbox'. Carrying out instructions, Baron Thomas Fairfax's Parliament soldiers overran the farm and its defences. Later, on entering the farmyard with his retinue, Fairfax observed a number of men to the right of the farmhouse in the process of building a scaffold.

Enquiring what was afoot his officers informed him of the preparation being undertaken to hang the Royalist traitor, Richard Elmhirst. 'Cease your endeavours,' was Fairfax's order. 'I have a kindness for this man.' Thus the alleged traitor was spared and the offence was taken through the Civil Courts which had the effect of fining the then owner Richard Elmhirst on the assumption of guilt, later reinforced by a further fine of £100, but the farm and properties was allowed to be retained within the Elmhirst family. The current family head, Alfred Elmhirst, a former solicitor, is now in his early 90s. The farm and its associated properties has been retained within the Elmhirst Family ownership for over 350 years.

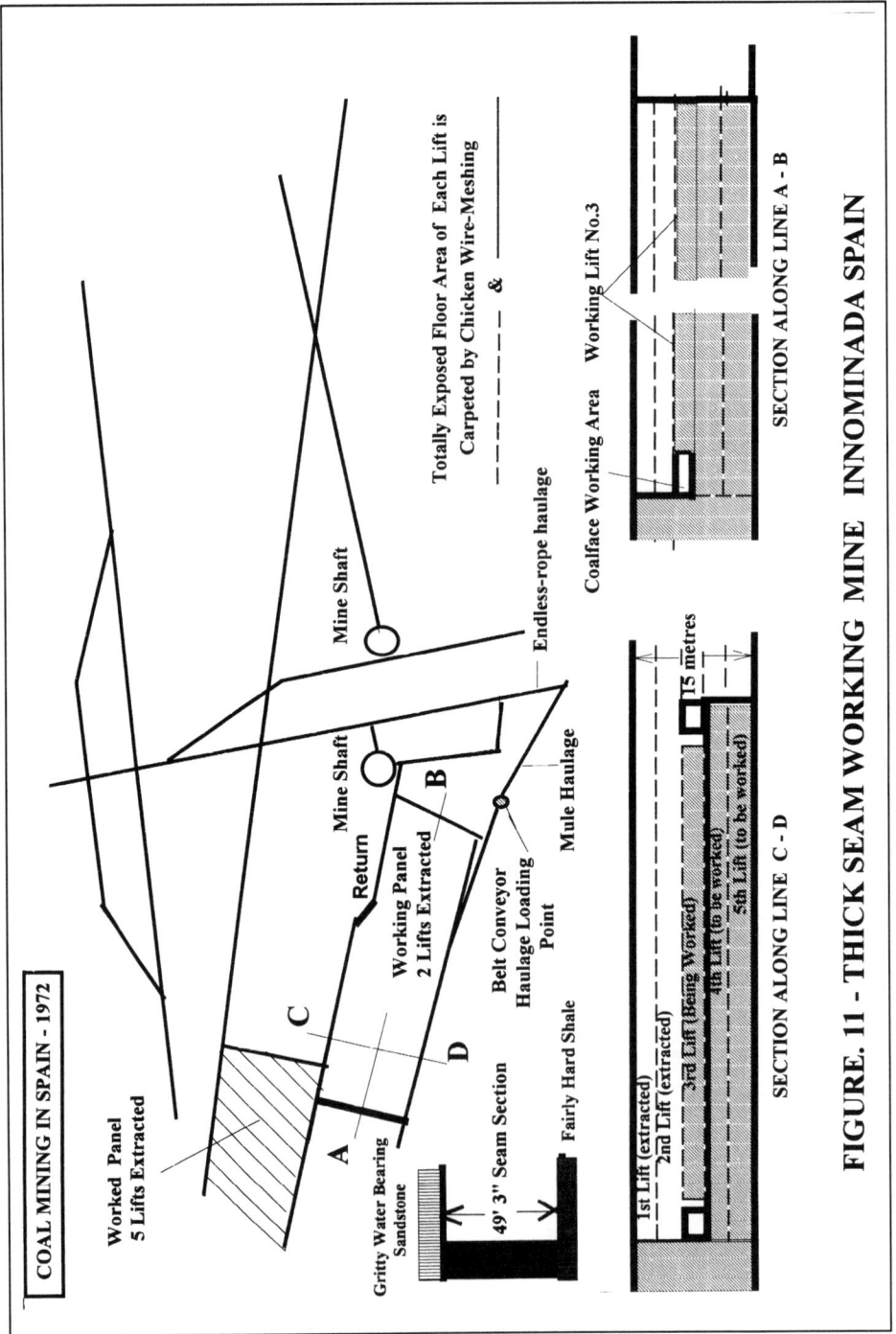

FIGURE. 11 - THICK SEAM WORKING MINE INNOMINADA SPAIN

(OPPOSITE PAGE) FIGURE II – THICK SEAM WORKING IN MINE
INNOMINADA, SPAIN

Some very remarkable achievements have occurred within Coal Mining Industries – the more spectacular these achievements are the less there seems to be known about them generally. The people responsible usually take them in their stride, happy in the knowledge that a difficult problem has been resolved or, where the problem involves a wide area of mining practice, such people will share their experiences in the form of visits or written papers presented before their Technical Associations. The general public appear to learn more of the negative side of coalmining involving partisan political issues, closures, strikes, or disasters than of the fantastic technical achievement and advances made within the coal industries of the world by mining engineers and the miners everywhere.

During an assignment in Spain early in 1972 two such examples came to my notice and gave me a tremendous boost of morale. The first was a method of working a 50 foot seam inclined at 18 degrees. The second was working a 4 foot seam inclined at 78 degrees to the horizontal. Both examples are worthy of short explanation and reference. Thicker seams have been worked in India, China and elsewhere for generations – but this to me I felt to be unique in the manner of practice and accomplishment particularly as alongside the feats being achieved, primitive application obtained in other directions.

The mine Innominada is the location of the first narrated example. Here the seam was 15 metres (49.25 feet) in thickness pitching about 18 degrees. The roof was a water bearing sandy grit whilst the floor at the foot of the seam was a fairly hard shale. The coal was extracted in lifts of 10 feet in descending order. It was considered that the first slice under the natural roof was the most difficult to take. To form a working roof after first lift had been taken a carpet of chicken wire mesh was placed at roof level above the supports with each incremental advance being connected to the previous ones. As the workings advanced the strata broke down and consolidated on top of the newly formed coal floor, such that when the next lift was taken it became quite a homogenous mass forming the roof with the chicken wire mesh binding it together and presenting a roof skin. This sketch indicates part of the mine's layout showing a worked out and a working unit. From the Section A–B and C–D it will be noted that two 'lifts' have been developed. There was quite a good technique in laying the mesh attached to a machine mounted on top of a panzer conveyor: rolls of mesh were paid out and secured above the supports supplementing normal production operations. At the time the men were operating with pneumatic picks with the management contemplating the use of power loading machines. With modern face equipment this could be regarded as a suitable subject for the extraction of 15 feet lifts still using the technique described and modern longwall face equipment in which case performance could be expected to be well in excess of 10,000 tons per day.

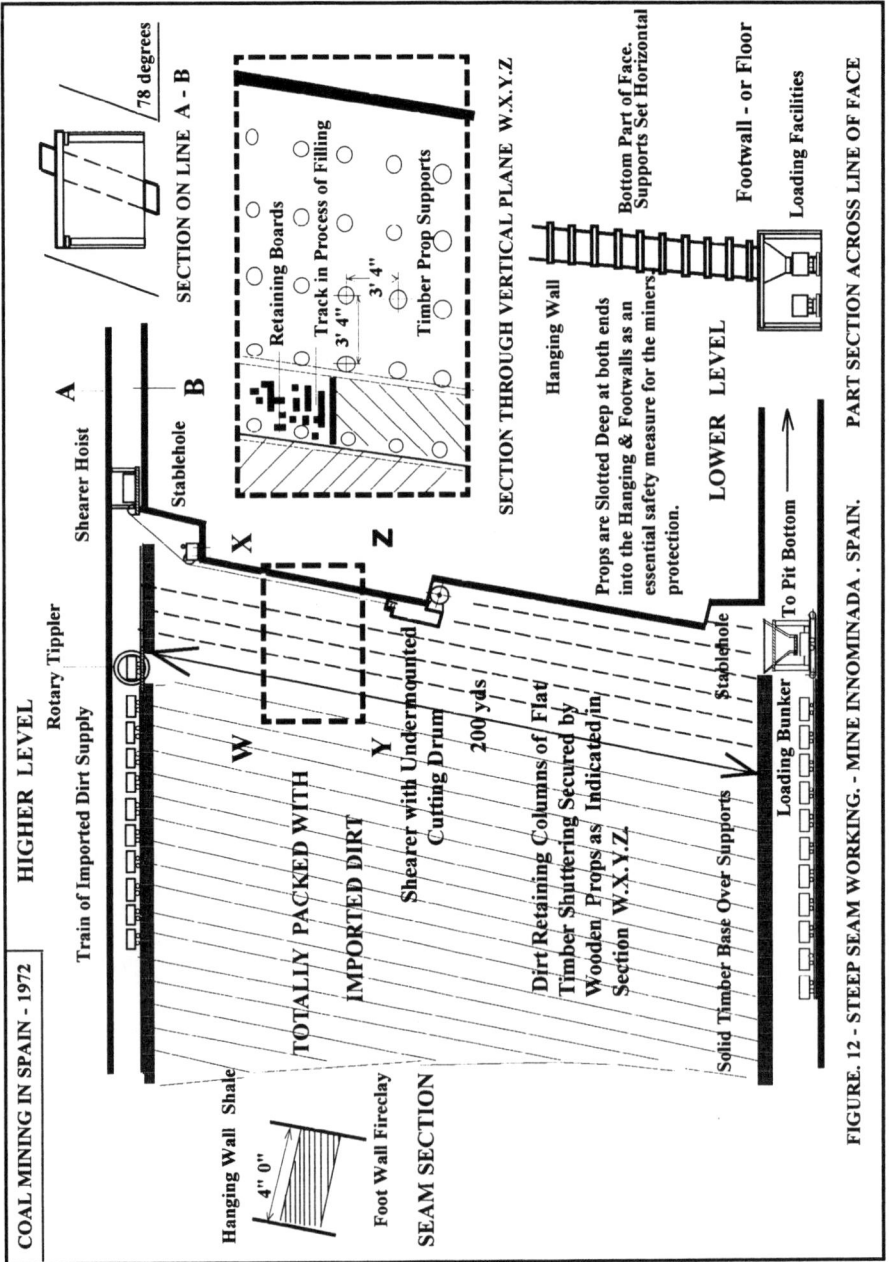

COAL MINING IN SPAIN - 1972

HIGHER LEVEL

Train of Imported Dirt Supply

Rotary Tippler

Shearer Hoist

Stablehole

A

B

SECTION ON LINE A - B

78 degrees

Retaining Boards

Track in Process of Filling

3' 4"

3' 4"

Timber Prop Supports

SECTION THROUGH VERTICAL PLANE W.X.Y.Z

Hanging Wall

Props are Slotted Deep at both ends into the Hanging & Footwalls as an essential safety measure for the miners protection.

Bottom Part of Face. Supports Set Horizontal

Footwall - or Floor

Loading Facilities

LOWER LEVEL

To Pit Bottom

Stablehole

Loading Bunker

PART SECTION ACROSS LINE OF FACE

TOTALLY PACKED WITH IMPORTED DIRT

W

X

Z

Y

Shearer with Undermounted Cutting Drum

200 yds

Dirt Retaining Columns of Flat Timber Shuttering Secured by Wooden Props as Indicated in Section W.X.Y.Z

Solid Timber Base Over Supports

Hanging Wall Shale

4' 0"

Foot Wall Fireclay

SEAM SECTION

FIGURE. 12 - STEEP SEAM WORKING. - MINE INNOMINADA . SPAIN.

– 348 –

(OPPOSITE PAGE) FIGURE 12 — STEEP SEAM WORKING
IN INNOMINADA MINE, SPAIN

This second example represents quite a unique mining system for a number of reasons, the heavy gradient at 78 degrees (almost vertical), the length of the coalface on this inclination, and the fact that mechanised production obtained using an Anderton shearer specially modified for the application by having its cutting drum mounted under the machine base. The application and use of powered supports was under consideration: quite an imaginative situation with the absence of the panzer conveyor and the supports were set horizontally. The principles developed in Figure 10 could well offer a possible contributory solution although some of the techniques used in mountain climbing, e.g. safety chains inter-connecting the units and anchor stations for security, might offer further possibilities.

The Russian-built unidirectional shearer was suspended by a steel rope secured to a haulage winch positioned adjacent to the face in the Upper Level. Cut coal fell into a bunker type loading media at the lower level from which the mine tubs were filled. The support system was quite simple and consisted of up to six rows of wooden props similar to the rungs of a ladder. For the men's security, both ends of the horizontal supports were ledged into the hanging and foot walls. The hanging wall (roof) was a weak form of shale, whilst the foot wall (floor) was a fairly strong fireclay. Dirt was imported into the mine, the face being packed by tipping the dirt at the 'upper level' and allowing it to fall into the track being filled at a prescribed distance from the coalface, timber shuttering being provided to ensure its containment. The vertical section WX-YZ shows detail of the support system and the manner in which the dirt or filling is stowed in the waste void after the coal has been extracted. The dirt is tipped mechanically and dirt falls into place under the influence of gravity being restrained within its track by the shuttered wall. Stableholes were formed at the top and bottom of the face, the machine operating unidirectionally. The arrangements worked very well. However, the underground haulage was most outdated at the time: haulage was undertaken by Spanish mules, rope haulage and narrow slow belt systems. The mules were magnificent animals, being bigger and stronger than the pit ponies of my early days in mining.

Loading facilities were simple – a portable bunker type collecting medium which enabled controlled discharge into the 2 ton or so mine-cars.

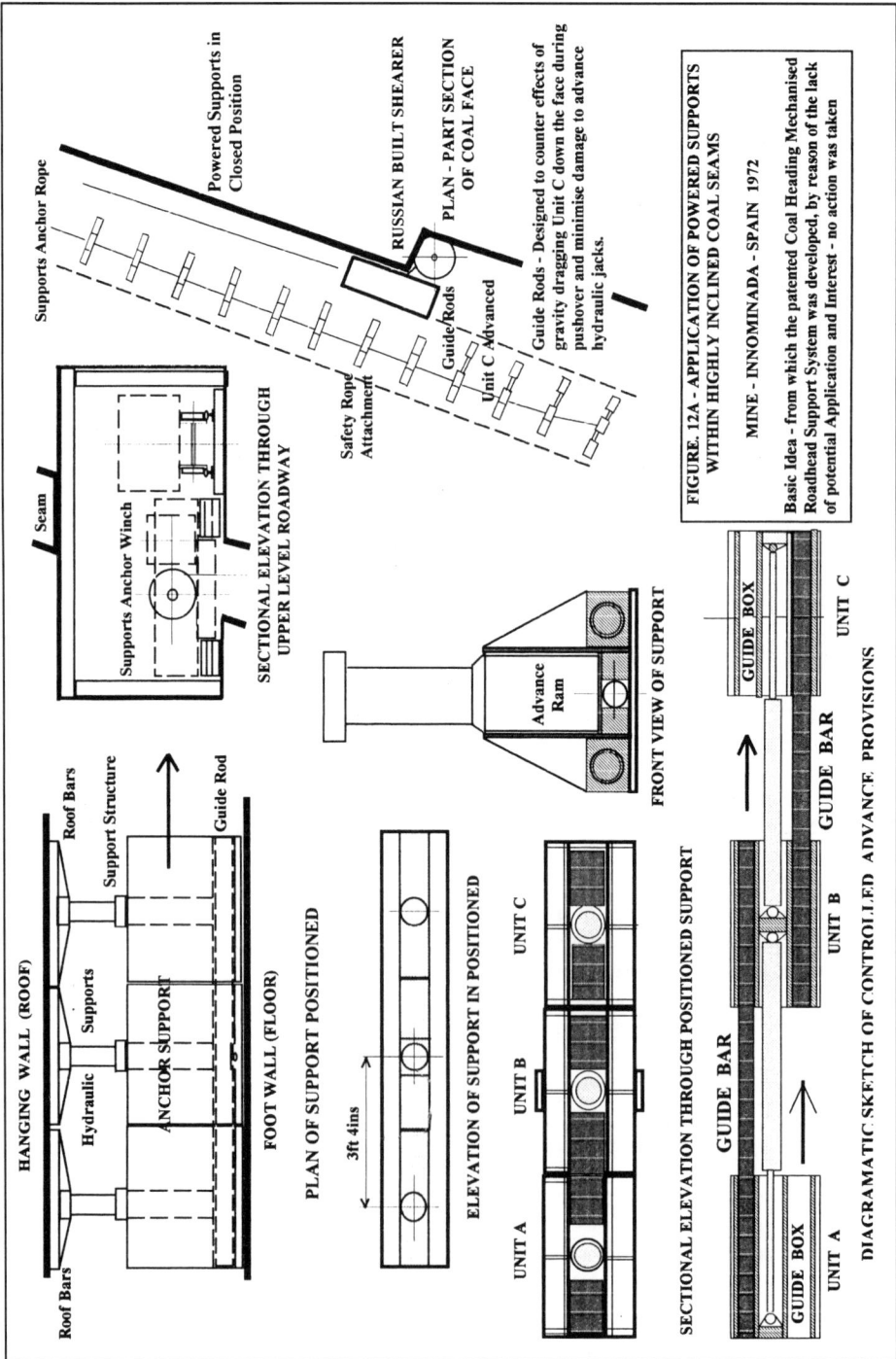

HANGING WALL (ROOF)

Roof Bars
Hydraulic
Supports
Support Structure
ANCHOR SUPPORT
Guide Rod
Roof Bars

FOOT WALL (FLOOR)

PLAN OF SUPPORT POSITIONED

3ft 4ins

UNIT A   UNIT B   UNIT C

ELEVATION OF SUPPORT IN POSITIONED

Advance Ram

FRONT VIEW OF SUPPORT

Seam

Supports Anchor Winch

SECTIONAL ELEVATION THROUGH
UPPER LEVEL ROADWAY

Supports Anchor Rope

Powered Supports in
Closed Position

RUSSIAN BUILT SHEARER

PLAN - PART SECTION
OF COAL FACE

Safety Rope
Attachment

Guide Rods
Unit C Advanced

Guide Rods - Designed to counter effects of
gravity dragging Unit C down the face during
pushover and minimise damage to advance
hydraulic jacks.

GUIDE BOX

GUIDE BAR

UNIT C

GUIDE BAR

UNIT B

GUIDE BOX

UNIT A

SECTIONAL ELEVATION THROUGH POSITIONED SUPPORT

DIAGRAMATIC SKETCH OF CONTROLLED ADVANCE PROVISIONS

FIGURE. 12A - APPLICATION OF POWERED SUPPORTS
WITHIN HIGHLY INCLINED COAL SEAMS

MINE - INNOMINADA - SPAIN 1972

Basic Idea - from which the patented Coal Heading Mechanised
Roadhead Support System was developed, by reason of the lack
of potential Application and Interest - no action was taken

(OPPOSITE PAGE) FIGURE 12A – STEEPLY INCLINED WORKINGS:
INNOMENADA MINE, SPAIN

The Spanish mining engineers indicated that within their projections for mechanisation powered supports would be part of the infrastructure. To the best of my knowledge, at the time no powered supports had been designed, constructed or installed on such heavy gradients anywhere in the world. Returning from my visit to Spain, I spent a great deal of time working on the problems of applying such supports on steep gradients. The basic difficulty was that of controlling each sectional unit as it was being advanced forward. Once released from the roof for incremental advance it would simply slide down the face under the influence of gravity.

The idea I came up with is embodied in Sketch 12A, in which a three element frame unit, A,B and C, incorporating guide rods was envisaged. Each element had its own support jack which enabled it to be advanced in section with some support to the roof maintained at all times. The sketches show how the system was projected to work.

Despite this, however, the situation called for a safety backup, this being provided in the form of an anchor winch situated in the top level with each support being linked individually to the one below and to the winch haulage. Moreover the winch was there to provide for the installation and withdrawal on and off the highly inclined coalface.

It was from this basic idea that the element principle was transferred to the support of coal heading roadheads (shown in Figure 10) which had a wider and more urgent need for application within the UK. Although patents were taken out for this latter system it never attracted interest and was never developed.

(ABOVE) PHOTOGRAPH 63. DARWIN ENSIGN AND PARTY FROM CHARLESTON,
WEST VIRGINIA, USA, VISIT UK

This photograph was taken during the visit of Darwin and Janet Ensign of Charleston, West Virginia to the United Kingdom. Davis (Derby) Ltd arranged sightseeing tours which took in Windsor Castle and Runnymede Meadows on the south bank of the Thames near Windsor, where King John met his rebellious Barons in 1215 and accepted Magna Carta.

The location of this photograph was in front of the Compleat Angler alongside the River Thames. L–R: Janet Ensign, Mary Round, Darwin Ensign, Charles Round. Mrs Jim Nash, daughter Kay Nash and the Davis (Derby) chauffeur Jim Nash, who took the picture.

Recreational and Other Activities

Despite the fact that my mining activities absorbed practically all my available time, somehow I found myself involved in limited leisure activities from time to time, a situation which extended my range of contacts and friendships to Ireland, Tasmania, Canada and the USA.

As a youngster, before I started work, I enjoyed cycling and together with a group of friends so inclined, we explored the Derbyshire Dales during weekends and on summer vacations made several trips to Scarborough some eighty-seven miles distant from Hoyland Common. My first cycle was built by Grave's of Sheffield. It was a 'Speed King' model with oil bath, Sturmey Archer three-speed and half-drop handle bars and the cost was £5 19s. 6d. It was a very sturdy, heavy model which gave many years of reliable service.

Occasional attendance at the Barnsley (Oakwell) football ground during a Saturday home match was a highlight topic of conversation for many weeks. Those were the days of Brough Fletcher (Manager), John Charles, of later Italian football fame, like Stanley Matthews one of the game's real gentlemen, Clifford Bastin, Tilson and Brooke (later transferred to Manchester City) who were all players of note in those days. Incidentally, Barnsley football team are reputed to be the only club ever to win the Football Association's Cup in Yorkshire by beating West Bromwich Albion on a Yorkshire football pitch; all the other Yorkshire clubs to win this trophy did so at Wembley or outside the county.

On starting work and studying I had little time for recreation although I did manage to read a number of books. Charles Dickens, R.L. Stevenson, H.G. Wells, Alexandre Dumas, Zane Grey, H. Rider Haggard and R.M. Ballantine were the authors who appealed to me in those early days; Emile Zola, Franz Kafka, Jack London, Rudyard Kipling, Koestler and others came along much later.

Whilst working under N.R. Smith, Production Manager, I was introduced to golf and along with Ron Swindall turned out on occasional Sunday mornings at one of the Nottinghamshire Clubs. I never really got seriously interested in the game: moreover, unlike anyone else who claimed that 'a round of golf obliterated all thoughts of work' it never applied to me. However I do have a small shield given to me by Gilbert Winder, former Captain of the Alfreton Golf Club, which testifies to my having 'holed out in one stroke' whilst playing with my son John. He started the game at the age of thirteen years or so and became an excellent player, winning many prizes over the years, in addition to having 'holed out in one' on three occasions. In one competition over four holes, he took but seven strokes (including a hole in one) and in that particular year he and Sue were awarded a week's golfing tuition under the direction of Christie O'Conner, the former Ryder Cup player, in the Canary Isles. In 1993 my daughter-in-law

Sue was made Lady Captain of Dun Laoghaire Golf Club, in Eglinton Park, Dun Laoghaire. As what I thought would be light relief, I produced a fifteen page satire, 'The Challenge of Golf', under the pseudonym Carlos Rotundi, which dealt with the DAUGS (The Disillusioned and Unwanted Golfers Society), complete with Articles and Memoranda of Association. This I sent along to her with the Society's Research Bulletin No.1 'Golf Clubs and Club Swing Characteristics'. This incorporated tabular and graphical detail of 'club size/distance relationships' etc., as produced by the Society's Research Director Sir Oswald Twitt. Golfers take their golfing pursuits seriously and it was quite a while before she realised the nature of the document.

Record and Tape – Audio Collection

Over the years I have built up a large record collection of old 78s and LPs covering world famous artists and a wide range of music: classical, light instrumental, vocal, brass and military band, and indeed popular music of the day. Among my favourites are Beethoven's piano sonatas played by Arthur Schnabel; violin and piano sonatas by Beethoven played by Fritz Kreisler and Franz Rupp; Haydn string quartets with the Pro Arte String Quartet; Beethoven symphonies and string quartets with Toscanini. The range of old artists and recordings I have really is something and has given me and others great pleasure over the years. Currently Gregorian chant is featured strongly in response to public demand on Classic FM Radio. I possessed copies of this on old 78s for a large number of years.

In 1983, together with a North Countryman Robert F. Wray, I set out to recover from old 78 recordings the best of the brass bands under the title 'Heritage in Brass'. Unfortunately we were only able to make two conversions from these old recordings; the first was the St Hilda Colliery Band, and the second featured Jack Macintosh, the cornet virtuoso.

Heritage in Brass – St Hilda Colliery Band

The St Hilda Colliery Band (1869-1927) has a unique place within the historical accounts of the brass band world, in that it achieved standards of musicianship akin to those of the professional players of the time.

The name St Hilda has its origin way back in the distant past in AD 648 when St Aidan, 'Apostle of the North' prevailed upon the Princess Hilda, daughter of King Edwin, to establish a religious house at Caer Urfa to be the centre for the evangelisation of an area which is now County Durham.

In 1905, under the patronage of Mr James Kirtley, Agent for the Harton Coal Company (which operated four collieries in the South Shields Area) and the St Hilda Miners Lodge (they provided financial support through the miners of the colliery), the St Hilda Colliery Band, under its Bandmaster James Oliver, became the leading brass combination in the UK within four years. In 1921 at the Crystal Palace contest with their winning performance of their test piece 'Life Divine' by Cyril Jenkins, the band created a national sensation. This was considered to be the band's most emphatic win of its contesting career.

Obviously, I wasn't the first mining engineer to have an interest in brass bands and as the St Hilda Band formerly was, the Grimethorpe Colliery Band today ranks amongst the best there is in the concert, contesting and recording fields.

The St Hilda Colliery Band's successes sustained in the contest and concert arenas established levels of musicianship, presentation, and performance equal to the best orchestral combinations of the time and has the unique distinction of being the only brass band within the brass band movement ever to take professional status. During its career, the band had the honour of five Royal Command performances, together with a long series of engagements at the leading seaside resorts, in music halls and in radio broadcasts, including the series production of 160 78 r.p.m. gramophone records.

In 1922 the band undertook its first continuous tour, which involved a colossal list of engagements, administered by the Secretary and Business Manager, James Southern. In that year 182 appointments, involving concerts with an average of fourteen items per programme, were successfully undertaken. The band's stationery proclaimed it was 'the most successful band in the country, the pride of the North, engagements booked anywhere, any time, distance no object; you have tried the rest – now try the best.'

With the retirement of James Southern in 1937 the famous St Hilda Band was disbanded. The recordings were converted to LP albums featuring the following items:

Heritage in Brass Volume 1
St Hilda Colliery Band

SIDE 1		SIDE 2,	
1. March: St Hilda March	Hawkins	1. March: Rubinstein	Bidgood
2. Grand Selection:		2. Overture:	
The Beggar's Opera	Pepusch	Poet and Peasant	F.V. Suppe
3. Cornet Solo (Harold		3. Cornet Polka:	
Laycock): Titania	Rimmer	The Bostonian	Rimmer
4. Valse: Allah	Nichols	4. Humoresque:	
		Pat in America	Hiram Eden
5. Air: Andante in G	Batiste	5. Sacred Air: Varie	
		Simeon	Rimmer
6. Symphonic Poem:		6. Grand Selection:	
Coriolanus	Jenkins	William Tell	Rossini

The challenge by the colliery and works brass bands to the professional musicians was seriously considered by the musical hierarchy in those early days. Derogatory slogans such as 'beer and baccy' bands were freely expressed. However today that no longer holds in that some of the finest brass players accepted within the orchestral world started out with their village colliery works and Salvation Army bands.

JACK MACINTOSH, CORNET VIRTUOSO (1891–1979)

Jack is featured on the second of the Heritage in Brass volumes. He was born in Sunderland, County Durham on 22 September 1891. His formal education commenced at Barnes School, Sunderland, following which he attended Skerry's College, Newcastle. He received his musical training under his father, a very competent cornetist in his own right and a very skilled musician. He started out at the age of six and at eight was playing second cornet with the East End Band, being regarded at the time as a child prodigy. In 1907, at the age of seventeen, he was struck low with rheumatic fever for over a year. By dint of tenacity, fortitude and exacting exercises he fought his way through his illness which left his limb joints distorted. His grip and finger positions were permanently affected and unorthodox. With rigid self discipline, sustained habits and physical exercise he not only overcame these problems but achieved a fine military bearing which he carried throughout the rest of his life.

Fit again, Jack went for a while to a business college, but almost immediately found himself a regular professional job playing with a band of five for the 'silent films' at the Sunderland Palace Theatre in 1909. In addition to his professional activities he was asked to play for the Hetton Silver Band for the Grand Shield, Second Section Contest at Crystal Palace. In 1913 he joined the St Hilda Colliery Band as a professional assistant to Arthur Laycock during a busy concert season. Later in 1916 Arthur Laycock was called up for military duties and for the next three years Jack became principal cornet for the St Hilda Colliery Band, whilst at the same time playing professionally with a band of thirty-five at the Sunderland Empire Theatre. His orchestral career began in July 1930 with his appointment as founder member with the BBC Symphony Orchestra under Sir Adrian Boult with whom he stayed for twenty-one years until 1952.

Heritage in Brass Volume 2
Jack Macintosh – cornet virtuoso

SIDE I		SIDE 2	
1. I Hear You Calling Me	Marshall	1. Bird of Love Divine	Wood
2. Alpine Echoes	Windsor	2. Zelda Caprice	Code
3. Columbine Caprice	Wright	3. Sounds From The Hudson: Valse Brilliante	Clarke
4. *Una Voce Poco Fa* Barber of Seville	Rossini	4. Bride of the Waves: Polka Brilliant	Clarke
5. Facilita: Air & Variations	Hartmann	5. Fascination: Air and Variations	Hawkins
6. Silver Showers	Rimmer		

7. Until	Sanderson	7. Lucille Caprice	Code
8. Cleopatra Fantasie Polka	Demare	8. *Ill Bacio* Waltz	Arditi
9. Carnival of Venice		9. Showers of Gold	
Air & Variations	Arr: Arban.	Scherzo	Clarke

Between 1952 and 1970 he became a member of the Philharmonia and New Philharmonia Orchestras, touring Europe and South America with Herbert Von Karajan and Sir John Barborolli respectively. He died in November 1979 at the age of eighty-eight.

We had 1,000 copies of each Volume made. Unfortunately the demand was below that expected, and on 1000 copies (500 of each) I lost about £300. This wasn't too bad and what was rewarding was the number (some thirty or forty) of the old bandsmen's sons and daughters who wrote to me expressing their thanks for the pleasure their fathers (usually in their late seventies and eighties) had received from both recordings. I suppose old bands and artists are like old systems and machines. They just get misplaced in antiquity but are never totally forgotten.

Radio WFDTS-FM. Columbia Union College, Maryland

Reference at intervals throughout this chronicle have been made directly to my interest in brass bands and in all kinds of music generally. Over several years from 1970 onwards this took a practical form in the shipping and receiving of band and brass instrumental recordings in the form of reel-tapes, cassettes and LP records to and fro across the Atlantic. One such activity was with the broadcast service of the Columbia Union College WGTS-FM in Takoma Park, Maryland, and my friends there, Alvin Schwab and Walter Dorn. In the early and mid 80s I was introduced to Alvin Schwab, a Consulting Engineer from Wheaton, Maryland, by Ronald F. Stokes of Norfolk, Virginia, with whom for quite a long time I had been exchanging tapes of trumpet players ancient and modern.

Alvin Schwab had been running for many years 'Band Concert of the Air' from this radio station on Sunday nights at 7.00 p.m. local time, featuring bands from Germany and other countries. I recall shipping out ten or so LP albums of brass bands (including Grimethorpe Colliery) in the early stages which I prepared as a programme, announcing the items and applying appropriate local comment for the enlightenment of the American listeners.

The radio station engineers were later able to make use of tape cassettes from which they were able to modify and prepare final tapes of the material submitted for direct broadcast. In this latter format, I made some twenty-five programmes covering most of our First and Second Class bands which were put out over the air covering a period of some two years or so. I enjoyed doing them. Walter Dorn had been running 'Musical Memories' on Saturday nights over a long period, sometimes featuring musical biographies of composers and artists. Other programmes presented popular musical memories of very many years.

Having received from my two friends and colleagues some fifty or more of their programmes, I found them to be of an excellent standard, tremendously well researched and having an all-embracing informative and entertainment value. I recall the programme of Stephen Foster most vividly, for it had a great impact on me emotionally. Some of the compositions I had never heard before, whilst the sadness which beset this fine composer's life was really moving. I often wondered if the pressures of life which beset him contributed to the beauty of his compositions. Other programmes included Strauss, Waldteufel, Paul Linke, Offenbach together with German 'light operetta' programmes.

I met both Alvin and Anita Schwab a couple of years or so ago in London. Unfortunately they were on a scheduled run which limited the time for off-programme contacts or visits. Fortunately on a second occasion we were able to have Walter Dorn over for a little longer; he was able to stay overnight and on the following day Mary and I took him down to Dawlish to savour his great interest in matters concerning railways. The first time I met Walter over here was in Bristol. He too was tied down to a rigid pre-timed and rigid schedule but we did have a pleasant night together.

Unfortunately the programmes 'Band Concert on the Air' and 'Musical Memories' were both closed towards the end of 1993 after twenty-one and twenty-eight years continuous broadcasting respectively.

ROBERT HOE JR. POUGHKEEPSIE, NY

Robert Hoe (1922–1983) of Poughkeepsie, New York State was a highly successful businessman. He was a man of great purpose and ideals, with a dream to preserve for posterity a huge reservoir of little heard music for bands, which he had collected over many years in both Europe and America, to educate and to raise the standard of the band to the level and status of the symphony orchestra. Of great practical vision, he promoted more than 180 LP albums as a series 'Heritage of the March', which featured a great deal of the music he had collected. This was often arranged from piano copy and manuscript by very competent musicians amongst whom our own Gay Corrie of Barrow-in-Furness took a leading role. Bands from the United States included those from universities, colleges, high schools and the armed forces, whilst those from the United Kingdom featured Black Dyke Mills, Besses o' the Barn and Stanshawe Bands, all under Roy Newsome, the then BBC presenter of 'Listen to The Band'. Certain of our military and service bands together with those from other European countries were also featured on the associated LPs to great effect.

Over the years the United States Marines Band (The President's Own) under the direction of Col. Schoepper made mono tape recordings of a large number of varied works and arrangements from the classics. From these old tapes he cut and pressed some twenty further volumes, a number of which are really classics since the band at the time was considered to be the US Marines Band at its best. He also had the band record the whole of John Philip Sousa's works including some 127 marches, fantasies, operettas and exhibition pieces. John Philip Sousa, trained as a violinist, wrote a single solo for the violin, 'Nymphalin', a very beautiful and pleasing solo which I feel to be one of his best compositions. In all, Robert Hoe produced well over 250 LPs which he shipped out to universities, colleges and high schools and to selected persons throughout Europe, America and Australia, gratis. He was a kind and generous person with a tremendous love for the euphonium which he sometimes played with the Allentown Band directed by Bert Myers.

In recent weeks, from late December 1993, I made contact with one of the UK's current leading trumpet virtuosi, John Wallace, a most brilliant instrumentalist and player who is regularly featured by Radio Classic FM in baroque items or with his 'John Wallace Connection'. In his latter capacity he was in the process of researching material and information for a projected exclusive Sousa and Scott Joplin concert in Queen Elizabeth II Concert Hall, London on 29 March 1994. He was having difficulty obtaining the manuscript of Sousa's violin solo 'Nymphalin'. I was able to contact, after an interval of some twenty-five years, an old American musician, Art Lehman, former principal euphoniumist of the US Marines Band for many years. He made contact with the US Marines Music Library (which is fully computerised). In a little less than a week the manuscript for piano and band were made available to him. I'm glad John Wallace and his Wallace Connection featured it in his project, for it is a most delightful work and Sousa's daughter considered it to be his best composition. I attended the concert which was a huge success with Joshua Rifkin playing the Scott Joplin compositions. Biographical notes prepared by Tony Stavecar about both John Philip Sousa and Scott Joplin were narrated by David Healy between the items featured.

Art Lehman was, and is, well known to many military and brass band euphoniumists and other musicians throughout the United Kingdom, and in a large number of cases he was most generous in fulfilling their wants and meeting the special difficulties they referred to him over a period of some thirty years or so.

PHOTOGRAPH 64 is a picture of Art and Freida his wife at their home in Camp Springs, Maryland. Special LP albums were put out by Robert Hoe Jr, featuring Art playing with great skill and virtuosity all the standard cornet and euphonium solos which bring out the artistry of this great instrumentalist and musician.

Following my short experience as a merchant seaman I had a very strong desire to sail with the *Queen Elizabeth II*, but for many reasons this was not possible until 1971. Having flown out to New York by air in August 1971 to meet friends and colleagues including Mr and Mrs Jim Nash from Charleston and Darwin and Janet Ensign in Fort Lauderdale and others in various other parts of the United States, I decided to return home via New York and Southampton First Class by the *Queen Elizabeth II*, and what a fantastic experience that proved to be. The accommodation, service and treatment was perfection. It was probably the only real luxury I have ever partaken of in the whole of my career but it is something I have savoured many times since. As a form of travel I found it to be the most restful and relaxing of all. Comfort, food and accommodation match the best hotels anywhere within my experience. Whatever one wants to do there is an available outlet.

The passage back home between 16 to 22 September 1971 was quite smooth and very pleasant. Whilst sitting dining in the Columbia Restaurant one had the impression of being in an exclusive London hotel with no sense of movement whatever. At our table we had the ship's hotel manager's wife and the ship's chef's wife, a fine couple from Connecticut, New Jersey and Mr and Mrs William Little of Ada, Oklahoma. I seem to recall him to have been the proprietor of a newspaper in the city. They were two unassuming people very kindly disposed, always cheerful and happy. For a very long time after we had each settled back home we were in correspondence. In fact Penny Little introduced Mary and her family to pecan pie, sending over the ingredients and recipes.

PHOTOGRAPH 65 shows Mr and Mrs William Little and myself and was taken on board the *Queen Elizabeth II* prior to going to dinner one evening. On one occasion at dinner I recall the subject discussed getting round to collective nouns and a very large number were quoted throughout the dinner. However the lady from Conneticut swept the board with:

'What is the collective noun for a group of 'ladies of the evening' blocking a sidewalk in New York?'

'A jamb of tarts,' was her answer.

The food throughout all meals for the duration of the journey was really something and the service immaculate. Discussions at dinner on one occasion centred on 'take-away'. However, reference was made to the English form which had been available very much longer, 'fish and chips'. Our American friends were very interested, particularly when the hotel manager's wife remarked how she got her best gastronomic pleasure in eating such fare.

Much depended upon the actual newspaper in which the 'fish and chips' were supplied. *The Times* was no good; it was much too 'uppish'. The food should be wrapped in one or other of the 'down to-earth' northern sports editions to obtain the maximum gastronomic value from this form of quick food.

The following day at lunch the ship's hotel chef provided our American friends and ourselves with a sample of this instantly sustaining meal, naturally without the newspaper. From their comments afterwards, the intent of our American visitors was to try the real thing during their stay in the UK.

Naturally, having been and being involved in technical and engineering matters, the *QE2* boiler plant, turbines and engines had to be seen. The ship's designers had certainly packed away tremendous power in a comparatively confined space whilst the screw propeller shafts which ran along the sides of the ship in themselves were quite something, particularly in relation to their size and alignment accommodation.

PHOTOGRAPH 64. ART & FREIDA LEHMAN

Art Lehman was for very many years solo euphoniumist of the United States Marines Band. In addition he was the band's recording engineer in which capacity he committed to tape some rare and fantastic Col Schoepper recordings of the classics, at the time when it was considered the band was at its best ever. He established a wide range of contacts with UK and European musicians to whom he readily and freely provided musical and playing advice to all who sought it. Robert Hoe Jr, business man of Poughkeepsie, USA, committed to posterity two private LP albums with a very wide range of cornet and euphonium solos played by this great artist.

PHOTOGRAPH 65. WILLIAM AND PENNY LITTLE,
QE2 PASSENGERS FROM ADA, OKLAHOMA, USA

During a return passage from New York to Southampton by the QE2 in September 1971, William and Penny Little were two fellow travellers. Throughout the whole voyage we shared the same hotel table reservation along with the ship's hotel manager and chef's wives and a further couple from Connecticut. They were a most excellent and interesting group of people and made a great contribution to what is probably the most enjoyable travelling experience anyone could have.

My memories of QE2 travel are really something special – everything about this great ship, its crew and services, ranks with the best available. After a hard period of visiting mines and spending much time underground it was a tremendous relaxation to return home in this manner. William and Penny went out of their way to avoid myself as a single traveller ever feeling outside whatever activities we were involved in. For quite some time after the voyage and on their return home we were in correspondence. Penny introduced to Mary and the family the pleasures of pecan pie, not only sending the recipes but also the ingredients. They were two of the most gracious and charming of all the people I have ever met in the United States or anywhere for that matter.

I 2

Decline of the UK Mining Industry

When I retired from the National Coal Board's service in 1971 the Coal Industry's future looked quite promising. Collieries at the end of their economic life were being closed from time to time but that was the normal situation.

The North Yorkshire Selby complex with its thick seams and 'state of the art' technology in which the Riddings Project played its part was coming on stream. The Industry had been stream-lined. Things looked quite good and augured well for the future.

However, with the advent of projected nuclear power generation during the 1960s, embraced by the Civil Service and politicians alike with the promise of low cost electrical power generation costs that alternative power station fuels could not hope to reach, I believe in hindsight that the early seeds were set for a decline in the Coal Industry's fortunes.

Lord Robens in his book *My Ten Year Stint* outlined in great depth the political situation at the time, making strong valid political and economic points, all of which have been factually borne out in the intervening years. With regard to the advent of nuclear energy he stated,

> Neither Coal Board as a whole nor I were personally against the building of nuclear power stations. Clearly the country where the atom was first split and which had in many ways originated this new hope for humanity must attempt to develop and reap the rewards for being early in the field with a power station building programme. Our criticism was that the programme was too big for a new process and that before proceeding to a large (and as it turned out disastrous) programme operating experience should first be gained. The fatal error of too big a programme was made and stemmed from there being several consortia for power station building. There never was really room for more than one.

During his Chairmanship he made strong attempts to have an independent enquiry into the true capital and generating costs of nuclear energy. Questions in Parliament relative to such costs elicited some detail of the escalating costs of Dungeness B and revealed that favourable coal cost comparisons could be made relative to the contemporary costs of nuclear fuel for electric power generation (much to the embarrassment of the CEGB).

With a change of Government, MPs were told in reply to their questions that such matters were for the Chairman of the Central Electricity Generating Board who would write to the Member. The answers to such future questions so received gave no information at all. In fact, about the only figures they contained were the dates. This conspiracy of silence was considered a sinister development in that Parliament was being denied information of the greatest national importance, information that was essential for seeing whether the vast sums of money were being wisely spent.

Probably one of the earlier blows struck against the Coal Mining Industry arose from the decision to set up a nuclear power station within the Durham Coalfield at Seaton Carew. It was instigated by the then Minister of Fuel and Power, Roy Mason, formerly a Barnsley miner. Strong efforts to institute an inquiry into the true costs involved he rejected. His reasoning to the Yorkshire miners was that he had refused the inquiry because he was satisfied that nuclear power was cheaper and that any examination would prove that and shatter the morale of the miners and that the result of any inquiry would be to kill coal. It was the miners' morale that mattered most.

Such reference certainly does not apply to the Yorkshire miners I worked amongst and later managed. The inherent fibre of the Yorkshire mining fraternity is probably stronger than that of any other coalfield with the possible exception of the Welsh, a situation which has been proved many times.

Winter of Discontent

Towards the end of Edward Heath's period as Prime Minister (17 December 1973) we had a miners' overtime ban which was skilfully led by Joe Gormley, the then President of the National Union of Mineworkers, such that industry throughout the country was obliged to work a three-day week. This was due to a shortage of deep mined coalstocks which had fallen to a level of some 40% of normal output. The miners won the confrontation and returned to work following which Edward Heath lost the backing of the Conservative Party and was replaced by Margaret Thatcher as Prime Minister.

It could well have been at this point that the Government's attitude to the Mining Industry hardened towards reducing the nation's dependency on using coal for power generation in view of the alternative fuels readily becoming available.

Joe Gormley retired shortly afterwards through ill-health and age and was replaced by Arthur Scargill, then President of the Yorkshire Miners Union. Arthur I knew quite well through my association with Woolley Colliery in the North Yorkshire Area. Arthur was and is a truly remarkable man of purpose, a highly intelligent, articulate, and natural inspiring leader and deeply left-wing politically motivated, set on improving the lots of his fellow men and the miners he represents. I found him to be an excellent negotiator and willing to listen to and accept convincing reasons on matters under discussion.

I mentioned, at the beginning of this narrative, the premature death of my elder sister Doris's husband Albert Sykes by contracting pneumoconiosis whilst working as a miner at Cortonwood Colliery, South Yorkshire. Many years after, during the time Arthur Scargill was dealing with matters of compensation on behalf of his miners and their widows in such and other cases, I referred Doris's situation to him. He applied himself on her behalf with deep understanding, care, unstinted effort and diligence, as with countless other similar situations, but in the circumstances was unable to secure redress for my sister. He did some excellent work with great sympathetic feeling in that field on behalf of the miners and their widows who had been shabbily treated by their former employers.

There were many times during media coverage in many situations over the past few years that I didn't recognise Arthur as the man I formerly knew. On the last occasion we met, many years ago, we were both paying tribute to my old colleague and friend Tommy Wright during his funeral service. Arthur I believe will go down in history as a leader who fought 'tough and hard' for what he believed to be the best interests of the miners he represented.

Orwellian Shadows – 1984

Shortly after taking up his appointment as the NUM President, Arthur Scargill obtained what was alleged to be a secret list of collieries scheduled for closure. He made the details public issue which elicited denials and the usual explanations as to the official situation. I recall similar situations on a minor scale during my employment in higher management. Later events however did prove the authenticity of his discoveries. Even though the country had passed quietly through the previous winter and coal stocks were high, apparently he considered the situation to be a strong subject with which to confront the Conservative Government led by Margaret Thatcher.

Prior to 1984 Arthur Scargill sought and obtained by a national ballot of all mineworkers a mandate for strike action. In this connection I believe he was motivated by two objectives: the first, to secure the future of the coalmining industry including the collieries listed for closure; the second, to bring down the Conservative Government led by Margaret Thatcher.

Early in 1984 the miners' strike was called amidst splits in the Miners Union ranks. The miners of Nottinghamshire had become disenchanted with Arthur Scargill's leadership by reason, amongst other things, of his refusal to call a second strike ballot of NUM members for entering into an actual strike. They broke away from the National Union to form their own organisation, 'The Union of Democratic Mineworkers'. They undertook separate negotiations with the National Coal Board and sustained several large Midland collieries in production throughout the strike period.

The strike took an Orwellian twist in terms of Big Brother and violence in that the Prime Minister appointed Ian McGregor, a very strong American industrialist and a former President of the United Steel Corporation USA, to take over as Chairman of the National Coal Board.

Arthur Scargill organised mass picketing on an unprecedented scale, during which violence flared time after time, involving a fatality and many injuries to both the miners and the police. The situation really did get ugly. In the early days the public had great sympathy for the miners but a visit to Libya by one of the NUM top officials and a meeting with Colonel Gadaffi was not at all helpful.

The Power Industry played no small part in breaking the strike and the defeat of the miners. Oil purchases dramatically increased with certain coal power stations being converted to burn oil and gas; coal imports were surreptitiously increased and supplemented the stocked fuel at coal-burning stations.

Once coal-fired power stations had converted to burning oil or gas, the likelihood of them turning back to coal was very remote. In such a situation the demand for coal following the miners' strike could then be expected to show a further decline. This took place and resulted in the pit closure list being extended.

Arthur Scargill's and the miners' strategies were out-manoeuvred by a series of events:

1. The combined efforts of the CEGB management and the power workers led by Lord Marshall, a Chairman who considered the situation as an all-out war. They certainly kept power stations operating, being more than imaginative in the process.

2. The loss of support from the Nottinghamshire miners and the breakaway of the Nottinghamshire Miners Union, together with the continued operation of several large Midland collieries.

3. A crucial second National Ballot decision taken by the junior officials, members of the National Association of Colliery Overmen, Deputies and Shotfirers, not to support the miners in their confrontation with the National Coal Board and the Government, which reversed their former decision to support the miners' case. This really overnight turned events in the Government's favour. It would appear that the Government came to within twenty-four hours of capitulation, being saved in the final analysis by the National Coal Board's junior officials in the decision they had just taken.

Ian McGregor proved to be the man for the occasion, piloting the industry through the strike, and after a period of upwards of twelve months the men's confrontation collapsed. I know of no other strike which inflicted greater damage to the Coal Industry than this did. The mineworkers paid heavily for losing this strike both during its pursuance and subsequently after its collapse.

Having lost the strike Arthur Scargill and the power and influence of the Miners Union were severely weakened and more or less impotent. The Government and the National Coal Board had *carte blanche* for their list of colliery closures to be implemented, but now, more importantly, also the freedom to add to that list at their discretion which they did in no small measure.

During the strike Mike Eaton, a former colleague referred to earlier, was called upon to act somewhat in the capacity of the Government spokesman, appearing on TV news bulletins and the like. This was a difficult situation for him, being caught up within great conflict. He acquitted himself quite well in the most unrewarding of circumstances.

Events have since indicated that quietly and surreptitiously a second approach detrimental to the Coal Mining Industry had been put into effect: that of strategically approving, constructing and commissioning a series of gas-fired power stations, greater in generation costs than the coal-fired stations they replaced. This had the further effect of reducing the electricity authority's demand for coal, creating further escalation of pithead coal stocks as they came on stream and replaced coal-fired power stations.

Subsequently, within a comparatively short time after the strike had finished, the pit closure rate escalated dramatically. As each individual colliery closed, the Industry's predicament deteriorated. The financial servicing of the increasing coal-stocks, the mounting costs of decommissioning the closed collieries, heavy redundancy payments and such financial write-off was levied against current falling production levels and artificially increased the average coal production costs to levels where it was argued coal was not competitive against gas, oil or nuclear energy. It difficult to regard this as nothing other than pure manipulation designed to eliminate the nation's former dependency upon coal as the source of electrical power generation and, in some respects, political revenge relative to past difficulties and confrontations with the mineworkers.

In contrast, the Nuclear Industry was cocooned against the decommissioning costs of its redundant power stations and the safe disposal of its nuclear waste, inestimable in value at the current time, although figures have been quoted such as £15.8 billion sterling ($23.8 billion) which is to be borne by the taxpayer and will take over a hundred years to implement. 'What a crazy world we live in.'

United Kingdom Mechanised Longwall Achievements

It is against the above background that the Coal Industry's potential in real terms should be considered, relative to what now follows.

Coal mining technology within the United Kingdom over the past forty years has undergone great revolutionary change which I have traced out with landmarks from the primitive situation to the system in which the current advanced mining technology and mind-boggling achievements have emerged.

Evolution has played its part but the real progress which has been made has been attained through the forward thinking of intellectual men at all levels from the humblest miner to the most brilliant technocrat, the latter often motivated by the observant employee innocently remarking in a specific situation,

'Gaffer, can't you do something about this . . . If you alter that by . . . and put in something like . . . we can more than double what we are doing now.'

Thus a need is established. The rest is pure intellectual and practical application.

What I have found distasteful in such circumstances is where the manager, technician, or any person for that matter, so inspired forgets or deliberately ignores the source of his or her later achievement. I shudder to think of how many people in the course of history have had their worthwhile ideas usurped for personal prestige and financial reward by other selfish people, or have been inadequately recognised by the enterprise they serve. Classification of people and individuals as workers often degrades the individual's true worth, particularly as they develop great skills and expertise in whatever activity they are engaged in, and individually have much to offer if one takes the trouble to refer to their experience and recognise the abilities they possess. Miners are particularly adept in this direction no matter what the situation, which to me has often been both stimulative and amazing. Quick to spot managerial or other people's mistakes but ever ready to applaud and support worthwhile endeavours, their contribution to the success of modern mining method and technical innovation has been immeasurable.

Worldwide, the UK longwall mining system has been accepted, the basic element of which is the actual coal producing machine, the shearer, with the transport, communications and support systems forming the matching infrastructure. Developments in the former affect the latter and, as such, an integrated entity progressively develops. This has been the case for many years from which the current 'advanced high technology longwall unit' has emerged, triggered by the insatiable desire on the part of mining engineers to have greater and greater power designed and built into this basic element.

The great stimulus built into the requirement specifications of American, Chinese and French engineers has contributed substantially to current electrical power levels of achievement. The Anderson Group Limited, Motherwell, Eichoff (GB) and others have responded to such demands admirably.

The particular responses of the Anderson Group Limited's design, mechanical, electrical and mining engineers to the emergence of more powerful shearers has been truly gratifying. Over the years they have come up with radical designs which have transformed the whole coal producing process. The introduction of the modular constructed shearer has been a 'moon landing' leap forward in terms both of power application and of intense reliability.

In simple terms, with modular construction the basic different functioning elements of the machine are designed and constructed as separate working units, later integrated onto a common base plate, or within a frame structure. This utilises the power more effectively and enables the overall power input to the machine to be substantially increased and more efficiently used than with the former single motor designs driving all operating facets of the machine through mechanical gears and the like.

I recall that in my early youth both at collieries and within factories in my locality, multiples of plant, shapers, lathes, drilling machines, drop hammers, borers, cutting saws, etc. located over a wide floor area were all driven by a single steam engine, belt driving systems of line shafting and belt pulleys which almost defies description. Later these individual items of plant were fitted with electric motors and became separate machines. This is a close analogy to the idea of modular construction of the complex sophisticated single machine.

Since this machine is the basic element of the longwall mining system, its further development sets the pace for the system as a whole, such that those responsible for the different remaining elements, armoured face conveyors, hydraulic roof supports, communications and the like where required to match any increases of production capacity need to subject their equipment to appropriate development. However they do not necessarily have to be so inspired; generally their designers and engineers are constantly seeking to extend their achievement frontiers in whatever manner enhances their equipment performances. This has been the case with many British and German manufacturers of mining equipment, and resulted in the UK advanced longwall mining technique becoming widely established on the world scene in the more enlightened countries of Australia, China, Poland, South Africa, and the United States of America. Other major coal-producing countries will follow if they wish to survive.

One of the major breakthroughs which has emerged is that of opening up thick seams of 6 metres or more to a highly efficient production of amazing proportions. I have indicated one such example on the Spanish scene. Currently one could extract such a seam of fifteen metres in thickness with two or three passes rather than as formerly with five, with spectacular high production low cost performance.

I have often reflected as to whether North Sea gas and oil production was a boon or disaster. What I find difficult to reconcile is the vast revenues which have been generated being more than offset by similar vast sums being used in confining skilled and willing personnel to enforced idleness, subsidised by 'state handouts' under the heading of unemployment benefits.

ANDERSON GROUP LIMITED ELECTRA SHEARER SERIES

Currently operating worldwide within the above series are the following models: E550, E600, E1000 and E2000, all on a step-up power and operational rating in terms of seam thicknesses and mining environments. They have been successfully tested and established throughout an extremely wide range of mining situations with examples of performance in a number of countries and installations to follow.

The range of photographs and sketches which now follow illustrate the current state of coalface mechanisation and feature the Electra 1000, in operation.

PHOTOGRAPH 66 shows the Electra 1000 in operation at Kerr McGee mine in Illinois, USA. Here the machine is producing at a rate of some 2,400 tons per hour. The hydraulic roof supports set 'skin to skin' give complete cover to the operator and men working the face.

Never in my sixty-six years association with the coal mining industry have I seen such immense production being delivered at the pick point, albeit mechanised pick point, a situation best described as fabulous, fantastic, miraculous, stupendous and voluminous, as the old cinema film trailers used to portray forthcoming films. These adjectives are not out of place; in fact other relevant ones can be added.

PHOTOGRAPH 67 provides a clearer view of the machine in all its majestic prowess and magnificence. The leading cutter drum, coal breaking assembly and deflecting cowl are all shown in the raised position. The following small cutting drum which completes the seam extraction has its scroll clearly defined. Every effort during design was made to obtain the maximum tunnel dimensions between the machine and the panzer, so to facilitate the passage of coal under the machine.

FIGURE 13 is a seam sectional drawing of the face with the machine approaching the LH Gate. The design flexibility of the Electra 1000 is such that it can readily accommodate a wide range of seam thickness or thickness variations.

PHOTOGRAPH 68: the Electra 550 shearer loader, designed for seams below two metres in thickness. It provides an excellent view of the leading cutting drum assembly and the rack-toothed track along which the machine travels backwards and forwards across the face. There is a good view of the panzer conveyor chain assembly showing the double centre strand chain and flights which move the coal, falling and being deflected onto the conveyor during production operations.

FIGURE 14 is an exploded sketch of the Electra 550 modular form of construction which together with the photographs explains this fabulous machine design far better than paragraphs of technical or even lay description. In detail it shows how various active elements are individually powered but integrated to great effect into a complete sophisticated entity. Its upgrading can be undertaken in terms of high-power, capacity and great reliability over a very wide range of mining conditions in the UK, USA, Australia, France, China, Poland and South Africa.

FIGURE 15 represents seam-sectional elevations relative to the Electra 1000 located within a six-metre seam. Note the matched design of the hydraulic roof supports substantially strengthened and simplified in terms of hydraulics and flexibility with which to accommodate a wide range of seam thickness.

THE FRENCH CONNECTION – LORRAINE COALFIELD

At this point, I would be most remiss not to make reference to the Electra 2000 as a specially designed machine for the Lorraine Coalfield as the circumstances are quite illuminating.

This is quite a difficult mining proposition. Seams vary between 1.5 and 5 metres in thickness, face gradients are from 0 to 45 degrees, the gradients on line of the face retreat are over 20 degrees, and generally the mining environments are most arduous. Normally the supply voltage for operating the existing equipment installed there was 3.3 Kv, but in these adverse mining conditions and from the experience of the French mining engineers something more was required. This led to the specification of a 5Kv electrical supply requirement subsequent to which a machine was developed following the construction principles of the Electra 1000. This has an installed cutting power of 900 Kw, haulage power of 112 Kw developing pulls of 100 tons with haulage speeds of 14 feet per minute, a 150 Kw coal sizer and a 100Kw power pack module giving a total power of 1,162 Kw.

Modifications to meet the increased supply voltage were quite involved. Moreover, the 85 ton machine working on the sort of gradients referred to involved intensive safety considerations. The electrical provisions had to meet both local and European standards for electrical acceptance and certification. Heavy and prolonged mechanical and electrical tests were necessary, following which the first Electra 2000 undertook field trials. Following the excellent results obtained by the first machine, another five have been supplied. I can visualise with great clarity what tremendous efforts must have been applied by the French mining engineers and the Anderson Group Limited to develop a successful application in the circumstances. In this I see something of great importance – the difference in attitude between the French Government and that of the United Kingdom. Whilst the former are prepared to husband and exploit successfully their natural assets (particularly their farming as is generally well known) and human resources, our Government are content to abandon our natural coal resources to oblivion and the skills developed there to the dole queue. In a career span of some sixty-six years actively and passively associated with the Coal Mining Industry, I find the above developments to be somewhat of a 'fairy tale' ending to what has been a very rewarding and happy calling, although one that in its past ten years is tinged with great sadness.

WORLD DISTRIBUTION OF ELECTRA SHEARER INSTALLATIONS

The first Electra shearer was supplied to the Old Ben mine (now Zeigler) in the USA in October 1988, when since fifty machines have been supplied into all the major coal producing countries throughout the world. The machine has great versatility and enormous potential in that it can be applied within a seam range of 1.8 to 4.5 metres in thickness and is achieving excellent results in a wide variety of mining environments.

Currently the disposition of machines worldwide is set out in the following table which refers to type and numbers in the countries specified. This does not take into account the similar machines manufactured in the USA, Germany and China.

DISTRIBUTION OF ELECTRA SHEARER INSTALLATIONS

COUNTRY	ELECTRA MACHINE TYPE				INSTALLATIONS OPERATING
	E550	E600	E1000	E2000	
UK	3		2		5
USA			19		19
Australia			11		11
China			7		7
France				6	6
Poland	1		4		5
South Africa		1			1
Totals:	4	1	43	6	54

To summarise, the American mining engineers are leading the field with their established and projected longwall mining system performances. Ever restless in that direction they continue to push out further and further the boundaries of achievement with each generation of mining technique development. The rest of the world looks upon this as the example to be followed.

In Australia the Electra 1000 is establishing a reputation as the workhorse of their coal industry. With seven machines currently in operation and two more for delivery in 1995 the machine is firmly established in the market place.

With the current rapid expansion in the Chinese economy, fuelled by a huge coal industry leaping in performance and efficiency from the benefits of imported equipment and technique, together with advancements of their own, the scene is being set for a major impact on the world scene with China becoming a superpower within the next 25 to 50 years and exerting such pressures on world markets as to make it more and more difficult for the West and the Americas to sustain competitive capability.

It is understood that Germany, USA and China each have the design and machine manufacturing capabilities and capacity for extending the number and types of the machines and systems described above, a further situation which compounds the artificial disaster which has been inflicted on our Coalmining Industry.

At this point I would like to disclose the source of my updated technical information which has been derived from a series of excellent technical papers written by Mr Ken Mackie, Engineering Director, Anderson Group Limited, Motherwell to whom I express my thanks for permission to include same and for regenerating the former excitement and pleasure of past years in the early machine and system developments described.

The observations and opinions expressed throughout are my own and are based on thought and analysis arising from both experience and event, as disclosed by the media and by technical sources of information. We now apparently live in an age in which the 'big lie' can be accepted as creditable currency under the blanket of being 'economical with the truth' and deceit is permissible under the Civil Service faceless determination of 'against the public interest'. In such a situation how does the ordinary person assess the credibility of official statements and understand the basic public issues of living in the 21st century? Or understand what those issues are and what the public interest really is particularly with reference to the billions of tons of untapped coal reserve assets of incalculable value available and their true place in the future economic life of the nation?

EXAMPLES OF ELECTRA SHEARER PERFORMANCE

Performance detail with reference to the application of UK advanced longwall mining technology now follow.

During the year 1969/70 the National Coal Board's No.3 (Rotherham) Area 19 collieries produced a total tonnage of 10,099,000 tons of which 9,796,000 tons was from mechanised coal faces. Overall productivity was 2.31 tons per manshift or approx 560 tons per man year. The number of longwall shearer installations was 73. The average tonnage per installation per machine year was 134,200 tons per installation.

The same number of current installations provided with Electras and matching infrastructure with overall colliery capacities designed at 4,000,000 tons per annum could be expected to produce, as a conservative estimate 73,000,000 saleable tons, i.e. 745% greater under the same mining conditions. With individual pit overall capacities of about 2,000,000 tons per year, one could still be looking at 57,000,000 tons from 38 advanced mechanised (installations) in that particular situation. I refer to colliery capacities in that at the time of the National Coal Board's 1967 Reorganisation Colin Rudge (Area Mechanisation Engineer), Ben Barraclough (Area Chief Mechanical Engineer) and myself were actively involved in upgrading the same by the introduction of better coal transport facilities and the build up of intense coal bunkerage. To the best of my knowledge these activities ceased on my transfer into the adjacent Barnsley Area.

AUSTRALIA

Six mines are considered:

> Appin
> Gordonstown
> Capcoal Southern Mine
> West Wallsend Mine
> Wyke Mine
> Newstan Mine

The performance detail which follows can be taken as the current representative performance situation at the present time.

APPIN MINE
LW25 UNIT

Machine commenced production	July 1992
Production to 10 October 1994	4,948,084 tons
Machine restarted	December 1994
Estimated panel operation + downtime	30 months
Machine utilisation (Reliability)	98%
Actual average annual rate of production	1,980,000 tons

GORDONSTONE MINE

The following represents the extraction of three panels over a period of 22 months during which the longwall mining system had to be withdrawn and reinstalled on two additional units between April 1993 and December 1994. In the absence of confirmatory detail, it is assumed both months were full months, though this may not necessarily have been the case.

LW101 + LW102 + LW103 PANELS

Commenced production	April 1993
Finished production	December 1994
Period of extraction	21 months
Total panels production	4,453,872 tons
Machine utilisation	97.84%
Average saleable tonnage per annum	2,450,000 tons

RECORD PERFORMANCES

Best Day	30,400 tons
Best Week	117,502 tons
Best 4 Weeks	418,688 tons
Best 5 Weeks	508,136 tons

These performances are apparently inclusive of the downtime taken to switch the full installation between the respective panels, a tremendous achievement by any standards.

CAPCOAL SOUTHERN MINE
LW605 PANEL

Commenced production	June 1993
Finished production	April 1994
Period of extraction	11 months
Total panels production	1,918,989 tons
Machine utilisation	95.27%
Average rate of saleable tonnage per annum	2,093,340 tons

LW606 PANEL

Commenced production	June 1994
Finished production	November 1994
Period of extraction	6 months
Total panels production	1,402,654 tons
Machine utilisation	97.07%
Average rate of saleable tonnage per annum	2,805,300 tons

Record Performances

Best Day	28,727 tons
Best Week	124,178 tons
Best 4 Weeks	339,106 tons
Best 5 Weeks	463,198 tons

LW605 + L606

Commenced production	June 1993
Finished production	November 1994
Period of extraction	18 months
Total panels production	3,332,640 tons
Machine utilisation	96.17 %
Average saleable tonnage per annum	2,214,430 tons

WEST WALLSEND MINE
LW 8 PANEL

Commenced production	September 1993
Finished production	May 1994
Period of extraction	9 months
Total panels production	1,150,058 tons
Machine utilisation	93.81%
Average rate saleable tonnage per annum	1,533,400 tons

WYEE MINE

Commenced production	June 1994
Finished production	December 1994
Period of extraction	6 months
Total panels production	858,148 tons
Machine utilisation	97.76 %
Average rate saleable tonnage per annum	1,716,290 tons

NEWSTAN MINE

Commenced production	May 1994
Finished production	December 1994
Period of extraction	8 months
Total panels production	1,077,225 tons
Machine utilisation	98.04%
Average rate saleable tonnage per annum	1,436,250

The foregoing has been processed from material I have been able to obtain. In the event of any variation with the actual detail at site level, this can be attributed to communication difficulties, together with my interpretation and processing of the statistics. I am, however, completely satisfied that it is sufficiently accurate (particularly the record achievements) to indicate the current situation and the potential of present productive systems and for comparison with back dated performance so avoiding any political or other misjudgments.

It will be seen that our mining colleagues 'down-under' have established the +2,000,000 tons per annum production unit, with the Electra as the workhorse of their activities.

UNITED STATES OF AMERICA

The current levels of coal production – longwall, bord and pillar or opencast oriented – provide an inspiration to coal producers anywhere. Having introduced the UK advanced longwall mining technique their mining technocrat's contribution to its development has been immeasurable and this dates back to the early days of John Todhunter Jr., President Barnes and Tucker and others already referred to. His modern counterpart has been no less committed so that today their achievements really are of astounding dimensions. Maybe they have an advantage in their superb mining environments with their immense, widely geographically spread, reserves of comparatively flat shallow seams, relatively free from heavy geological disruption and disturbance. Be that as it may, what has been and still continues to be accomplished, is the extraction of those reserves at low levels of cost with high levels of productivity.

In this respect two collieries are examined: Cyprus Twentymile Mine and Consol's Enlow Fork Mine.

CYPRUS AMAX TWENTYMILE MINE – OAK CREEK COLORADO

This is located eighteen miles from Steamboat Springs, Routt County, Colorado. Reserves of low sulphur (0.41%) exceed 200,000,000 tons. Current workings are in the Wadge Seam, 9 feet thick at a depth of 1,100 feet. The longwall panels are 840 feet wide and panel depths 11,000 (plans call for doubling panel depths to 22,000 feet).

The Electra 1000, operating on a 2,370 volt electrical system, traverses the face at a speed of 84 feet per minute extracting coal at the rate of 3,000 tons per hour, piling a continuous river of coal two feet in depth on the Westfalia PD 4 armoured face conveyor. Hydraulic roof shield supports each with its own sophisticated onboard computer automatically advance the roof supports and the face conveyor in sequence with the Electra's passage along the face. Rates of face advance are 55 feet per ten-hour shift which with a coalface of that length is really something. With our Riddings Project in 1970 we set standards of face advance of 140 feet per twelve hour two-shift period, but then our face was only 220 feet in length compared with 840 in this case. However, we did establish rapid face retreat as a practical possibility.

PERFORMANCE LEVELS:

Longwall unit: Annual production 1994 3,820,000 (run of mine) tons
Development – 2 continuous miners 1,160,000 (run of mine) tons
Total Mine Output 4,980,000 (run of mine) tons

MONTHLY RECORDS

Month	Shifts	Longwall Tons	CM Tons	Totals
Sept 94	34	415,058 r.o.m	76,202 r.o.m	491,594 r.o.m
Oct 94	35	434,058 r.o.m	63,500 r.o.m	497,558 r.o.m

Best longwall ten-hour shift 17,141 r.o.m
Productivity tons per man-year. 4,980,000/310 r.o.m
(Staff 310) 16,000 tons per man-year

CONSOL ENLOW FORK MINE

In 1994 this mine worked two longwall Electra mechanised units together with three continuous miners involved in development drivages, blocking out the longwall production panels in a seam 5ft. 10ins. in thickness. Face lengths 177 metres.

PERFORMANCE DETAIL FOR 1994

Two longwall mechanised units 6,290,000 saleable tons
Continuous miners 31,080,000 saleable tons
Total mine production 7,370,000 saleable tons
Total mine production 8,700,000 r.o.m tons
Saleable coal percentage 7,370,000/8,700,000
= 84.71%

RECORD PERFORMANCES

Best eight-hour shift – 21 Shears 11,237 r.o.m tons

MONTHLY RECORD SEPTEMBER 1994

Unit	Run of mine tons	Saleable tons
Longwall A	431,732	349,678
Longwall B	375,981	370,688
Total:	807,713	720,366
Continuous Miners	141,492	97,519
Grand Total:	949,205	817,885

The average saleable output per longwall unit 1994 3,245,000 saleable tons.

In 1989 total production from 95 longwall faces exceeded 106,000,000 tons roughly averaging 1.12 million tons per installation. I have little doubt in the light of the later performances (from the updated inductions achieved in 1992, 93, 94 and 95) but that the current situation has substantially improved and will continue to do so.

Between 1980 and 1989 longwall average face lengths have increased from 160 to 210 metres; the average seam thickness has increased from 1.65 to 1.95 metres over the same period. With the Electra's successful breakthrough into seams up to 6 metres in thickness one can foresee that the trend of increasing average seam thickness will continue.

CHINA

Longwall mining is established in China as the leading coal mining system. There are some 2022 such units of which 913 are mechanised, producing about 10% of the total present output of 1,034 billion tons. Current output is built up as:

Ministry Controlled	350,000,000 tons
Province Controlled	500,000,000 tons
Commune Mines	184,000,000 tons
Total Production	1,034,000,000 tons

Longwall faces: average face length	340 feet
Average production (all faces)	28,879 tons per annum
(Mechanised faces)	47,501 tons per annum
Best recorded monthly performance	228,000 tons
Number of longwall 1 million ton units	48
Best Performance	2,000,000 from one unit

Between 1970 and the current time, China has kept abreast of shearer design by both imports from Europe and domestic manufactured machines. Those for thin seams (under 1.3 metres) have progressed from 70Kw to 300Kw and those working seams up to 4.2 metres have had their power increased from 200Kw to 750Kw. At the same time design of their machines has developed from an installed power of 70/100Kw to 750Kw and are of similar design to imported machines. Looking at these figures I am struck by the potential degree of coal output expansion open to the Chinese economy. At the bottom end of the scale there are 1,100 non mechanised longwall faces producing at the rate of approximately 25,000 tons per annum, a little further along 913 mechanised faces producing 47,500 tons per annum (almost double) and at the end of the scale some 48 faces mining over 1,000,000 tons per annum almost wholly from domestic constructed equipment. What a base they have for supporting an emergent highly competitive economy in the current world situation and what problems it poses for the Western societies!

France

The French Coal Industry, not unlike that of Germany and our own, has been subject to intense difficulties. Nuclear power has been accepted and developed substantially and plays an important part in the French scheme of things. I have referred to the difficulties of mining in the Lorraine Basin (Houillères Du Basin – HBL) in which there are four collieries producing a total of 8 million tons from seams varying between 5ft. 3ins. and 16ft. 5ins.

However the French engineers and mining engineers, together with the Anderson Group Ltd., have gone a long way with the design of the Electra 2000 towards securing a viable and successful future for the coalfield, in view of its excellent performance.

LORRAINE BASIN – ELECTRA 2000 PERFORMANCE

Best shift	5,300 tons r.o.m	4,500 tons saleable
Best month	247,048 tons r.o.m	207,043 tons saleable
Total panel	1,380,000 tons r.o.m	1,180,000 tons saleable

In the field of communications great strides have been made with their Data Transmission – Telsafe System. HBL Central Headquarters and the local Area network are able to access data from, and send back to, each operating machine by analysing parameters such as machine position, haulage speed, and power consumption of the electric motors. A central core of expertise is available at Headquarters which can provide 'telemaintenance' and assist with identifying breakdown causes.

South Africa

The current rate of South African coal production is 180,000,000 tons per annum of which 109,000,000 is obtained from deep mining operations.

Longwall mining, because of the prevailing geological conditions, embracing low seam depths with massive sandstone cover, has resulted in high support costs and is considered not to be cost effective, although during the 1970s, 13 longwall installations gave excellent results. However short faces of 210 to 420 feet in length with panel depths have proved a very effective means of coal extraction – particularly with reference to the recovery of remnant coal pillars and awkward panels.

The current overall approach is one of simplicity. Roadways are restricted to 23 feet wide in seams varying between 5.25 and 8.0 feet in thickness. Outputs vary between 125,000 and 150,000 tons per month with the best performance being 204,137 tons over a fifty shift duration. This too mirrors a great deal the Riddings Project of the 1970s since we were obtaining somewhat similar results under like logistic parameters.

Over panel changeovers were often accomplished in a weekend with the new panel starting up on the following Monday morning. Even in those early days of 1970 with comparative primitive equipment they were producing 500,000 tons from a shearer installation at production rates of some 2,400 tons per man year.

UNITED KINGDOM

Over the past twenty-five years our later generation of mining engineers have acquitted themselves alongside the best anywhere, their levels of achievements having been obscured by the nature of political events.

The first Electra 550 was installed in the Fenton Seam, 6 feet in thickness, at Houghton Main Colliery (North Yorkshire Group). Although the mining conditions were adverse and difficult, by 1989 the machine had produced 1,800,000 tons with great reliability.

The First Electra 1000 was introduced at Castlebridge Colliery (Longannet Complex) Scotland in April 1993, under British field trial conditions. The results of those field trials were most encouraging. During them in one single week 38,417 tons were produced with great reliability.

In the Selby Complex (North Yorkshire), 600 miners at Riccall Colliery produced over 2,000,000 tons in 36 weeks. Their close neighbours at the Wistow Mine produced a similar tonnage in 48 weeks during the same year.

In January 1994 the miners at Daw Mill Colliery (Midlands) produced 114,021 tons in a single week using an Eichoff (GB) shearer (similar in design to the Electra 1000) with matching infrastructure from a face 985 feet long in a seam of about 14 feet in thickness. The extraction rate for that particular week was 107 strips, establishing a European record, prior to which the face had consistently averaged a weekly rate of 60,000 tons.

Privatisation of 'British Coal'

Immediately following the 1984 miners' strike British Coal (formerly the National Coal Board) coal sales to the Central Electricity Generating Board started progressively to decline, but between the years 1991/92 and 1994/95 the fall in sales from 73,000,000 tons to 36,000,000 was most dramatic. This was a reduction of 36,8000,000 tons, roughly the equivalent of thirty-six deep-mined collieries. Undoubtedly this situation was influenced by new gas, and maybe additional oil fired, electricity generating stations coming on stream. Nuclear power stations had a guaranteed share of the market irrespective of their competitive costs.

During the first nine months of 1994 the sixteen remaining collieries were producing at the combined rate of 28,500,000 tons per annum with a labour force of 7,500 men on colliery books.

Productivity was running at about 3,800 compared with the former average of 2,020 tons per man-year. The effective result of this on 1993/94 prices was to reduce the cash operating costs from £2.39 to below £1.30 per gigajoule ($3.728 to $2.028 @ $1.56 per pound sterling). From a base of this nature the

Coal Industry, not unlike the proverbial phoenix, can and should emerge from its ashes to ascend to even greater heights than formerly.

Although the government is committed to privatisation it still retains state ownership of the coal reserves, management of which is now vested in the Coal Authority which is empowered to grant licences for exploration and the projected working of deep-mine and opencast operations. It has the power conditionally to lease 'the right to work a particularly designated area for a designated period of time'. Thankfully this would seem to indicate that the coal reserves have not been written off, although the circumstances under which they can be exploited in the future in the absence of a publicised policy statement is still obscure. Unless the future 'competitive energy fields' are open to free and fair competition as opposed to the restrictions and manipulations described, disposal and purchase of collieries and the development of new mines will have difficulties in attracting City capital or capital from foreign sources. Maybe future possibilities may envisage foreign capital extracting our reserves for its own profit and use.

In subjecting British Coal to privatisation the Government decided that its production operations were to be divided into 'regional packages' based on the geographical location of the coalfields and opencast sites. Moreover it provided for prospective bidders to have wide flexibility in their choice of regions from one to five or any combination they chose to submit.

The two main packages are located in England with the three smaller packages being located in the North-East, Scotland and Wales. On the basis of media reports I followed at the time two major consortia elected to place bids.

The first consortium, led by Richard J. Budge, had its bid of £900,000,000 for the two major packages accepted. His bid was on the basis of markets for power generation, at 27,000,000 tons, with a further 8 million tons of industrial and housecoal, with a weighted cost of £34 per ton. Within that figure the 8 million tons industrial and housecoal are expected to yield revenues at over £60 per ton.

The second consortium was formed from the National Union of Democratic Mineworkers and Management led by Malcolm Edwards (Commercial Director of British Coal) with a bid of some £500/600,000,000 on the major English packages. Their bid was unsuccessful. Eventually the business was sold to RJB Mining, Mining Scotland and Celtic Mining. There were two successful 'buy-outs' from management sources; Bettws (West Wales) and Hadfield Colliery (Yorkshire).

The current situation in which the remnants of the Coal Industry is to flourish and emerge is totally different from that in 1947 when the collieries were nationalised. Coal markets then were extensive with a buoyant market for exports. Traditional markets at the time were:

Steam Coals   industrial boilers, factories, maritime & naval
                                    shipping, railways, exports.

Electrical power heat and light
    generation   large & small private electrical generating
                                    stations

Gas and coke generating plants	town, city & urban private generating plants
Coke ovens and steel plants	special coking coal for steel coke manufacture and chemical and other by-products
Housecoal and central heating fuels	including anthracite, household, hotels, commercial greenhouses.

During the past forty-nine years of national mines ownership, substantial changes have taken place. With the nationalisation of steel, gas, railways and the electricity generation industry, centrally organised monolithic boards were formed to run these large enterprises, a situation which often gave rise to conflict of interests. This was particularly so with the breaking into traditional coal markets such as shipping, railways and central heating.

This situation was later compounded by the discovery and exploitation of North Sea gas and oil, the advent of nuclear power generation and decline of the steel industry, such that certain coal markets disappeared or were drastically cut back.

Both the National Coal Board and the Government failed to recognise the permanence of these changes and to effect compensating policies, adjustments and changes in the interest of achieving the best economic strategies for the overall national interest. Maybe it was inevitable that the difficulties and upheavals such as occurred were bound to arise, this being the price levied against the lack of foresight and action appropriate to the situation at the time.

It is felt there is much to be learned from these past experiences to good effect in the light of what is now emerging on the world scene in which intensive change is taking place at an ever-accelerating rate.

The Current World Order

Recently I have read two books about the current and projected order of things world-wide which bear out some of the points already made.

The World in 2020 (Power, Culture and Prosperity – A Vision of The Future) by Hamish McRae (Harper Collins, London) is described by the publishers as:

> . . . a radical new version of the way the world is changing. It shows how clearly the cutting edge will be cultural rather than technological and why originality and intellect will prove an important discipline in the struggle for dominance. Using the analytical tools which reveal the potential for growth and social harmony in each country, Hamish McRae shows how demographic change, the costs of social disintegration in terms of crime, violence and disorder, and rapidly developing new resources are bringing about profound shifts in world order. Drawing on the best research available in Europe, Japan and America in long term growth prospects McRae expounds a highly original version of the economic case for good behaviour.

As the developed nations are ever more able to imitate each other, innovations cross national boundaries within days and weeks rather than months or years - it will be those countries which combine ingenuity with social order that will emerge triumphant by the year 2020.

This is one of the most interesting books of its kind I have read in recent years. Dealing with the strengths of the North American economy McRae postulates that these are neither its vast resources nor its low cost manufacturing ability, but rather two interrelated different qualities: its culture and its intellect. These are human resources rather than physical attributes. Most people thinking of the USA would also admire its scientific skills – scientific advance is certainly brought about by clever well educated people, but in economic terms it is the culture of America and the flexibility of thought the US culture encourages which give the country its unique and enviable role in world affairs.

McRae quotes from 'A World Bank Paper 1993 – The East Asian Miracle – Economic Growth and Public Policy' by Konosuke Matsushita, head of one of Japan's largest electronics group, on attitudes within the Japanese industry. His paper was given before a group of visiting foreign managers when he referred to the outdated concepts of exclusive technocrat thinking with labour confined to working to instruction. An excerpt is reproduced on p.74.

Tom Peter's *Seminar (Crazy Times Call For Crazy Organisations)* is a most stimulating and radical exposition of the current world order and contains much with which I can identify. Peters welcomes his readers with:

Welcome to a world in which imagination is the source of value in the economy, an insane world, and in an insane world, sane organisations make no sense.

This statement is followed by a series of examples which include the following:

18 October 1993 *USA Today* revealed:

that the Golden State Warriors offered a $74.4 million contract to 20 year old Chris Walker, who had not played a minute's professional basketball at the time. On the same day the *Financial Times* headlined a strong threat to 400,000 auto part workers.

Three months later in January 1995 *Business Week* published:

a poll reporting that 90% of the Executive foresaw sales going up during the year. Yet despite this optimism over half the Executive planned to maintain their payroll as it is or lay off employees.

Further examples involving drastic changes from the old order:

In the span of two weeks during the summer of 1993, MCI and British Telecom do a huge deal, TimeWarner & US West team up, then Time-Warner & Silicon Graphics, then US West & Microsoft. Like a new solar system forming from galactic dust, forming a monstrous mulitrillion computer-software, telecommunications-electronics services mega industry is born in cyberspace and on the television screen.

Meanwhile a whole continent – whole civilisations break loose. China's economy grows at the rate of 14% in 1992 but is expected to 'slow' to 10%. However the First Quarter comes in at 15%. Sophisticated Singapore 'per capita' income passes that of Great Britain. Speaking of sophistication advanced electronic design work in Asia extends beyond Japan's borders to Taiwan, India, Korea, Malaysia and Thailand.

The several party delegates to Guangdong Communist party meeting arrive in gold plated Mercedes, and the Chinese are rioting to acquire stock and share allocations; is it any wonder that the Chairmen of General Motors, IBM, Westinghouse, American Express and Kodak are all fired within a few months of each other.

. . . since that day early in 1992 when the stock value of Microsoft with $2 billion in revenues surpassed that of General Motors, should it come as a surprise that Boeing, Digital Equipment, Compaq, Daimler-Benz and ICI, Phillips, Hyudai, Volkswagen and Bosch have announced restriction 'revolutions' which decimate the workforce, starting with the whitest of white collars.

. . . if Hewlett-Packhard can run its 9,000 person ink-jet business with a headquarters staff of four people and companies are flattening out a 'warp' speed, is it difficult to believe that Manpower Inc., America's largest provider of its temporary worker population, has now more than 500,000 people on its rolls? Or that the temporary worker population has risen 150% since 1982, whilst the total work force has increased by only 20%? Or that some firms are growing like Topsy by supplying transient CEOs, VIPs and Senior Managers to 'networked companies' that viewed performance as a mortal sin?

When we confront an average of 300 programmed electronic micro-controllers each day and my new Minalota Camera has more intelligence than my Apple II Computer should we be nonplussed with Nintendo with only 892 workers racking up 85.5 billion in sales ($6 million per employee) ranking 3rd in all Japan.

The United Kingdom can add to the above situations:

. . . the collapse of Barings Bank during the First Quarter of 1995 – due to a loss of £800,000,000 as the result of diversified speculation in Japanese Contracts on the Singapore Stock Exchange in which the Baring Bank Officials pressed for, and were awarded bonuses of £100,000.000 during take-over negotiations with the Dutch Bank ING.

Chairman and Directors of the Water Utilities awarding themselves substantial salary increases and share-options on the grounds of their achieved excellent performance – weeks later followed by a near breakdown of essential water supplies to its consumers. [Unusually dry summer of 1993].

The recent October 1995 proposed substantial increase of certain 'rail fares' with the apparent object of reducing commuter demand.

Such is the background to the future, providing an indication of the lessons to be learned by the modern entrepreneurs and managers from which it would appear that the monolithic structures which grew out of the Industrial Revolution and subsequent commercial growth are moving towards being relegated to the role of the 'administrative dinosaur', in that the future cannot accept the associated delays and wasteful costs embodied in these outdated structures.

In the new market world environment, design, design changes, manufacture and delivery will need to be met in a matter of hours, days and weeks, according to the situation, as opposed to the months and years formerly existing with the Western monopolised market situation.

Non-direct revenue producing, time and cost consuming, staff departmentalism, adversely delaying and killing radical thought and other decisions, will be swept away by the 'contracting out of required skills and informational know-how'. In the new 'age of rapid, almost instantaneous communication and rapid transport', the location of such skills nationally or internationally will have less meaning.

Prestige titles, privileged position and nepotism inadequately backed by performance cannot be expected to survive in a situation in which the concept of 'corporate imagination, intellect and intelligence is the order of things' and everyone is a business person in their own right.

Tom Peters further draws upon the examples of 'flattening' out current organisational structures in a number of fields, stressing the need for greater speed and flexibility and freedom in decision making. In line with such thinking and expression the National Coal Board's organisational restructuring of 1967 might well have been served better by flattening out at the top instead of cutting away at the centre, leaving the Divisional Boards free to operate in competition, somewhat in the manner of the former large mining companies prior to nationalisation. Whatever the short term gains were, subsequent events have shown that it could hardly have been considered successful in securing a future for the Coal Industry.

It is realised that such a step would have been fraught with political and industrial difficulties, together with amendments to the Nationalisation Act 1946. Such difficulties however were met and overcome in the late 1980s. Maybe such an act might have saved many collieries from closure – I don't know nor does anyone for that matter, but what I do feel strongly is that such momentous decisions and objective changes should have been the subject of considerable detailed thought and controlled application, which didn't seem evident to me at the time.

Reproduced in this book are a number of expressions and quotations from several American sources, some of which I can readily identify and are appropriate to my situation and personal views.

Do something – make things happen, it's inaction that kills you. (*Middle Manager MCI*)

The world of technology thrives when individuals are left alone to be different, creative and disobedient. (*Don Valentine*)

The lumbering bureaucracies of this century will be replaced by field, independent groups of problem solvers. (*Steve Truett and Tom Barrett*)

Job security is gone – The driving force must come from the individual. (*Homa Bahrami*)

Does your company have a 'clean desk policy'? If so, the company is 'nuts' and you are 'nuts' for staying there.

. . . on the virtual corporation – edgeless – permeable – constantly reforming according to need. (*Bill Davidson and Mike Malone*)

If we don't make fools of ourselves from time to time – we grow smug – that is we don't grow at all.

Never use the excuse of following orders as a rationale for a poor course of action (*Rodger Meade*)

We are trying to sell more intellect and less and less materials. (*3M Planner*)

Microsoft's only asset is human imagination. (*Fred Moody New York Times*)

The conversion of an 'organisation' into a 'business' ALWAYS strengthens performance. (*Stan Davis & Bill David*)

Career and Family Reflections

Throughout the whole of my life I have been driven from within by a strong desire to seek out and develop new experiences and to accomplish things in the face of challenge. Generally I have found such challenge in whatever form it appeared, although often accompanied by mental and physical stress, highly stimulating.

Maybe by 'flowing with the stream of event or attempting to lead from behind' things could have been easier. Thankfully this was not in my make-up; had it been so I certainly would not have had such an interesting and worthwhile career. Although I worked hard with great dedication I had pleasure in doing so, although there were many times when matters were at a low ebb. On these occasions I would recall the advice of Sir Noel Holmes, the first Chairman of the National Coal Board's NE (Yorkshire) Division:

'Charles – what you are doing is all right but you must learn to live.'

This often gave me food for thought but at this late hour, looking back over many years, I have had my full share of accomplishment and real happiness. I remember the short poem of my late school days:

Happy the Man

Happy the man and happy he alone,
He who can call today's his own,
He who, secure within, can say,
Tomorrow do thy worst, for
I have lived today.
Be fair or foul or rain or shine,
The joys I have possessed,
In spite of fate are mine.
Not Heaven itself has power,
But what has been, has been, and
I have had my hour.

John Dryden (1631–1700)

People at their best can be truly inspiring, at their worst irksome and annoying. Thankfully I have neither reason nor desire to bear ill will, ill feeling or hatred to any person. Should any impressions have been created in that direction from any of the information or opinions disclosed, I would quickly deny them. No human is infallible in all his or her endeavours, actions and behaviour, and I myself am no different from the ordinary person.

Many of life's disappointments and conflicts are often engineered by events over which no-one has complete control, but in many such instances they do have a useful part to play in the scheme of things when considered positively, whether it be in correcting one's weakness, changing one's approach or altering direction.

In June 1984 Dorothy, her husband John and our grandson Paul, then aged five, moved from Ecclesfield (Yorkshire) to Taunton (Somerset). John, a qualified accountant, had joined a large south-western firm of chartered accountants, Butterworth Jones & Partners, based in the county town. Some two years later Dorothy, formerly trained by Lloyds Bank over many years, joined Marks & Spencers (Taunton Branch) staff on a part time basis and has worked happily for them for over nine years.

In 1985 Mary and I wound up our affairs in Wickersley, Rotherham and moved south-west to join them. We wanted to be close to our young grandson, having lost out on seeing our other three grandsons, Michael, Peter and David (John and Sue's three boys), develop over in Ireland.

Mary and I quickly settled down in Taunton, life having been very rewarding to us both. Looking back over the years we had an extremely happy life together as a family, largely shaped by Mary whose uncomplicated view of living and simple but homely desires created an atmosphere of harmony and happiness which I am more than grateful to have enjoyed. Throughout our life together, her love and devotion, unstintedly given to us all, was immeasurable. During my darkest hours of which there were a few during my career her simple statement to the effect, 'We managed before you met this situation – we'll manage whatever happens,' never failed to give the supplementary strength I needed at the time I needed it.

John and Dorothy developed well and were not too badly disturbed by the many early school changes they were subjected to in their early life, and as teenagers they never gave us cause to worry. Both went on to develop as responsible young adults, a situation which they have carried through into their married and family lives.

On the Saturday night of 11 July 1987 Mary and I spent upwards of three hours together on the Quantock Hills (a local unspoilt area of great beauty). It really was exquisite. Blueberries were there to be picked in great profusion. After picking about three pounds we returned home. During the early morning of 12 July 1987 Mary quietly passed away in her sleep around 5.00 a.m. without pain or suffering, thus leaving a gap in my life that can never be filled. Being the great and wonderful person she was hardly a day goes by but that I remember and relive memories of our great life together.

My son John and his partner Eddie Robinson over a period of some twenty-seven years or so built and nurtured their business, Industrial Detergents, against fierce competition, to become one of the leading Irish manufacturers in its field with an ultimate turnover in excess of £3,000,000. At this point it attracted the interest of the multi-nationals, of whom Henkel (Germany), one of the largest privately owned chemical companies in Europe, started direct negotiations for takeover.

The ultimate outcome was that they were bought out. Eddie Robinson retired at the time of the takeover, with John being retained on a three-year contract to run the company on behalf of the new owners. Later Henkel (Germany) teamed up with EKOLAB, the large American multi-national, in a joint European venture. At the end of John's contract, however, he was retained for a further two years. During his tenure with the Company the business has expanded, currently having a turnover of upwards of £4,500,000. Officially he retired at the end of 1995.

Michael Round his son, my eldest grandson, operates as a production engineer. He married a charming Irish girl, Lesley Briggs of Dublin, who runs her own jewellery business. They have an infant son, James L. Round, born on 7 September 1995.

Peter Round, our second grandson, operates as a salesman in the Dublin area. He too married a delightful young Irish girl, from Kilkenny, County Kilkenny. They too have a young child, Katie L. Round, born on the 7 August 1995, our first great-grandchild.

David C. Round, our third grandson, is currently taking a business course whilst seeking an employment outlet. He like his father is an excellent golf enthusiast playing down to a 5 handicap.

On Dorothy and John Fordham's side, our fourth grandson Paul in his sixteenth year just completed his schooling at Wellington College. He secured nine passes at GCSE and is currently attending the Somerset College of Art & Technology. (SCAT) taking courses in business studies and computers, in preparation for taking his Advanced Level examinations. Currently he is gaining experience in a variety of spare time jobs: the traditional 'paper round'; part time shop assistant; car cleaning; lawn cutting etc., balanced by an interest in golf, football, badminton, darts and ten-pin bowling, and as a young leader of the Church Lads Brigade.

A deep regret I have lies with the inability of Mary to have been present at the weddings of her grandchildren and later to see her great-grandchildren. It would have been a crowning glory to the useful and creative life we jointly shared, particularly in view of her love and devotion to young infants.

As for myself, I have just moved onto the Information Highway – Internet, which has opened up vast panoramic horizons of interest and activity which will absorb all the remaining days Providence in its kindness has allocated to me.

Having spent more than two enjoyable years writing up and completing this account of my personal historic and coal mining experiences, before and after nationalisation, and embracing the current privatisation of the Coalmining Industry, common experiences of many former and working miners, mine officials, colliery managers and higher management at all levels, I hope that like myself they have been able to recall some of the close comradeship and pleasures of their experiences in what was and is a noble industry which has served this nation well over a great number of years.

To those who have been more closely introduced to coalmining for the first time, I hope you have found something of interest in the matters I have tried simply to explain, being motivated throughout in a manner consistent with having your individual presence alongside me at all stages of its development. The sort of difficulties and problems which beset those attempting to emerge from the lowest to the highest levels of operation whatever the vocation or calling, I hope has clearly emerged. Moreover, it is felt that the human characteristics of 'intellect and imagination' are not necessarily confined to the birth accidents of privilege and academic prowess nor are the exclusive province of the aristocracy or ruling classes, as demonstrated throughout history over countless generations in the affairs of ordinary working people.

PHOTOGRAPH 66. ELECTRA 1000 SHEARER LOADER
ANDERSON GROUP LIMITED, MOTHERWELL, SCOTLAND

The current 'state of the art' longwall mining system embracing the ELECTRA 1000 shearer loader, skin to skin shield type powered supports and the high capacity panzer face conveyor. The ELECTRA 1000 is controlled remotely. Here it is operating in the Kerr McGee Mine, Illinois, USA, producing at the rate of about 2500 tons per hour, in a seam 15 ft. in thickness. The reliability factor is over 98% and maintenance at the face is greatly eased by the designed modular exchange characteristic.

The original Anderton shearer loader development shown in Figures 34, 35 and 36 gave us about 100 tons per hour with a probable limit of about 150 tons per hour. Reliability was a problem, as can be imagined in an 'embryo' development situation. The panzer face conveyor and the powered supports show somewhat similar comparative rates of design advance in attainment.

PHOTOGRAPH 67 – ELECTRA 1000 SHEARER LOADER

Here the leading scrolled drum is raised to the selected shear cutting height. This assembly incorporates a cowl assembly which assists in loading the shearer coal onto the panzer face conveyor. The cowl can be rotated through 180 degrees for bi-directional working. Mounted in front of the machine over the face conveyor is the pick breaker drum. This crushes large material to a size which will pass under the tunnel formed between the underside of the machine and the face conveyor. The four machine conveyor mounting blocks can clearly be seen at the corners. These incorporate the haulage 'rack and pinion' provisions, the rack being integral with the face conveyor structure, more clearly seen in Figure 14 – Exploded View. Each ranging arm is equipped with a cutter motor of 350 Kw providing a total cutting power of 750 Kw (1000 h.p. hence the machine's designation). The pick drum crusher (coal sizer) is driven by a 100 h.p. motor.

Special links couple each section of the conveyor together and permit a degree of horizontal and vertical misalignment without detriment to its operation. The conveying media takes the form of two centre chains and heavy duty steel flights attached. Chain speed, compared with the early former conveyors, has been increased together with the horsepower of the leading and tail end drive units.

PHOTOGRAPH 68. ELECTRA 550 SHEARER LOADER

This is an exploded sketch view of the machine with the various independently powered elements carried on a basic 'base plate'. This represents to me the culmination of some 40 to 50 years painstaking development and application by many capable people, a situation which has placed mechanised longwall mining practice as the most productive deep mined mining system anywhere. There are so many advanced features about this design compared with the single motor driven format of yesteryear; a separate book could be written about them.

Referring to Photograph No. 68 one cannot but be struck by the symmetry and apparent simplicity of the format. The leading cutting drum does give the impression of an efficient powerful tool. The haulage arrangements are clearly shown. The 'rack' built into the panzer conveyor has certainly eliminated many of the former dangerous aspects of the early rope haulage arrangements and permits machine transit backwards and forwards across the face at speeds we would have previously found too dangerous to contemplate.

Note the heavy double standard and flighted nature of the panzer conveying element, a feature well in advance of former arrangements. Integrated into modern and equally developed infrastructure the potential of the mechanised longwall mining system has been greatly extended in terms both of general application and of high production and low cost performance.

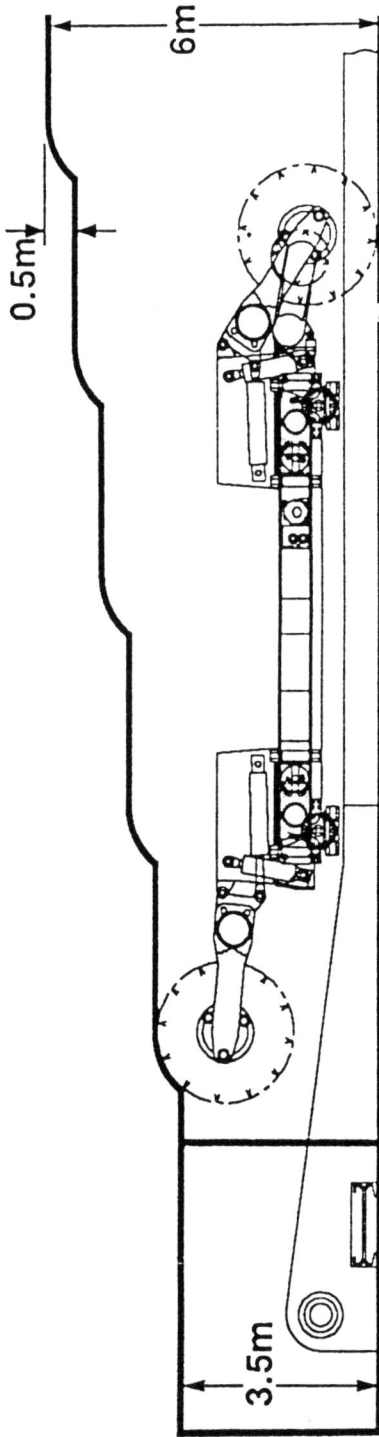

Electra 1000. variable geometry – cutting horizon approaching face end

FIGURE 13. FACE SECTION: CUTTING HORIZON VARIABILITY

This sketch indicates the flexibility of the machine with regard to its application in seams between 5 and 20 ft in thickness, without the need for any special modification. Moreover, it indicates how accommodation is made for the reduced height taken in the gate roadways, up to about 12 ft. in situations of this kind.

The machine is remotely controlled and a great deal of further sophistication is involved with the development of automatic steering control and data transmission monitoring individual aspects of the machine elements.

Electra 550 – modular construction

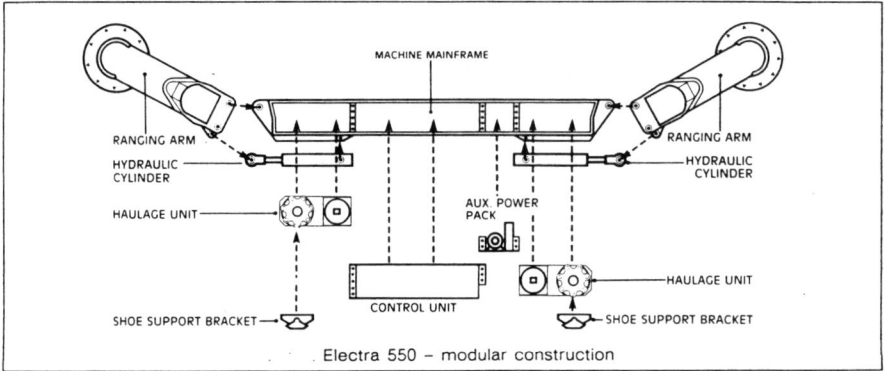

(ABOVE) FIGURE 14 – ELECTRA SHEARER LOADER – MODULAR

(BELOW) FIGURE 15 – ELECTRA 1000 SKETCHES OF SIDE AND END
ELEVATION

I get quite a repeated 'kick' out of the photographs of the ELECTRA 550 and 1000 which
have all the beauty and symmetry of the Rolls Royce car and aircrafts Concorde and
Boeing 707, embracing as they all do a range of characteristics dreams are made of.
Reference back to Photographs 33 to 38 inclusive (at the beginning, when James Anderton
& Anderton Boyes Ltd, Motherwell, introduced and developed the most successful coal
producing power loading machine of its time, my datum of comparison) shows the
tremendous development and sophistication that has evolved within those intervening
years. Similar comparisons can be made with the hydraulic powered road supports. Note
the primitive nature of those featured in Photographs Nos 41 and 42, which were pioneers
in every sense of the term and the design developments. Here the sketches show the
integration of the machine into the powered support infrastructure and the degree of
security provided with their operation.

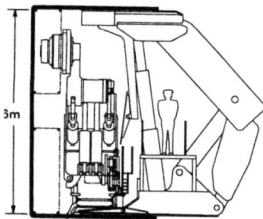

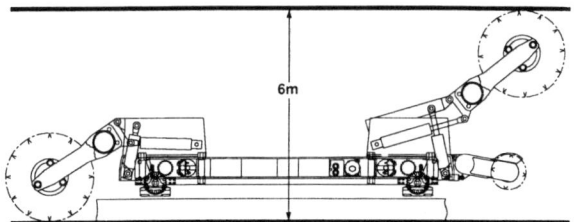

Appendix

NATIONAL COAL BOARD PERFORMANCE – YEAR 1947

National Coal Board – Division and Coalfield	Tonnage Long tons 2240 lbs	Proceeds Revenue Per ton £ Sterling	Revenue Total Gross £s	Profit £, Total	Loss £, Total
SCOTTISH					
Fife and Clackmannan	6,468,821	1.978	12,795,328	1,509,392	
Lothians	3,283,051	1.96	6,434,780	410,381	
Central West	4,499,656	2.088	9,395,282		-772,441
Central East	3,686,391	1.792	6,606,031		-116,736
Ayr and Dumfries	4,247,144	2.039	8,659,927	539,741	
Divisional Totals:	22,185,063	2.039	43,891,329	2,459,514	-889,177
				1,570,338	
NORTHERN					
North East Durham	3,509,048	2.066	7,249,693		-688,651
Mid East Durham	4,904,998	1.998	9,800,186		-1,265,080
South East Durham	3,478,885	2.018	7,020,390		-407,319
South West Durham	3,639,653	2.03	7,388,496		-1,445,246
Mid West Durham	3,157,533	2.002	6,321,381		-1,439,309
North West Durham	5,450,249	2.049	11,167,560		-1,921,213
Southern Northumberland	2,866,976	1.993	5,713,883		-475,440
Central Northumberland	4,372,787	1.948	8,518,189		-129,362
Northern Northumberland	3,436,803	1.966	6,756,755		-514,088
Cumberland	1,082,540	2.338	2,530,979		-514,658
Divisional Totals:	35,899,472	2.019	72,467,511		-8,285,707

National Coal Board – Division and Coalfield	Tonnage Long tons 2240 lbs	Proceeds Revenue Per ton £ Sterling	Revenue Total Gross £s	Profit £, Total	Loss £, Total
NORTH-EASTERN (YORKSHIRE)					
Worksop	5,488,865	1.921	10,544,110	226,416	
Doncaster	7,488,358	1.946	14,572,345	1,104,533	
Rotherham	5,704,415	1.917	10,935,364		-164,002
Carlton	4,728,502	1.97	9,315,149		-673,812
South Barnsley	3,192,573	1.917	6,120,162		-216,829
North Barnsley	3,323,028	1.951	6,483,228	186,920	
Wakefield	2,835,866	2.008	5,694,419		-224,506
Castleford	5,548,683	1.975	10,958,649	381,472	
Divisional Totals:	38,310,290	1.948	74,623,425	1,899,341	1,279,148
				620,193	
NORTH WESTERN					
Manchester Area	5,371,178	2.253	12,101,264		-568,450
Wigan	2,543,227	2.221	5,648,507		-308,449
St Helens	2,434,133	2.285	5,561,994		-271,812
Burnley	846,026	2.214	1,873,102		-130,782
North Wales	2,038,446	2.078	4,235,891		-291,328
Divisional Totals:	13,233,010	2.223	29,416,981		1,570,820
EAST MIDLANDS					
Chesterfield	4,930,829	1.866	9,200,927	638,953	
Mansfield	5,118,694	1.875	9,597,551	906,435	
Edwinstowe	5,879,822	1.832	10,771,833.9	1,765,552	
Alfreton	4,473,345	1.868	8,356,208	1,545,389	
Ilkeston	4,946,561	1.889	9,344,054	95,354	
Nottingham	3,333,337	1.922	6,406,674	20,806	
South Derby and Leicestershire	6,001,395	1.825	10,952,546	2,037,974	
Divisional Totals:	34,683,983	1.718	64,629,794	7,010,463	

National Coal Board – Division and Coalfield	Tonnage Long tons 2240 lbs	Proceeds Revenue Per ton £ Sterling	Revenue Total Gross £s	Profit £, Total	Loss £, Total
WEST MIDLANDS					
North Staffordshire	6,210,847	2.076	12,893,718	1,180,060	
Cannock Chase	4,317,903	2.039	8,804,204	95,354	
South Staffs and Shropshire	1,314,075	2.044	2,685,969	20,806	
Warwickshire	4,799,345	2.03	9,742,670	1,517,793	
Divisional Totals:	16,642,170	2.05	34,126,562	2,814,013	
SOUTH WESTERN					
Swansea	3,828,965	2.299	8,802,791		-4,307,586
Maesteg	2,568,157	2.11	5,418,811		-1,348,282
Rhondda	3,683,281	2.16	7,955,887		-1,368,954
Aberdare	2,743,942	2.18	5,981,794		-703,135
Rhymney	3,496,315	2.634	9,209,294		-148,593
Monmouth	4,756,014	2.132	10,139,822		-2,358,190
Bristol and Somersset	589,214	2.221	1,308,644		-245,998
Forest of Dean	752,976	2.089	1,572,967		-261,659
Divisional Totals:	22,418,864	2.17	48,648,935		-10,742,397
KENT COALFIELD	1,375,359	2.293	3,153,698		-490,526
UNITED KINGDOM	184,748,211	2.013	385,754,265		-9,203,905

NATIONAL AVERAGES:

Proceeds	40s. 3d/ton
Total Costs:	41s 3d/ton
Profit/Loss	1s 0d/ton loss
Manpower	711,400 Average
Productivity	1.075 tons/ms
Absentee	12.43%
Number of Collieries Year Ending 31 December 1947	980
Capital Expenditure (Auuthorised @ Vesting Date) 1 January 1947 Authorised:	£37,003,190
Expenditure Year 1947	£ 2,892,344

PRODUCTION COSTS NATIONAL AVERAGES COST ELEMENT

	s	d
Wages and allowances	25	6.9
Holiday pay	—	10.7
Supplementary injuries	—	—
Men's compensation	1	0.5
National Insurance		6.1
Roof supports	2	6.4
General Stores	2	8.1
Repairs and renewals	—	8.4
Power, heat and light	2	11.6
Salaries	—	11.8
General expenses	1	5.5
Administration expenses	—	6.7
Depreciation	1	4.3
Total costs:	41	3.0
Average earnings/ms	27	8.2
Tons per man year:	260 (Estimated)	

Bibliography

Lord Robens, former National Coal Board Chairman, *My Ten Year Stint*.

John Threlkeld, *Pits* (Wharncliffe Woodmoor Investments Ltd., Barnsley).

Tom Peters, *Tom Peters Seminar*, (New York).

Hamish McRae, *The World in 2020* (Collins, London).